MEDICAL PHYSICS

Medical Physics

John R. Cameron
University of Wisconsin, Madison
and
James G. Skofronick
Florida State University, Tallahassee

A Wiley-Interscience Publication
JOHN WILEY & SONS
New York · Chichester · Brisbane · Toronto

Library of Congress Cataloging in Publication Data

Cameron, John R.
  Medical physics.

  Bibliography: p.
  Includes index.
  1. Medical physics. I. Skofronick, James G.,
joint author. II. Title.

R895.C28   612'.014   77-26909
ISBN 0-471-13131-8

Printed in the United States of America

10 9 8 7 6 5 4 3 2 1

To Raymond G. Herb
friend and dedicated scientist

How far even then mathematics will suffice to describe, and physics to explain, the fabric of the body, no man can foresee. It may be that all the laws of energy, and all the properties of matter, and all the chemistry of all the colloids are as powerless to explain the body as they are impotent to comprehend the soul. For my part, I think it is not so. Of how it is that the soul informs the body, physical science teaches me nothing; and that living matter influences and is influenced by mind is a mystery without a clue. Consciousness is not explained to my comprehension by all the nerve-paths and neurones of the physiologist; nor do I ask of physics how goodness shines in one man's face, and evil betrays itself in another. But of the construction and growth and working of the body, as of all else that is of the earth earthy, physical science is, in my humble opinion, our only teacher and guide.

D'Arcy Wentworth Thompson,
On Growth and Form, 1917

PREFACE

Have you ever wondered about the forces of your body during various activities, or how much useful work can be done by your body, or the relationship of an electrocardiogram to your heart's electrical activity, or how a medical x-ray works, or how much radiation you receive from an x-ray? These questions and many others like them involve applications of physics to medicine and are answered in this book.

While the roles of chemistry and biology in medicine are well accepted, the role of physics is usually not as obvious. Even though most medical and paramedical students take an introductory physics course, they often see little or no relationship between physics and medicine. This communication gap is due primarily to the large amount of material in a traditional year-long physics course that precludes an adequate treatment of physics applied to medicine.

This book is written primarily for students who plan to make a career in some field of medicine. We describe in a simple fashion the usefulness of physics in understanding the behavior of the body. The book covers two broad areas of medical physics: the application of physics such as mechanics, heat, light, sound, electricity, and magnetism to medicine and the physics of the various organ systems such as the eyes, ears, lungs, and the heart and circulatory system. Since at present the most important applications of physics in medicine involve the field of radiology, we have four chapters that deal with the various aspects of radiology. They cover diagnostic radiology, nuclear medicine, radiotherapy, and radiation protection. Most professional medical physicists work in one of these fields. The last chapter describes applications of computers in medicine.

Although this book was written primarily as a text for students who have some understanding of elementary physics, we believe much of it will be interesting and understandable to any person who is curious about how the body works and the physical principles of the instruments used in medical diagnosis and therapy. The mathematics is basically at the algebra level. Although logarithms and exponentials are used, they are reviewed in Appendix A. International System (SI) units are used most often, but occasionally units common to some particular branch of medicine are used. Blood pressure in pascals will take a while to catch on. Tables in Appendix B give information on the SI and other units. Appendix C gives physical data on the Standard Man. This book is not intended to be a reference book, but it does include a bibliography at the end of each chapter and a general bibliography at the end of the book.

Most of the chapters were written by the two authors; however, colleagues who are specialists in their particular fields either wrote or assisted in the writing of several of the chapters. In particular, C.R. Wilson assisted in the writing of Chapter 3, J.G. Webster wrote Chapter 10, J. Wagner assisted with Chapter 17, and R.B. Friedman wrote Chapter 20. Their cooperation is greatly appreciated.

The text has gone through several drafts, but it is certain to contain a number of errors still. These errors are of three general types: (1) errors introduced in the production of the text and not caught during the proofreading (typographic errors), (2) errors due to the inadequate knowledge of the authors, and (3) errors due to inadequate information even by the experts in a given field. We apologize for the first two types of errors and hope that the readers will call these to our attention. If you detect the third type of error, you should publish your results in an appropriate journal.

The authors express their deep gratitude to Barbara Sandrik (see Fig. 8.7, p. 162), our editorial assistant, who worked with dedication on the many aspects of this manuscript even though she had many other commitments. We also thank Nancy Clark (see Fig. 3.3, p. 43), who typed and retyped this manuscript in addition to carrying out her many regular secretarial duties.

One of the authors (JGS) thanks and recognizes the help of the many colleagues, students, and staff at Florida State University who made contributions and provided encouragement as the text proceeded through the various drafts. In particular, Dexter Easton, Ed Deloge, and Kurt Hofer were always available with helpful comments on the manuscript. Some draft versions were typed by Sandy Crum, and some of the illustrations were done by Ken Ford and Wally Thorner.

The other author (JRC) apologizes for his failure as a humorist. The reader will be relieved to know that the corniest jokes have been edited out. He thanks his friends, colleagues, and students for their help in the preparation of this book. In particular he thanks the following for their special contributions: Lee Baird, Nury Canto, Orlando Canto, Shannon Dolan, Mark Madsen, Charles Mistretta, Earl Nepple, Jerry Nickles, Linda Robbins, John Sandrik, Thomas J. Smith, Thomas Stevens, Anne Vancura, Jude Wiener, Al Wiley, John Wochos, and Jim Zagzebski.

We also thank Dick Walters at Duke University, Rod Grant at Denison University, and Jerry Wagner at Vanderbilt University, who contributed to this book by using it in draft form and offering many helpful suggestions. We also appreciate the patience and understanding of our editor, Bea Shube of Wiley-Interscience.

Finally, we thank Dot Skofronick and Von Cameron, our wives, for hanging in there during the difficult five-year gestation of this book.

John R. Cameron
James G. Skofronick

Madison, Wisconsin
Tallahassee, Florida
January 1978

CONTENTS

MEDICAL PHYSICS

CHAPTER 1

Terminology, Modeling, and Measurement

In this chapter we present three broad topics. First we examine what we mean by medical physics and describe some related and overlapping disciplines. In the next section we discuss modeling, a concept that is essential in science, engineering, and medicine. We discuss and give examples of feedback, an important feature of many models. In the last section we discuss measurement. Most of this last section is a review of material taught in an elementary physics course. As part of this last section we discuss the problem of accurately measuring a person's weight. As you will see, this measurement is far from simple.

1.1. TERMINOLOGY

The field of medical physics overlaps the two very large fields of medicine and physics. In this book the term *medical physics* refers to two major areas: the applications of physics to the function of the human body in health and disease and the applications of physics in the practice of medicine. The first of these could be called the physics of physiology; the second includes such things as the physics of the stethoscope, the tapping of the chest (percussion), and the medical applications of lasers, ultrasound, radiation, and so forth.

The word *physical* appears in a number of medical contexts. Only a generation ago in England a professor of physic was actually a professor of medicine. The words *physicist* and *physician* have a common root in

1

the Greek word *physikē* (science of nature). Today the first thing a physician does after taking a medical history of a patient is to give him a physical examination. During this examination he uses the stethoscope, taps the chest, measures the pulse rate, and in other ways applies physics. The branch of medicine referred to as *physical medicine* deals with the diagnosis and treatment of disease and injury by means of physical agents such as manipulation, massage, exercise, heat, and water. *Physical therapy* is the treatment of disease or bodily weakness by physical means such as massage and gymnastics rather than by drugs.

In principle, the field of biophysics should include medical physics as an important subspecialty. In fact, biophysics is a relatively well-defined field that has very little to do with medicine. It is primarily involved with the physics of large biomolecules, viruses, and so forth, although it does approach medical physics in the area of transport of materials across cell membranes. Biophysicists conduct basic research that may improve the practice of medicine in the next generation, while medical physicists generally engage in applied research that they hope will improve the practice of medicine in the current generation.

The field of medical physics has several subdivisions. Most medical physicists in the United States work in the field of radiological physics. This involves the applications of physics to radiological problems and includes the use of radiation in the diagnosis and treatment of disease as well as the use of radionuclides in medicine (nuclear medicine). Another major subdivision of medical physics involves radiation protection of patients, workers, and the general public. In the United States this field is often called *health physics;* this name was given to it during World War II by members of the Manhattan Project (the group responsible for the development of the atomic bomb). Health physics also includes radiation protection outside of the hospital such as around nuclear power plants and in industry. Health physics is discussed in detail in Chapter 19.

Very often an applied field of physics is called *engineering.* Thus, medical physics could be called *medical engineering.* However, for practical purposes if you meet an individual who refers to himself as a medical physicist it is highly probable that he is working in the area of radiological physics; a person who refers to himself as a medical engineer or biomedical engineer is likely to be working on medical instrumentation, usually of an electronic nature. In addition, the medical physicist usually has a bachelor's degree in physics, while the medical engineer usually has a bachelor's degree in some field of engineering—usually electrical engineering. In some areas, such as the applications of ultrasound in medicine and the use of computers in medicine, you are likely to find medical physicists and medical engineers in nearly equal numbers. (The word *medical* is sometimes replaced with the word *clinical* if the job is closely connected with patient problems in hospitals, i.e., clinical engineering or clinical physics.)

In the United States relatively few hospitals or medical schools have departments of medical physics or medical engineering, although they are quite common in medical schools in the United Kingdom and in the Scandinavian countries. The largest department of this type is in Glasgow, Scotland; in 1974 this government-supported Department of Clinical Physics and Engineering of Western Scotland had over 150 professional staff members serving the hospitals in the area.

Although the terms *medical engineering* and *biomedical engineering* are essentially synonymous, the word *bioengineering* has a much broader meaning. Bioengineering involves the application of any engineering to any biological area. Bioengineering includes medical engineering as an important category, but it also includes other fields such as agricultural engineering. Designing cow barns and manure spreaders are bioengineering problems!

1.2. MODELING

Even though physicists believe that the physical world obeys the laws of physics, they are also aware that the mathematical descriptions of some physical situations are too complex to permit solutions. For example, if you tore a small corner off this page and let it fall to the floor, it would go through various gyrations. Its path would be determined by the laws of physics, but it would be almost impossible to write the equation describing this path. Physicists would agree that the force of gravity would cause it to go in the general direction of the floor if some other force did not interfere. Air currents and static electricity would affect its path.

Similarly, while the laws of physics are involved in all aspects of body function, each situation is so complex that it is almost impossible to predict the exact behavior from our knowledge of physics. Nevertheless, a knowledge of the laws of physics will help our understanding of physiology in health and disease.

Sometimes in trying to understand a physical phenomenon we simplify it by selecting its main features and ignoring those that we believe are less important. Our description may be only partially correct, but it is probably better than none at all. In trying to understand the physical aspects of the body, we often resort to analogies. Just as Christ taught his followers by parables, physicists often teach and think by analogy. Keep in mind that analogies are never perfect. For example, in many ways the eye is analogous to a camera; however, the analogy is poor when the film, which must be developed and replaced, is compared to the retina, the light detector of the eye. In this book we often use analogies to help explain some aspect of the physics of the body. We hope that we are successful, but please remember that all explanations are incomplete to some degree. The real situation is always more complex than the one we describe.

Many of the analogies used by physicists employ models. Model making is common in scientific activities. A famous nineteenth century physicist, Lord Kelvin, said, "I never satisfy myself until I can make a mechanical model of a thing. If I can make a mechanical model I can understand it!" Some models involve physical phenomena that appear to be completely unrelated to the subject being studied; for example, a model in which the flow of blood is represented by the flow of electricity is often used in the study of the body's circulatory system. This electrical model can simulate very well many phenomena of the cardiovascular system. Of course, if you do not understand electrical phenomena the model does not help much. Also, as mentioned before, all analogies have their limitations. Blood is made up of red blood cells and plasma, and the percentage of the blood occupied by the red blood cells (the hematocrit) changes as the blood flows toward the extremities. This phenomenon (discussed in Chapter 8) is difficult to simulate with the electrical model.

Other models are mathematical; equations are mathematical models that can be used to describe and predict the physical behavior of some systems. In the everyday world of physics we have many such equations. Some are of such general use that they are referred to as *laws*. For example, the relationship between force F, mass m, and acceleration a, usually written as $F = ma$, is known as Newton's second law. There are other mathematical expressions of this law that may look quite different to a lay person but are recognized by a physicist as other ways of saying the same thing. Newton's second law is used in Chapter 2 in the form $F = \Delta mv/\Delta t$, where v is the velocity, t is the time, and Δ indicates a small change of the quantity. The quantity mv is the momentum, and the part of the equation $\Delta/\Delta t$ means rate of change (of momentum) with time.

One of the physicist's favorite words is *function*. The symbol for function (f) should not be confused with the symbol for force (F). The equation $W = f(H)$ means the weight W is a function of the height H. It does not tell you how weight and height are related or what other factors are involved. It is sort of a mathematical shorthand. In the medical field we could write $R = f(P)$ to indicate that the heart rate R is a function of the power produced by the body P. The next step—to leave out the f and write an equation that tells how the things are related to each other—is the hard one.

A medical researcher may use a model of some function of the body to predict properties that were not originally suspected. On the other hand, some models are so crude that they are only useful for serving as guides to improved models.

Many functions of the body are controlled by *homeostasis,* which is analogous to *feedback control* in engineering. An engineer who wants to control some quantity that changes with time will take a sample of what is being produced and use this sample as a signal to control the production to some desired level. That is, some of the output is fed back to the source to regulate the production. If the system is designed so that an increase in the amount that is fed back decreases the production and a decrease in the sample increases the production, the feedback is negative. Negative feedback produces a stable control, while positive feedback, in which a change in the sample fed back causes a change in the same direction, produces an unstable control.

A simple example of negative feedback is the control of house temperature by a thermostat. The furnace produces heat, and the thermostat samples the heat output via a thermometer. When the temperature rises above a fixed point the thermostat sends a signal to the furnace to shut off the production of heat. As heat is lost from the house, the temperature falls until the thermostat reaches a preset point; it then sends a signal to turn on the heat again.

Negative feedback control is common in the body. For example, one important function of the body is to control the level of the calcium in the blood. If the level drops too low, the body releases some calcium from the bones to increase the level in the blood. If too much calcium is released, the body lowers the level in the blood by removing some via the kidneys.

While many of the control mechanisms of the body are not yet understood, various diseases have been found to be directly related to the failure of these mechanisms. For example, as the body grows, its cells keep increasing in number until it reaches adult size, and then the body remains more or less constant in size under some type of feedback control. Occasionally some cells do not respond to this control and become tumors.

1.3. MEASUREMENT

One of the main characteristics of science is its ability to reproducibly measure quantities of interest. The growth of science is closely related to the growth of the ability to measure. In the practice of medicine, early efforts to measure quantities of clinical interest were often scorned as detracting from the skill of the physician. For example, even though body temperature and pulse rate could be measured during the seventeenth century, these measurements were not routinely made until the nineteenth century. In this century there has been a steady growth of science in medicine as the number and accuracy of quantitative measurements used in clinical practice have increased.

Figure 1.1 illustrates a few of the common measurements used in the practice of medicine. Some of these measurements are more reproducible than others. For example, an x-ray gives only qualitative information about the inside of the body; a repeat x-ray taken with a different machine may look quite different to the ordinary observer.

In an introductory physics course many different types of measurement are studied. In general, International System (SI) units, or metric units, are used to measure various quantities (see Appendix B). Unfortunately, they are not yet in common use in the United States as they are in most of the world. The basic SI units we use in this book are the meter (m) for length, the kilogram (kg) for mass, and the second (sec) for time. For

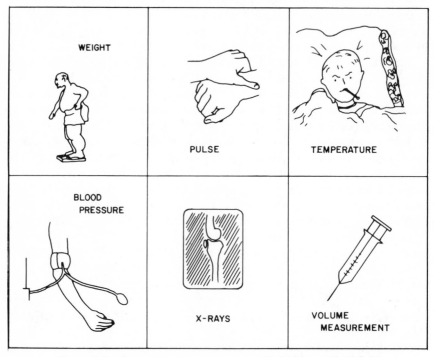

WEIGHT

PULSE

TEMPERATURE

BLOOD PRESSURE

X-RAYS

VOLUME MEASUREMENT

Figure 1.1. Some common measurements in the medical field.

convenience we often use larger or smaller units such as the centimeter (cm), one-hundredth of a meter; the millimeter (mm), one-thousandth of a meter; the gram (g), one-thousandth of a kilogram; and the milligram (mg), one-thousandth of a gram or one-millionth of a kilogram. We sometimes give quantities in English units such as feet, pounds, and gallons for easy visualization, and we sometimes use the units common to a particular branch of medicine (see Appendix B).

The use of a unit implies that there is a "true" or standard unit to which measurements with that unit can be compared. Most modern countries have laboratories that specialize in the standardization of measuring systems. In the United States, the National Bureau of Standards (NBS) located near Washington, D.C., performs this function. The NBS has little contact with medicine, but it has played an important role in standardizing the measurement of ionizing radiation in the treatment of cancer.

In medicine it is often convenient to measure quantities in nonstandard units. For example, while the correct units for pressure involve force per unit area such as newtons per square meter, dynes per square centimeter, and pounds per square inch, blood pressure is generally expressed in millimeters of mercury (Hg), a length of liquid! This length is the height of a column of mercury that has a pressure at its base equal to the blood pressure. We discuss pressure further in Chapter 6.

Let us give you an example of a nonstandard unit that is no longer used but that is of historical interest. You recall (we hope) that Galileo discovered the pendulum principle in church while watching a chandelier swing. He used his pulse to time the period of the swing and found that the time for one large swing was the same as the time for one small swing. This led to his development of the pendulum clock. Shortly thereafter (in 1602) Sanctorius, a medical friend of Galileo, invented the pulsologium (a simple pendulum) to measure the pulse rate of his patients. An assistant adjusted the length of the pendulum until it swung in time to the pulse rate as called out by Sanctorius. The pulse rate was then recorded as the length of the pendulum.

There are many other physical measurements involving the body and time. We can divide them into two groups: measurements of repetitive processes, such as the pulse, and measurements of nonrepetitive processes, such as how long it takes the kidneys to remove a foreign substance from the blood. Nonrepetitive time processes in the body range from the action potential of a nerve cell (1 msec) to the lifespan of an individual.

Measurements of the repetitive processes usually involve the number of repetitions per second, minute, hour, and so forth. For example, the pulse rate is about 70/min and the breathing rate is about 15/min. Different frequencies in the electrical signals from the brain, caused by mental activity, range from less than 1/sec to more than 20/sec. [A special unit, the hertz (Hz), is used for cycles per second.]

Many subtle processes in man are repetitive. For example, the body has a circadian (approximately 24 hr) rhythm that is present even when a person is living in a deep cave without any way of telling the time. Another biologically controlled clock is the very slow running clock that governs a growing person's mass. This clock causes the rate of growth to fluctuate with a period of 14 to 18 months.

In science *accuracy* and *precision* have different meanings. Accuracy refers to how close a given measurement is to an accepted standard. For example, a person's height measured as 1.765 m may be accurate to 0.003 m (3 mm) compared to the standard meter. Precision refers to the reproducibility of a measurement and is not necessarily related to the accuracy of the measurement. For example, an ill person measured her temperature ten times in a row and got the following values in degrees Celsius (formerly called degrees centigrade): 36.1, 36.0, 36.1, 36.2, 36.4, 36.0, 36.3, 36.3, 36.4, and 36.2.* The precision was fairly good, with a variation of 0.2°C from the average value of 36.2°C. However, when compared to a recently calibrated standard thermometer the thermometer was found to be defective, reading 3°C low. This inaccurate thermometer would not be satisfactory for clinical use.

In general, it is desirable to have both good accuracy and good precision. However, sometimes an accurate measurement cannot be obtained even with a measuring technique that has good precision, and in such cases the precise but inaccurate measurement may be useful. For example, the defective thermometer discussed previously could be used to determine if a patient's temperature was stable, rising, or falling. Sometimes the accuracy is limited by uncontrollable factors; for example, it is difficult to accurately measure internal parts of the body, such as the amount of mineral in bones. One technique for measuring the bone mineral has a precision of about 1% but an accuracy of 3 or 4% (see Chapter 3, p. 59). The technique is nonetheless very useful for showing changes in the amount of bone mineral.

It is an accepted fact in science that the process of measurement may significantly alter the quantity being measured. This is especially true in medicine. For example, the process of measuring the blood pressure may introduce errors (uncertainties). Although the data are scarce, it is generally believed that when an attractive woman is performing the measurement, the blood pressure of a young man will increase. Similarly, a

*The normal body temperature is about 37°C or 98.6°F.

handsome man may affect the blood pressure measurement of a female patient. This type of error may also be introduced in taking a patient's history. The patient may not answer factually questions dealing with personal matters, such as sex habits. Using computers for patient interviews is thought to reduce this type of error (see Chapter 20, p. 555).

A clinical measurement in itself does not necessarily determine whether a patient is well or ill. For each medical test there is usually a well-established range of normal values. In addition, the values just above and below the normal range are usually considered equivocal; no decision can be made with certainty from these values. Also, the results of many clinical laboratory tests can be affected by outside factors such as medications. In Chapter 20 (p. 561) we discuss how computers are used to sort out such drug-test interactions.

After a physician has reviewed a patient's medical history, the findings of the physical examination, and the results of clinical laboratory measurements, he or she must decide if the patient is ill and if so, what the illness is. It is not surprising that sometimes wrong decisions are made. These wrong decisions are of two types: false positives and false negatives. A false positive error occurs when a patient is diagnosed to have a particular disease when he or she does not have it; a false negative error occurs when a patient is diagnosed to be free of a particular disease when he or she does have it. In some situations a diagnostic error can have a great impact on a patient's life. For example, a young woman was thought to have a rheumatic heart condition and spent several years in complete bed rest before it was discovered that a false positive diagnosis had been made—she really had arthritis, a disease in which activity should be maintained to avoid joint stiffening. In the early stages of many types of cancer it is easy to make a false negative diagnostic error because the tumor is small. Since the probability of cure depends on early detection of the cancer, a false negative diagnosis can greatly reduce the patient's chance of survival.

Diagnostic errors (false positives and false negatives) can be reduced by research into the causes of misleading laboratory test values and by development of new clinical tests and better instrumentation. Errors or uncertainties from measurements can be reduced by using care in taking the measurement, repeating measurements, using reliable instruments, and properly calibrating the instruments. To illustrate how measurement errors can be reduced, let us consider the problem of determining a person's weight accurately and precisely. (All characters in the following discussion are fictitious.)

We began by having Sam, a willing subject, stand on a bathroom scale while we carefully read the scale with a magnifying glass. Although ideally he should have been naked so that we would not later have to weigh his clothes and correct for their weight, we decided to let him remain clothed. We had him step off and on the scale several times to see if we got the same weight each time, and we found small variations due to his positions on the scale. The results are given in column 2 of Table 1.1. His average (mean) weight was 156.3 lb, with two-thirds of all the measurements between 156.0 and 156.6 lb. This result can be written as 156.3 ± 0.3 lb. The 0.3 lb is often referred to as the uncertainty or the *standard deviation* (SD) of the measurements. (If you want to learn more about standard deviations and related matters see Chapter 13 of the book by Stibitz listed in the bibliography.)

It occurred to us that the bathroom scale might not be very accurate as it was quite old. Someone suggested that we go to the local drugstore where there was a coin-operated scale. He felt that surely it would be accurate or the owners would not charge good money to use it! The financial investment needed for the study seemed exorbitant. Since this was unfunded research we decided to use several of the new scales at the gymnasium instead. The results obtained on two of these scales are given in columns 3 and 4 of Table 1.1. The means and uncertainties were 155.2 ± 0.2 and 155.5 ± 0.2. As we were about to comment on the error of the bathroom scale we recalled that Sam had eaten lunch and used the toilet between weighings. It seemed best not to draw any conclusions since we had not weighed either his food intake or his excreta.

As Sam stood on the scales at the gym we noticed that his weight jiggled no matter how carefully he tried to stand still. Then someone suggested that the jiggling might be due to the beating of Sam's heart. He pointed out

Table 1.1.   Results of Several Weighings on Different Scales

Weighing	Bathroom Scale	Gym Scale #1	Gym Scale #2	Hospital Scale
1	156.3	155.0	155.6	156.0
2	155.9	155.2	155.4	156.2
3	156.0	155.3	155.3	155.9
4	156.7	155.4	155.5	155.8
5	156.8	155.1	155.7	156.1
6	156.1	155.5	155.2	156.0
7	156.3	154.9	155.8	155.9
8	156.2	155.3	155.4	156.1
9	156.4	155.1	155.7	156.1
Mean	156.3	155.2	155.5	156.0
SD	0.3	0.2	0.2	0.1

that each time the heart beats it forces a mass of blood upward which forces the body downward (Newton's third law), resulting in an apparent increase in weight with each heartbeat. As the bulk of the rushing blood reaches the first bend in the aorta and heads toward the feet it has the opposite effect and causes an apparent decrease in weight. When we felt Sam's pulse we found it was synchronized with the jiggles.

It occurred to us that we could eliminate the jiggle if we had a scale that responded much more slowly and would thus average the variations caused by the heartbeat. Someone suggested that we stop at the local hospital and use the beam balances, which are operated by sliding a weight on a metal arm to counterbalance through a lever system the weight on the scale. Sure enough, we could no longer see the jiggle due to the heartbeat. It was noticed that one scale reader was doing much better than the others and getting the same value each time. However, instead of looking for a balance while tapping the slider, he repositioned the slider to the same point each time and then decided whether the scale was balanced, and his data were thrown out as they did not represent independent weighings. The results for one scale are given in column 5 of Table 1.1. The average weight was 156.0 ± 0.1 lb.

We asked one of the nurses how accurate the scales were and how often they were checked. He told us that accuracy was not a great concern since an error of 1 lb is of no clinical importance. He suggested that we talk to the hospital physicist who should know about such things, and we found him in the radiotherapy area having a cup of tea. (He seemed to have more interest in the technician he was chatting with than in our scientific inquiry.) After we discussed our problem, he explained that all measuring devices have some error. He said that if we wanted to pursue this matter further we should contact the state's Weights and Measures Laboratory since it has calibrated weights for checking meat scales, and so forth. We called this office. Although the people there were reluctant to participate in our research project as it fell outside their job area, they did tell us that they had special weights that were kept only to check the weights they used in their daily work. They said that these weights, called secondary standards, had been compared to the primary standards at NBS and that they had a certificate that they could show us to prove it.

We decided to drop the problem of calibrating the hospital scales as it appeared to us that the nurse was right—an error of 1 lb would not make much difference. However, all this research aroused our interest in other sources of error. Someone asked about the effect of breathing on body weight. Does a person weigh more when his lungs are full than when they are empty? We decided to try the experiment. Sam weighed the same when he held a deep breath and after a forced expiration, but we noticed that as he took a deep breath and as he let out his breath his weight appeared to jiggle. Someone remembered that air weighs about 1 g/liter and that the lungs hold about 5 liters. Since there are 454 g in 1 lb, the weight of the air in the lungs would be about 0.01 lb, much less than we could hope to see with our uncertainty of 0.1 lb. Then someone else remembered Archimedes' principle. Since the body is in a "sea" of air it is buoyed up by the weight of the air displaced by the body. Thus it does not make any difference whether the air is inside or outside the lungs. We decided to calculate the buoyant effect for a 154 lb (70 kg) person. The body is about as dense as fresh water (that is why it just floats in fresh water), and 1 kg of water has a volume of 1 liter. Thus a 70 kg person has a volume of about 70 liters. Since a liter of air has a mass of about 1 g, the buoyant effect on the body would be about 70 g or about 0.15 lb.

This exercise still did not explain why Sam's weight appeared to change when he breathed. We recalled that when we breathe in the diaphragm lowers, causing the center of gravity to lower slightly; when we breathe out, it rises slightly. The acceleration of the center of gravity downward at the beginning of an inspiration causes a momentary apparent reduction in weight; the acceleration of the center of gravity upward during expiration

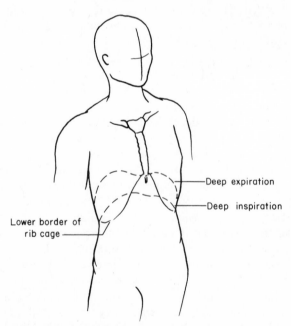

Figure 1.2. Levels of diaphragm during deep inspiration and deep expiration. Upon inspiration, the center of gravity is lowered slightly.

causes a similar increase (Fig. 1.2). We found that a larger effect was produced when Sam breathed in and out more rapidly.

We decided to adjourn to a nearby bar to discuss other errors in getting an accurate weight. Someone recalled that the moisture evaporating continuously from the body surface as well as from the lungs reduces the weight slightly with time. Someone else remembered reading in Chapter 7 of this book that for each breath the amount of carbon dioxide expired is slightly less than the amount of oxygen absorbed, although he could not remember what it amounted to. After another beer we dropped the problem of accurate weighing and someone brought up the medical physics problem of accurately measuring bladder pressure.

In summary, (1) all measurements are uncertain and inaccurate, (2) with special effort we can reduce the error and the uncertainty, (3) in many cases there is no need to improve the measurement because the quantity being measured is variable, and (4) drinking beer increases bladder pressure.

BIBLIOGRAPHY

Camac, C. N. B. (compiler), *Classics of Medicine and Surgery*, Dover, New York, 1959.
Clendening, L. (compiler), *Source Book of Medical History*, Hoeber, New York, 1942.
Stibitz, G. R., *Mathematics in Medicine and the Life Sciences*, Year Book, Chicago, 1966.
Thompson, D. W., *On Growth and Form*, Cambridge U.P., London, 1961.
Tustin, A., "Feedback," *Sci. Amer.*, **186–187**, 48–55 (1952).

REVIEW QUESTIONS

1. Give three examples of the use of the word *physical* in medicine.
2. Can a medical engineer always be called a clinical engineer?
3. What would be a more descriptive title for a health physicist?
4. Explain in what sense alcoholism is a disease that involves positive feedback.
5. List three physical measurements usually made during a physical examination.
6. Determine your age in minutes. Include an estimate of the error.
7. (a) Describe a method you might use for measuring the height of a person who is still growing. What are the sources of error?
   (b) Is measuring the height the best way of determining growth changes? Can you think of other methods?
8. Measure your pulse 10 times for 15 sec periods.
   (a) What is your average pulse rate per minute?

(b) Estimate the accuracy and the precision of this measurement. How can the accuracy be improved?

9. The period of a simple pendulum is given by $T = 2\pi\sqrt{l/g}$, where l is the length of the pendulum and g is the acceleration of gravity. Use the pulsologium idea of Sanctorius to determine the "length" of your pulse rate.

10. What is the ratio of your pulse rate to your breathing rate? (It is usually about four.)

11. The following systolic blood pressures (in millimeters of mercury) were recorded for one individual over a period of several days.

112	128	110	117	133
127	118	117	124	112
123	127	114	115	125
132	133	132	126	136
123	119	132	134	131

(a) Find the mean pressure $\bar{P} = \Sigma P_i/n$, where ΣP_i is the sum of all ($n = 25$) values of pressure.

(b) Find the standard deviation

$$\sigma = \sqrt{\frac{\Sigma(P_i - \bar{P})^2}{n - 1}}$$

(c) What are some general sources of error in these measurements?

CHAPTER 2

Forces on and in the Body

Force is such a common concept that unless we are physicists or engineers we just use our intuitive feeling about it. Force controls all motion in the world. We are often aware of forces on the body such as the force involved when we bump into objects. We are usually unaware of important forces in the body, for example, the muscular forces that cause the blood to circulate and the lungs to take in air. A more subtle example is the force that determines if a particular atom or molecule will stay at a given place in the body. For example, in the bones there are many crystals of bone mineral (calcium hydroxyapatite) that require calcium. A calcium atom will become part of the crystal if it gets close to a natural place for calcium where the electrical forces are great enough to trap it. It will stay in that place until local conditions have changed and the electrical forces can no longer hold it in place. This might happen if the bone crystal is destroyed by cancer. We do not attempt to consider all the various forces in the body in this chapter; it would be an impossible task.

Physicists like to consider the very fundamental origins of force. The first fundamental force described was *gravitational* force. Newton formulated the law of universal gravitation. This law states that there is a force of attraction between any two objects; our weight is due to the attraction between the earth and our bodies. The gravitational force is much smaller on the moon.

One of the important medical effects of gravitational force is the formation of varicose veins in the legs as the venous blood travels against the force of gravity on its way to the heart. We discuss varicose veins more in Chapter 8. Another medical effect of gravity is on the bones. Gravitational force on the skeleton in some way contributes to "healthy bones." If a

15

person becomes "weightless," such as in an orbiting satellite, he may lose some bone mineral, and this may be a serious problem on very long space journeys. Long-term bed rest removes much of the force of body weight from the bones and can lead to serious bone loss. In Chapter 3 we describe how bone mineral mass is measured.

The second fundamental force described by physicists was *electrical* force. This force is more complicated than gravity since it involves attractive and repulsive forces between static electrical charges as well as magnetic forces produced by moving electrical charges (electric currents). Electrical forces are immense compared to gravitational force. For example, the electrical force between an electron and a proton in a hydrogen atom is about 10^{39} times greater than the gravitational force between them.

Our bodies are basically electrical machines. The forces produced by the muscles are caused by electrical charges attracting or repelling other electrical charges. Control of the muscles is primarily electrical. Each of the billions of living cells in the body has an electrical potential difference across the cell membrane because of a difference in charge between the inside and outside of the cell (see Chapter 9). This amounts to less than 0.1 V, but because of the very thin cell wall it may produce a field as large as 10^5 V/cm. Electric eels and some other animals are able to add the potential from many cells to produce a stunning voltage of several hundred volts. This special "cell battery" occupies up to 80% of an eel's body length! Since the eel is essentially weightless in the water, it can afford this luxury. Land animals have not developed electrical weapons for defense or attack. In Chapter 9 we discuss the way we get information from the body by observing the electrical potentials generated by the various organs and tissues.

We are compelled (by our physics colleagues) to mention the two other known fundamental forces, which involve the nucleus of the atom. One of these, the strong nuclear force, is much larger than the other; it acts as the "glue" to hold the nucleus together against the repulsive forces produced by the protons on each other. The second, a weaker nuclear force, is involved with electron (beta) decay from the nucleus. Our current thinking is that the weak force may be related to the electrical force; thus, we may find we have only three fundamental forces. Although nuclear forces are indirectly involved with nuclear medicine, for the purposes of this book we do not need to consider them further.

There are two types of problems involving forces on the body: those where the body is in equilibrium (statics) and those where the body is accelerated (dynamics). Statics is considered in Section 2.1, and dynamics is discussed in Section 2.3. Friction (Section 2.2) is involved in both statics and dynamics. Forces and velocities are vector quantities; however, to simplify matters in this chapter we usually work in only one dimension. In the figures the vectors representing forces are not always drawn to scale.

2.1. STATICS

In this section we discuss forces involved with muscles, bones, and tendons.

When objects are stationary (static) they are in a state of equilibrium— the sum of the forces in any direction is equal to zero, and the sum of the torques about any axis also equals zero.

Many of the muscle and bone systems of the body act as levers. Levers are classified as first-, second-, and third-class systems (Fig. 2.1). Third-class levers are most common in the body, second-class levers are next in number, and first-class levers are least common.

A simple example of a lever system in the body is the case of the biceps muscle and the radius bone acting to support a weight *W* in the hand (Fig. 2.2*a*). Figure 2.2*b* shows the forces and dimensions of a typical arm. We

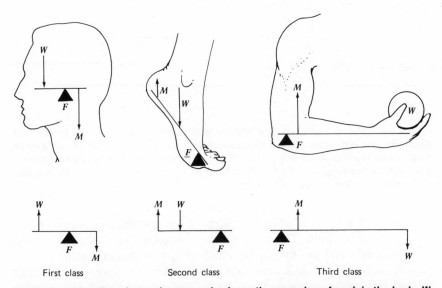

First class   Second class   Third class

Figure 2.1. The three lever classes and schematic examples of each in the body. *W* is a force that could be the weight, *F* is the force at the fulcrum point, and *M* is the muscle force.

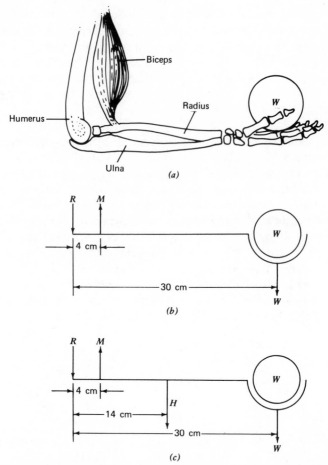

Figure 2.2. **The forearm. (*a*) The muscle and bone system. (*b*) The forces and dimensions: *R* is the reaction force of the humerus on the ulna, *M* is the muscle force supplied by the biceps, and *W* is the weight in the hand. (*c*) The forces and dimensions where the weight of the tissue and bones of the hand and arm *H* is included and located at their center of gravity.**

can find the force supplied by the biceps if we sum the torques about the pivot point at the joint. There are only two torques: that due to the weight *W*, which is equal to 30*W* acting clockwise, and that produced by the muscle force *M*, which is counterclockwise and of magnitude 4*M*. With the arm in equilibrium we find that $4M - 30W = 0$ and $M = 7.5W$ or that a muscle force 7.5 times the weight is needed. For a 100 N (~22 lb) weight the force needed is 750 N (~165 lb).

In our simplification of the problem we neglected the weight of the forearm and hand. This weight is not present at a particular point but is nonuniformly distributed over the whole forearm and hand. We can imagine this contribution as broken up into small segments and include the torque from each of the segments. A better method is to find the center of gravity for the weight of the forearm and hand and consider all the weight at that point. Figure 2.2c shows a more correct representation of the problem with the weight of the forearm and hand H included. A typical value of H is 15 N (~3.3 lb). By summing the torques about the joint we obtain $M = 3.5H + 7.5W$, which simply means that the force supplied by the biceps muscle must be larger than that indicated by our first calculation by an amount $3.5H = 52.5$ N (~11.8 lb).

Let us now consider the effect on the muscle force needed as the arm changes its angle as shown in Fig. 2.3a. Figure 2.3b shows the forces we

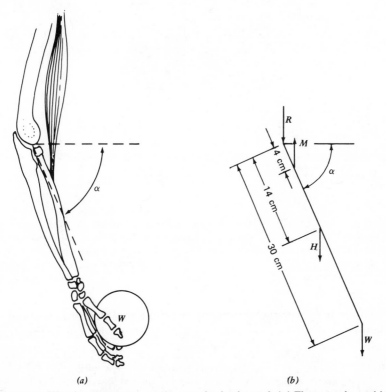

(a)   (b)

Figure 2.3. The forearm at an angle α to the horizontal. (a) The muscle and bone system. (b) The forces and dimensions.

must consider for an arbitrary angle α. If we take the torques about the joint we find that M remains constant as α changes! However, the length of the biceps muscle changes with the angle. In general, each muscle has a minimum length to which it can be contracted and a maximum length to which it can be stretched and still function. At these two extremes the force the muscle can exert is essentially zero. At some point in between, the muscle can produce its maximum force (Fig. 2.4). If the biceps pulls vertically the angle of the forearm does not affect the force required but it does affect the length of the biceps muscle, which affects the ability of the muscle to provide the needed force. Most of us become aware of the limitations of the biceps if we try to chin ourselves. With our arms fully extended we have difficulty, and as the chin approaches the bar the shortened muscle loses its ability to produce force.

The arm can be raised and held out horizontally from the shoulder by the deltoid muscle (Fig. 2.5a); we can show the forces schematically (Fig. 2.5b). By taking the sum of the torques about the shoulder joint, the tension T can be calculated from

$$T = \frac{2W_1 + 4W_2}{\sin \alpha} \tag{2.1}$$

If $\alpha = 16°$, W_1 (the weight of the arm) = 68 N ($\sim$15 lb), and W_2 (the weight in the hand) = 45 N ($\sim$10 lb), then $T = 1145$ N ($\sim$250 lb). The force needed to hold up the arm is surprisingly large.

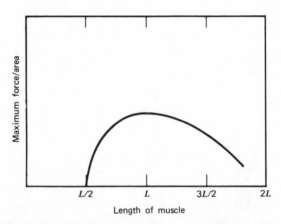

Figure 2.4. At its resting length L a muscle is close to its optimum length for producing force. At about half this length it cannot shorten further and the force it can produce drops to 0, whereas at a stretch of about 2L irreversible tearing of the muscle takes place. The maximum force of muscle at its optimum length is 3.1×10^7 N/m², or 4500 lb/in.²

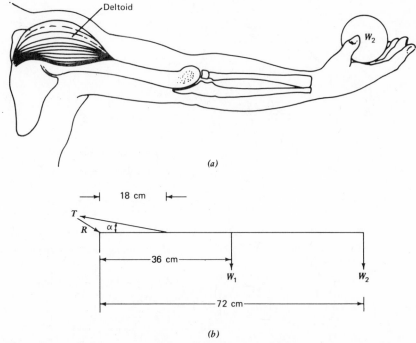

Figure 2.5.   Raising the arm. (a) The deltoid muscle and bone structure involved. (b) The forces on the arm. T **is the tension in the deltoid muscle fixed at the angle** α, R **is the reaction force on the shoulder joint,** W_1 **is the weight of the arm located at its center of gravity, and** W_2 **is the weight in the hand. (Adapted from L.A. Strait, V.T. Inman, and H.J. Ralston,** *Amer. J. Phys.***, 15, 1947, p. 379.)**

An often abused part of the body is the lumbar (lower back) region, shown schematically in Fig. 2.6*a*. The calculated force at the fifth lumbar vertebra (L5) with the body tipped forward at 60° to the vertical and with a weight of 225 N (~50 lb) in the hands can approach 3800 N (~850 lb) (Fig. 2.6*b*).

It is not surprising that lifting heavy objects from this incorrect position is thought to be a primary cause of low back pain. Since low back pain is rather serious and not too well understood, physicians are interested in finding out exactly how large the forces are in the lumbar region of the back. Measurements of the pressure in the discs separating the vertebrae have been made. A hollow needle connected to a calibrated pressure transducer is inserted into the gelatinous center of an intervertebral disc. It measures the pressure within the disc. The pressures in the third lumbar disc for an adult in different positions are shown in Fig. 2.7*a* and *b*. Even

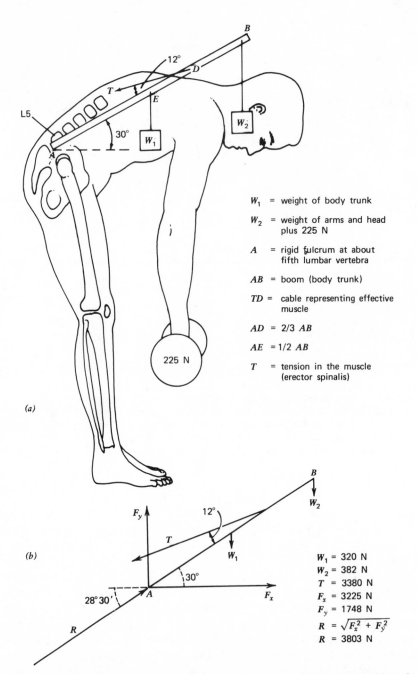

W_1 = weight of body trunk

W_2 = weight of arms and head plus 225 N

A = rigid fulcrum at about fifth lumbar vertebra

AB = boom (body trunk)

TD = cable representing effective muscle

AD = 2/3 AB

AE = 1/2 AB

T = tension in the muscle (erector spinalis)

W_1 = 320 N
W_2 = 382 N
T = 3380 N
F_x = 3225 N
F_y = 1748 N
$R = \sqrt{F_x^2 + F_y^2}$
R = 3803 N

Figure 2.6. Lifting a weight. (a) Schematic of forces used. (b) The forces. Note that the reaction force R at the fifth lumbar vertebra is quite substantial. (Adapted from L.A. Strait, V.T. Inman, and H.J. Ralston, *Amer. J. Phys.*, 15, 1947, pp. 377–378.)

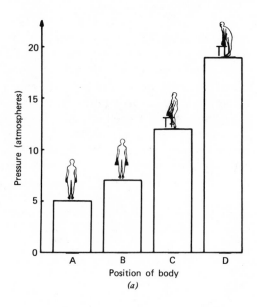

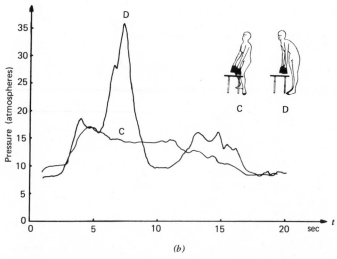

Figure 2.7. Pressure on the spinal column. (*a*) The pressure on the third lumbar disc for a subject (A) standing, (B) standing and holding 20 kg, (C) picking up 20 kg correctly by bending the knees, and (D) picking up 20 kg incorrectly without bending the knees. (*b*) The instantaneous pressure in the third lumbar disc while picking up and replacing 20 kg correctly and incorrectly. Note the much larger peak pressure during incorrect lifting. (Adapted from A. Nachemson and G. Elfstrom, *Scand. J. Rehab. Med.*, Suppl. 1, 1970, pp. 21–22.)

when standing erect there is a relatively large pressure in the disc due to the combined effects of weight and muscular tension. If the disc is over-loaded in a lifting accident it can rupture, causing pain either from the rupture or by allowing irritating materials from inside the disc to leak out.

It has been argued that the symptoms of low back pain are the price that man pays for being erect; however, veterinarians have shown that disc degeneration also occurs in four-legged animals. The symptoms for both animals and man occur at the regions with the greatest stress.

Sometimes vertebral bone collapse rather than disc damage occurs. This often happens in elderly women who suffer from weakened bones, or osteoporosis (see Chapter 3). Collapse of a vertebra can lead to a hunch-back stature.

Just as forces can be transmitted over distances and through angles by

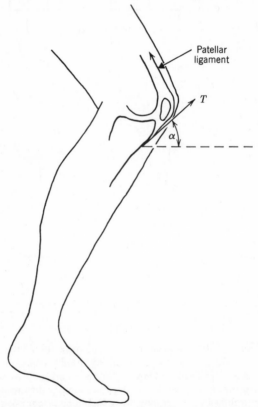

Figure 2.8. Diagram of the tensile force on the patellar ligament during squatting. The tension *T* may be quite large when a person is in a low squat.

cable and pulley systems, the forces of muscles in the body are transmitted by tendons. Tendons minimize the bulk at a joint. For example, the muscles that move the fingers to grip objects are located in the forearm, and long tendons are connected to appropriate places on the finger bones. Of course, the tendons have to remain in their proper locations to function properly. For example, in the leg, tendons pass over grooves in the knee cap (patella) and connect to the shin bone (tibia). With your leg extended you can move the patella with your hand but with your knee flexed you cannot—the patella is held rigidly in place. The patella also serves as a pulley for changing the direction of the force. One of the effects is to increase the mechanical advantage of the muscles that straighten the leg. Some of the greatest forces in the body occur at the patella. When a person is squatting, the tension in the tendons that pass over the patella may be more than two times his weight (Fig. 2.8).

2.2. FRICTIONAL FORCES

Friction and the energy loss due to friction appear everywhere in our everyday life. Friction limits the efficiency of most machines such as electrical generators and automobiles. On the other hand, we make use of friction in devices such as rubber tires and automobile brakes.

In the body, friction effects are often important. When a person is walking, as the heel of the foot touches the ground a force is transmitted from the foot to the ground (Fig. 2.9a). We can resolve this force into horizontal and vertical components. The vertical reaction force is supplied by the surface and is labeled N (a normal force). The horizontal reaction component must be supplied by frictional forces. The maximum force of friction f is usually described by

$$f = \mu N$$

where N is a normal force and μ is the coefficient of friction between the two surfaces. The value of μ depends upon the two materials in contact, and it is essentially independent of the surface area. Table 2.1 gives values of μ for a number of materials.

Measurements have been made of the horizontal force component of the heel as it strikes the ground when a person is walking (Fig. 2.9a), and it has been found to be approximately $0.15W$, where W is the person's weight. This is how large the frictional force must be in order to prevent the heel from slipping. If we let $N = W$ then we can apply a frictional force as large as $f = \mu W$. For a rubber heel on a dry concrete surface, the maximum frictional force can be as large as $f = W$, which is much larger

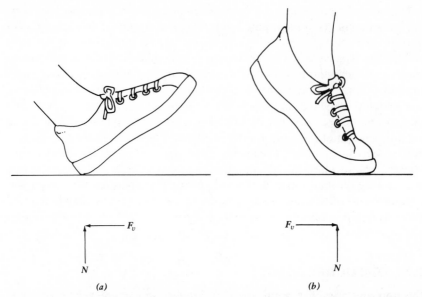

F_v

N

(a)

F_v

N

(b)

Figure 2.9.   Normal walking. (a) Both a horizontal frictional component of force F_v and a vertical (normal) component of force N exist on the heel as it strikes the ground. Friction between the heel and surface prevents the foot from slipping forward. (b) When the foot leaves the ground the frictional component of force F_v prevents the toe from slipping backward. (Adapted from Williams, M., and Lissner, H.R., *Biomechanics of Human Motion*, Philadelphia, W.B. Saunders Company, 1962, p. 122, by permission.)

than the needed horizontal force component ($0.15W$). In general, the frictional force is large enough both when the heel touches down and when the toe leaves the surface (Fig. 2.9b) to prevent a person from slipping. Occasionally, a person is on an icy, wet, or oily surface where μ is less than 0.15 and his foot slips. This is not only embarrassing; it may result in broken bones.

Table 2.1.   Example Values of Coefficient of Friction

Material	μ (Static Friction)
Steel on steel	0.15
Rubber tire on dry concrete road	1.00
Rubber tire on wet concrete road	0.7
Steel on ice	0.03
Between tendon and sheath	0.013
Lubricated bone joint	0.003

Friction must be overcome when joints move, but for normal joints it is very small. The coefficient of friction in bone joints is usually much lower than in engineering-type materials (Table 2.1). If a disease of the joint exists, the friction may become large. The synovial fluid in the joint is involved in the lubrication, but controversy still exists as to its exact behavior. Joint lubrication is considered further in Chapter 3.

The saliva we add when we chew food acts as a lubricant. If you swallow a piece of dry toast you become painfully aware of this lack of lubricant. Most of the large organs in the body are in more or less constant motion. Each time the heart beats it moves. The lungs move inside the chest with each breath, and the intestines have a slow rhythmic motion (peristalsis) as they move food toward its final destination. All of these organs are lubricated by a slippery mucus covering to minimize friction. Prior to coitus a woman secretes a "lubricant" in the vagina. The details are discussed in books on sex.

2.3. DYNAMICS

Let us now examine forces on the body where acceleration or deceleration is involved; for simplicity, we will usually consider cases in which the acceleration or deceleration is constant. If we limit ourselves to one-dimensional motion, then Newton's second law, force equals mass times acceleration, can be written without vector notation as

$$F = ma$$

This is not the way Newton originally wrote the law; he said force equals the change of momentum $\Delta(mv)$ over a short interval of time Δt or

$$F = \frac{\Delta(mv)}{\Delta t}$$

Examples 2.1 and 2.2 show how Newton's second law can be applied to some forces in the body.

Example 2.1
A 60 kg (~135 lb) person walking at 1 m/sec (~2 mph) bumps into a wall and stops in a distance of 2.5 cm in about 0.05 sec. What is the force developed on impact?

$$\Delta(mv) = (60 \text{ kg})(1 \text{ m/sec}) - (60 \text{ kg})(0 \text{ m/sec}) = 60 \text{ kg m/sec}$$

$$F = \frac{\Delta(mv)}{\Delta t} = \frac{60 \text{ kg m/sec}}{0.05 \text{ sec}} = 1200 \text{ kg m/sec}^2$$

or 1200 N (~270 lb, or ~2 times her weight)

Example 2.2

a. A person walking at 1 m/sec hits his head on a steel beam. Assume his head stops in 0.5 cm in about 0.01 sec. If the mass of his head is 4 kg, what is the force developed?

$$\Delta(mv) = (4 \text{ kg})(1 \text{ m/sec}) - (4 \text{ kg})(0 \text{ m/sec}) = 4 \text{ kg m/sec}$$

$$F = \frac{\Delta(mv)}{\Delta t} = \frac{4 \text{ kg m/sec}}{0.01} = 400 \text{ N } (\sim 90 \text{ lb})$$

b. If the steel beam has 2 cm of padding and Δt is increased to 0.04 sec, what is the force developed?

$$F = \frac{\Delta(mv)}{\Delta t} = \frac{4 \text{ kg m/sec}}{0.04 \text{ sec}} = 100 \text{ N } (\sim 22.5 \text{ lb})$$

An example of a dynamic force in the body is the apparent increase of weight when the heart beats (systole). About 60 g of blood is given a velocity of about 1 m/sec upward in about 0.1 sec. The upward momentum given to the mass of blood is (0.06 kg) (1 m/sec) or 0.06 kg m/sec; thus the downward reaction force (Newton's third law) produced on the rest of the body is (0.06 kg m/sec)/0.1 sec or 0.6 N ($\sim$2 oz). This is enough to produce a noticeable jiggle on a sensitive spring-type scale (see Chapter 1, p. 10).

If a person jumps from a height of 1 m and lands stiff-legged, he is in for a shock. Under these conditions, the deceleration of the body takes place mostly through compression of the padding of the feet. The body is traveling at 4.5 m/sec just prior to hitting, and if the padding collapses by 1 cm the body stops in about 0.005 sec. The force in the legs is almost 100 times the person's weight. If this person landed on a gym mat the deceleration time would be longer, and if he followed the normal body reaction he would land toes first and bend his knees to decelerate over a much longer time, thus decreasing the landing force.

Because of the large velocity of modern cars the riders have a larger momentum than when walking. In an accident the car often stops in a short time, producing very large forces. The results of these large forces on the passengers can be broken bones, internal injuries, and sometimes death.

Consider the case of "whiplash." A person sitting in an auto that is struck from behind will often suffer a whiplash injury of the neck (cervical region of the spine). When the car is struck, forces act through the seat forcing the trunk of the body ahead (Fig. 2.10a), while the inertia of the head causes it to stay in place, leading to severe stretching of the

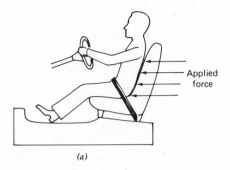

(a)

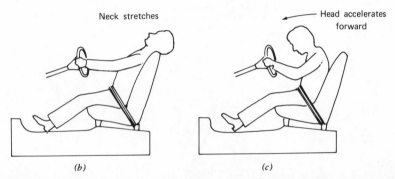

Neck stretches

Head accelerates forward

(b)   (c)

Figure 2.10. Whiplash injury. (a) The trunk of a person sitting in an automobile struck from behind is accelerated forward due to the forces acting through the seat. (b) The inertia of the head causes it to stay in place while the trunk of the body moves forward, leading to severe stretching in the neck region. (c) A moment later the head is accelerated forward.

neck (Fig. 2.10b). In milliseconds the head is forced to accelerate forward (Fig. 2.10c). Is it any wonder that severe damage to the neck results? The headrests currently installed in autos reduce the effects of this form of whiplash.

Although seat belts in automobiles have helped to reduce injuries from accidents, a person wearing a seat belt can still suffer acute head injury in an accident. Figures 2.11a and b show an auto traveling at 15 m/sec (~34 mph) that stops in 0.5 m due to a collision; the passenger's head and body are thrown against the dash and stopped (Fig. 2.11c). If the dash is padded, the decelerative effects may be minimized. If, however, the dash is not padded or if the head strikes a metal surface, forces far beyond that of human tolerance occur and severe head injuries or death can be

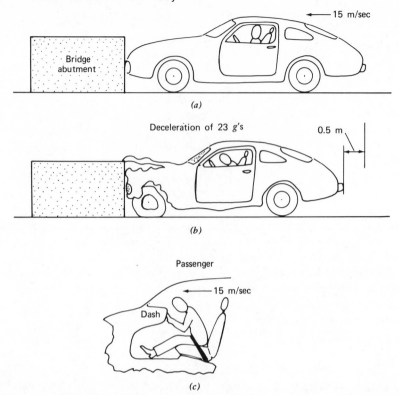

Figure 2.11. (*a*) An automobile traveling at 15 m/sec (~34 mph) is involved in a collision (*b*) and stops in 0.5 m. (*c*) A passenger wearing a seat belt is rotated forward and strikes the dash. 1 *g* equals the acceleration due to gravity.

expected. Air bags and shoulder belts effectively reduce the possibility of this type of injury.

Because of the hazards of uncontrolled auto collisions, safety devices are required by federal law. These include not only the headrests, seat belts, and shoulder belts that we have mentioned, but also energy-absorbing steering columns, penetration-resistant windshields, and side-door beams. The energy-absorbing steering column reduces the decelerative forces during a collision by increasing the time the trunk of the body takes to come to a stop. The old-style steering wheel sometimes impaled the driver. Most of the safety devices that have been installed have been evaluated in use and found to be effective in reducing body injury. These devices are designed for adults. If a child is in the car special safety devices are needed.

The behavior of the body under accelerative and decelerative forces has been an area of active interest for those concerned with space vehicles, aircraft, and automobiles. We are all aware of the effects of accelerations as they occur in the rides in an amusement park. The amount of acceleration the body can withstand depends on the orientation of the body and the time the acceleration (force) lasts. Figure 2.12 shows the tolerance to rearward acceleration of humans strapped in a seat as a function of the duration of the acceleration. The body can withstand large forces for short periods of time. Information is available for linear and rotary acceleration and deceleration for many different body positions.

Information such as this is used in connection with problems such as emergency escape from high-performance aircraft. If a pilot is to be shot upward through an escape hatch, we need to know the effects of acceleration in the seat-to-head direction. By knowing the limitations of the body, the accelerative force and its duration can be adjusted to minimize the probability of injury during emergency escape.

Accelerations can produce a number of effects such as (1) an apparent increase or decrease in body weight, (2) changes in internal hydrostatic pressure, (3) distortion of the elastic tissues of the body, and (4) the tendency of solids with different densities suspended in a liquid to separate. If the accelerations become sufficiently large the body loses control

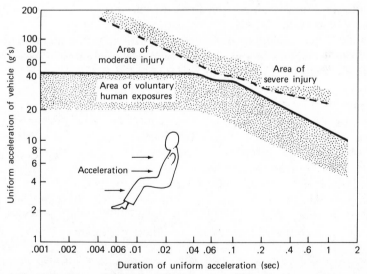

Figure 2.12. Tolerance of humans strapped in a seat due to rearward acceleration as a function of the time of the acceleration; 1 *g* equals the acceleration due to gravity. (Adapted from A.M. Eiband, NASA Memo 5-19-59E, 1959, Fig. 2.)

because it does not have adequate muscle force to work against the larger acceleration forces. Under certain conditions, the blood may pool in various regions of the body; the location of the pooling depends upon the direction of acceleration. If a person is accelerated head first the lack of blood flow to the brain can cause blackout and unconsciousness (see Chapter 8, p. 163).

Astronauts in an orbiting satellite are in a condition of "weightlessness." Prior to man's first space flights, many concerns were expressed about the physiological effects of weightlessness. Many of the effects predicted were based on the behavior of the body during extended periods of bed rest. We now have information about the effects on man of extended duration in space and have observed that some physiological changes do take place. However, none has proved to be incapacitating or permanent.

Tissue can be distorted by acceleration and, if the forces are sufficiently large, tearing or rupture can take place. Laboratory information is sparse, but some experiments in huge centrifuges have shown that tissue can be stretched by accelerative forces until it tears.

The tendency of suspended solids of different densities to separate when accelerated is not important in the body but it is utilized in the common laboratory centrifuge, which we discuss later in this chapter.

We have thus far concerned ourselves with linear acceleration and deceleration. If we subject the body to oscillatory motion, *resonance behavior* can occur. Each of our major organs has its own resonant frequency depending on its mass and the elastic forces that act on it. Pain or discomfort occurs in a particular organ if it is vibrated at its resonant frequency. Figure 2.13 shows the various sensations observed by humans subjected to vibrations of different frequencies.

We find that excessive vibration often occurs in motor trucks. It results in fatigue and discomfort and may cause visual disturbances. The vibratory frequency of motorized vehicles is usually 8 Hz or less. As might be expected, aircraft and space vehicles have higher vibratory frequencies. Sound pressure waves below 20 Hz (infrasound) can also cause fatigue and discomfort (see Chapter 12, p. 253).

We close this chapter by examining the centrifuge. The centrifuge is an artificial way to increase gravity. It is especially useful for separating a suspension in a liquid. It speeds up the sedimentation that occurs at a slow rate under the force of gravity.

When we drop pebbles into a pond they fall to the bottom at a constant speed much slower than their speed in air. Grains of sand fall through water at a still slower speed. Very fine particles (silt) appear to remain suspended. The speed at which small objects fall through a liquid depends

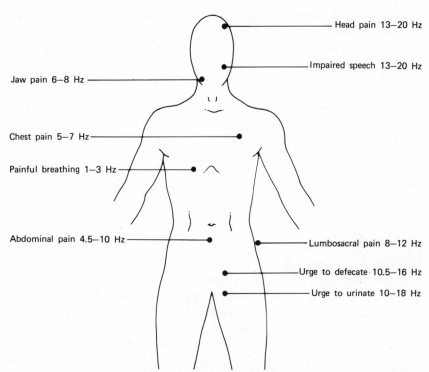

Head pain 13–20 Hz

Impaired speech 13–20 Hz

Jaw pain 6–8 Hz

Chest pain 5–7 Hz

Painful breathing 1–3 Hz

Abdominal pain 4.5–10 Hz

Lumbosacral pain 8–12 Hz

Urge to defecate 10.5–16 Hz

Urge to urinate 10–18 Hz

Figure 2.13. Symptoms of humans subjected to vibrations from 1 to 20 Hz. (Adapted from E.B. Magid, R.R. Coermann, and G.H. Ziegenruecker, "Human Tolerance to Whole Body Sinusoidal Vibration," *Aerospace Med.,* **31, 1960, p. 921.)**

on their size, the viscosity (consistency) of the fluid, and the acceleration due to gravity g. We can artificially increase g by spinning the fluid in a centrifuge. This will cause very fine particles to separate from the fluid.

Let us consider first sedimentation of small spherical objects of density ρ in a solution of density ρ_0 in a gravitational field g. We know that falling objects reach a maximum (terminal) velocity due to viscosity effects. Stokes has shown that for a spherical object of radius a, the retarding force F_d and terminal velocity v are related by

$$F_d = 6\pi a \eta v$$

where η is the viscosity of the liquid through which the sphere is passing. The SI unit for viscosity is the pascal second (Pas), which has the mks units of kg/m sec. The cgs unit for viscosity is the poise (1 Pas = 10 poises). The viscosity of water at 20°C is about 10^{-3} Pas (10^{-2} poise). When the particle is moving at a constant speed, the retarding force is in equilibrium with the difference between the downward gravitational force and the upward buoyant force (the weight of the liquid the particle displaces). Thus we have:

1. the force of gravity $F_g = \dfrac{4}{3}\,\pi a^3 \rho g$

2. the buoyant force $F_B = \dfrac{4}{3}\,\pi a^3 \rho_0 g$

3. the retarding force $F_d = 6\pi a \eta v$

F_g acts downward and F_B acts upward, and the difference is equal to F_d. From $F_g - F_B = F_d$ we obtain the expression for the terminal velocity (sedimentation velocity),

$$v = \frac{2a^2}{9\eta}\,g(\rho - \rho_0) \tag{2.2}$$

Equation 2.2 is valid only for spherical particles; however, we can use it as a guide to the behavior of particles with a more complicated shape.

In some forms of disease such as rheumatic fever, rheumatic heart disease, and gout, the red blood cells clump together and the effective radius increases; thus an increased sedimentation velocity occurs. In other diseases such as hemolytic jaundice and sickle cell anemia, the red blood cells change shape or break. The radius decreases; thus the rate of sedimentation of these cells is slower than normal. Determining the sedimentation rate of red blood cells is a simple and routine clinical laboratory test.

A related medical test that also depends on Eq. 2.2 is the determination of the hematocrit, the percent of red blood cells in the blood. Since the sedimentation velocity is proportional to the gravitational acceleration, it can be greatly enhanced if the acceleration is increased. We can increase g by means of a centrifuge, which provides an effective acceleration $g_{\text{eff}} = 4\pi^2 f^2 r$, where f is the rotation rate in revolutions per second and r is the position on the radius of the centrifuge where the solution is located.

Since the packing of the red blood cells takes place in the centrifuge, the hematocrit obviously depends upon the radius of the centrifuge and the speed and duration of centrifugation. The increase of any of these leads to more dense packing of the red blood cells or a smaller hematocrit. One standard method utilizes centrifugation for 30 min at 3000 rpm with $r = 22$ cm. A normal hematocrit is 40 to 60; a value lower than 40 indicates anemia, and a high value may indicate *polycythemia vera*.

The ultracentrifuge is a research tool used in the determination of the molecular weight of large macromolecules. Ultracentrifuges run at 40,000 to 100,000 rpm and have a g_{eff} of about 300,000 g. They have been very useful in protein research; however, they are not used in clinical medicine.

BIBLIOGRAPHY

Clynes, M., and J. H. Milsum, *Biomedical Engineering Systems,* McGraw-Hill, New York, 1970.
Frost, H. M., *Orthopaedic Biomechanics,* Thomas, Springfield, Ill., 1973.
Fung, Y. C., N. Perrone, and M. Anliker (Eds.), *Biomechanics: Its Foundations and Objectives,* Prentice-Hall, Englewood Cliffs, N.J., 1972.
Hardy, J. D., *Physiological Problems in Space Exploration,* Thomas, Springfield, Ill., 1964.
Kulowski, J., *Crash Injuries,* Thomas, Springfield, Ill., 1960.
Physiology in the Space Environment, Vol. 1, *Blood Circulation,* National Academy of Sciences, National Research Council, Washington, D.C., 1967–1968.
Physiology in the Space Environment, Vol. 3, *Respiration,* National Academy of Sciences, National Research Council, Washington, D.C., 1967–1968.
Tempest, W. (Ed.), *Infrasound and Low Frequency Vibration,* Academic, New York, 1976.
Williams, M., and H. R. Lissner, *Biomechanics of Human Motion,* Saunders, Philadelphia, 1962.
Yamada, H., *Strength of Biological Materials,* edited by F. G. Evans, Williams and Wilkins, Baltimore, 1970.

REVIEW QUESTIONS

1. What is the basic type of force that causes muscle contraction?
2. In the lever of the foot shown in Fig. 2.1, is M greater or smaller than the weight on the foot? (Hint: remember that the muscle that produces M is attached to the leg.)
3. Show that for Fig. 2.3, the muscle force is independent of the angle.
4. Derive Equation 2.1 for the arm and deltoid muscle system.
5. A muscle is capable of supplying a maximum force per unit area of 3.1×10^7 N/m² (see Fig. 2.4).
   (a) If the cross-sectional area of the muscle is 20 cm², what is the maximum force that can be supplied at the muscle's normal length?
   (b) Estimate the force that can be supplied by this muscle at 1.5 times its normal length.
6. The action of chewing involves a third-class lever system. Figure *A* shows the jaw and chewing (Masseter) muscle; Fig. *B* is the lever diagram. M is the force supplied by the chewing muscles that close the jaw about the fulcrum F. W is the force exerted by the front teeth.
   (a) If $l_2 = 3l_1$ and $W = 100$ N, find M.

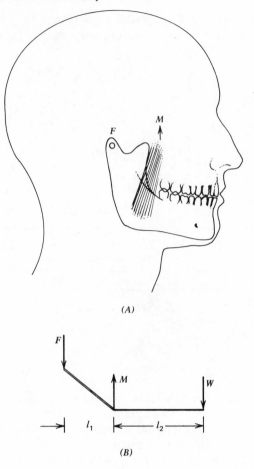

(A)

(B)

(b) If the front teeth have a surface area of 0.5 cm² in contact with an apple, find the force per unit area (N/m²) for part (a).

7. One first-class lever system involves the extensor muscle, which exerts a force M to hold the head erect: the force W of the weight of the head, located at its center of gravity (CG), lies forward of the force F exerted by the first cervical vertebra (see figure). The head has a mass of about 4 kg, or W is about 40 N.

(a) Find F and M.

(b) If the area of the first cervical vertebra, which the head rests on, is 5 cm², find the stress (force per unit area) on it.

(c) What is this stress for a 70 kg person standing on his head? How

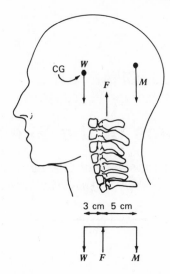

     does this stress compare with the maximum compression strength
     for bones ($\sim 1.7 \times 10^8$ N/m²)?

8. A 50 kg person jumping from a height of 1 m is traveling at 4.5 m/sec
just prior to landing. Suppose she lands on a pad and stops in 0.2 sec.
What maximum force will she experience?

9. Estimate the force on the forehead in Fig. 2.11 if the mass of the head
is 4 kg, its velocity is 15 m/sec, and the padded dash stops it in 0.002
sec.

10. Find the effective acceleration at a radius $r = 22$ cm for a centrifuge
rotating at 3000 rpm ($g = 9.8$ m/sec²).

CHAPTER 3

Physics of the Skeleton

With the assistance of Charles R. Wilson,
Medical College of Wisconsin, Milwaukee

Anthropologists have been interested in bones for a long time. Bone can last for centuries and in some cases for millions of years. Because of its strength, bone has been used by man for a wide variety of tools, weapons, and art objects. It provides the anthropologist with a means of tracing both the cultural and physical development of man.

Because of the importance of bone to the proper functioning of the body, a number of medical specialists are concerned with problems of bone. Two medical specialties, dentistry and orthopedic surgery, are completely devoted to this area. Other medical specialists who have considerable interest in bones are rheumatologists, M.D.'s who specialize in problems of rheumatism and arthritis, and radiologists, who base many diagnostic decisions on x-ray images of bony structures.

Bones also are of interest to medical physicists and engineers. Perhaps this organ system of the body appeals most to physical scientists because it has engineering type problems dealing with static and dynamic loading forces that occur during standing, walking, running, lifting, and so forth. Nature has solved these problems extremely well by varying the shapes of the various bones of the skeleton (Fig. 3.1) and the types of bony tissue of which they are made. In adapting bone for different functions nature has done a better "design job" than modern engineers are yet capable of doing. In fairness, it should be pointed out that nature has had millions of years to refine its design, while man has only recently attempted to duplicate the functions and properties of bone for bone replacement.

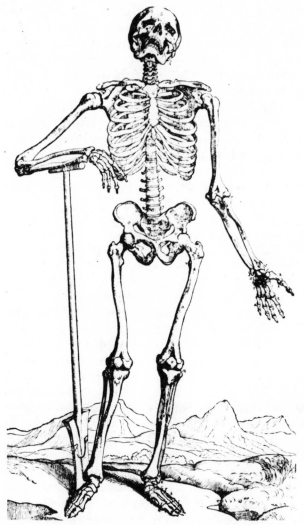

Figure 3.1. The skeleton of the body. (From A. Vesalius, *De Humani Corporis Fabrica*, Basle, 1543.)

Bone has at least six functions in the body: (1) support, (2) locomotion, (3) protection of various organs, (4) storage of chemicals, (5) nourishment, and (6) sound transmission (in the middle ear). In some animals bone is also involved in reproduction. The primates (except man, unfortunately), the walrus, and the raccoon all have penis bones.

The support function of bone is most obvious in the legs. The body's muscles are attached to the bones through tendons and ligaments and the system of bones plus muscles supports the body. In old age and in certain diseases some of this support structure deteriorates. If we lived in the sea where we would be "weightless" due to the buoyancy of the water, our need for a bony skeleton would be greatly reduced. Sharks do not have any bones; their skeleton is made of cartilage.

Bone joints permit movement of one bone with respect to another. These hinges, or *articulations,* are very important for walking as well as for many of the other motions of the body. We can manage with some loss of joint movement, but the destruction of joints by arthritis can seriously limit locomotion.

Protection of delicate body parts is an important function of some of the bones. The skull, which protects the brain and several of the most important sensory organs (eyes and ears), is an extremely strong container. The ribs form a protective cage for the heart and lungs. (The ribs and muscles of the chest also act as a bellows-like structure, which upon expansion and contraction allows the inhalation and expiration of air.) In addition to its support role, the spinal column acts much like an armored cable sheath to provide a flexible shield for the spinal cord.

The bones act as a chemical "bank" for storing elements for future use by the body. The body can withdraw these chemicals as needed. For example, a minimum level of calcium is needed in the blood; if the level falls too low, a "calcium sensor" causes the parathyroid glands to release more parathormone into the blood, and this in turn causes the bones to release the needed calcium.

The teeth are specialized bones that can cut food (incisors), tear it (canines), and grind it (molars) and thus serve in providing nourishment for the body. In man they come in two sets—deciduous (baby) teeth and permanent teeth (a third set is sometimes obtained from a dentist).

The smallest bones of the body are the ossicles in the middle ear. These three small bones act as levers and provide an impedance matching system for converting sound vibrations in air to sound vibrations in the fluid in the cochlea (see Chapter 13). They are the only bones that attain full adult size before birth!

It is sometimes thought that bone is a rather dead or inert part of the body and that once it has reached adult size it remains the same until death or some other calamity (such as a skiing accident) strikes. Actually, bone is a living tissue and has a blood supply as well as nerves. Most of the bone tissue is inert, but distributed through it are the *osteocytes,* cells that maintain the bone in a healthy condition. Cells make up about 2% of the volume of bone. If these cells die (e.g., due to a poor blood supply), the bone dies and it loses some of its strength. A serious hip problem is caused by a condition called *aseptic necrosis* in which the bone cells in the hip die due to lack of blood. The hip usually fails to function properly and sometimes has to be replaced with an artificial joint.

Since bone is a living tissue it undergoes change throughout life. A continuous process of destroying old bone and building new bone, called *bone remodeling,* is performed by specialized bone cells. *Osteoclasts* destroy the bone, and *osteoblasts* build it. Compared to many body processes, bone remodeling is slow work. We have the equivalent of a new skeleton about every seven years; each day the osteoclasts destroy bone containing about 0.5 g of calcium (the bones have about 1000 g of calcium), and the osteoblasts build new bone using about the same amount of calcium. While the body is young and growing the osteoblasts do more than the osteoclasts, but after the body is 35 to 40 years old the activity of the osteoclasts is greater than that of the osteoblasts, resulting in a gradual decrease in bone mass that continues until death.* This decrease is apparently faster in women than in men and leads to a serious problem of weak bones in older women. This condition, called *osteoporosis* (literally, porous bones), results in spontaneous fractures, especially in the spine and hips. In Section 3.4 we discuss how this disease can be diagnosed and studied through the use of a physical measurement.

3.1. WHAT IS BONE MADE OF?

The detailed chemical composition of bone is given in Table 3.1. Note the large percentage of calcium (Ca) in bone. Since calcium has a much heavier nucleus than most elements of the body, it absorbs x-rays much better than the surrounding soft tissue (see Chapter 16, p. 395). This is the reason x-rays show bones so well (Fig. 3.2).

Bone consists of two quite different materials plus water: *collagen,* the major organic fraction, which is about 40% of the weight of solid bone and 60% of its volume, and *bone mineral,* the so-called "inorganic" component of bone, which is about 60% of the weight of the bone and 40% of its volume. Either of these components may be removed from bone, and in each case the remainder, composed of only collagen or bone mineral, will look like the original bone. The collagen remainder is quite flexible, somewhat like a chunk of rubber, and can even be bent into a loop (Fig. 3.3). While it has a fair amount of tensile strength, it bends easily if it is

---

*As in other aspects of life, destruction is easier than construction. One osteoclast can destroy bone 100 times faster than one osteoblast can build new bone.

Table 3.1.   Composition of Compact Bone[a]

Element	Compact Bone, Femur (%)
H	3.4
C	15.5
N	4.0
O	44.0
Mg	0.2
P	10.2
S	0.3
Ca	22.2
Miscellaneous	0.2

[a]Adapted from H.Q. Woodard, *Health Physics,* **8,** 516 (1962), by permission of the Health Physics Society and the author.

compressed. When the collagen is removed from the bone, the bone mineral remainder is very fragile and can be crushed with the fingers! A simple way to remove the collagen is to put the bone in a furnace and "ash" it. Cremation is the ashing of the whole body; the bone mineral is the matter that is put in an urn.

Collagen is apparently produced by the osteoblastic cells; mineral is then formed on the collagen to produce bone. Bone collagen is not the

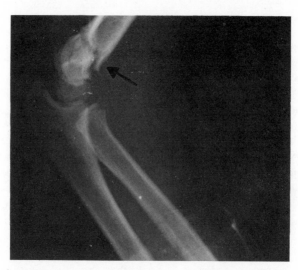

Figure 3.2.   An x-ray of the upper and lower arm showing a break just above the elbow (arrow). The lower part of the arm was broken earlier and is covered with a cast.

Figure 3.3. If the bone mineral in a bone is dissolved with a 5% acetic acid solution, the remaining collagen is quite flexible. Here Nancy Clark, typist of the manuscript copy of this book, easily bends an adult tibia (shin bone) that has been demineralized by this method.

same as the collagen found in many other parts of the body such as the skin. Its structure corresponds to the crucial dimensions of the crystals of bone mineral, and it forms a template onto which the bone mineral crystals fit snugly.

Bone mineral is believed to be made up of *calcium hydroxyapatite*—$Ca_{10}(PO_4)_6(OH)_2$. Similar crystals exist in nature; *fluorapatite*, a common rock, differs from calcium hydroxyapatite in that fluorine takes the place of the OH. Fluorine in drinking water may prevent *caries*, or cavities in the teeth, by turning microscopic areas of the teeth into the rock fluorapatite, which is more stable than bone mineral.

Studies using x-ray scattering have indicated that the bone mineral crystals are rod shaped with diameters of 20 to 70 Å and lengths of from 50 to 100 Å. (1 Å = 10^{-10} m. The angstrom is a convenient unit for measuring atomic dimensions since many atoms have diameters of about 1 Å.) Because of the small size of the crystals, bone mineral has a very large surface area. In a typical adult, it has a surface area of over 4×10^3 m^2 (~100 acres)—roughly the area of 12 city blocks! Around each crystal is a layer of water containing in solution many chemicals needed by the body. The large area of exposed bone mineral crystal permits the bones to interact rapidly with chemicals in the blood and other body fluids.

Within a few minutes after a small quantity of radioactive fluorine (^{18}F)

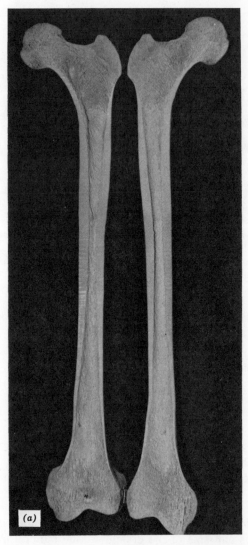

(a)

Figure 3.4. Cross-sections of (a) an adult femur (thigh bone), (b) a normal vertebra cut vertically, and (c) an osteoporotic vertebra (from an 80-year-old woman) cut vertically. Note the arrangement of compact and trabecular bone. (Vertebra figures courtesy of B.L. Riggs, M.D., Mayo Clinic, Rochester, Minn.)

(Figures 3.4b and 3.4c appear on page 45)

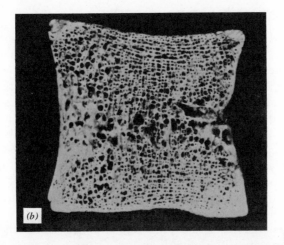

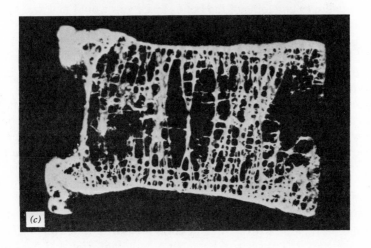

is injected into a patient, it will be distributed throughout the bones of his body. Bone tumors not yet visible on an x-ray can be identified by this method. Bone in a bone tumor is being destroyed somewhat like a brick house being torn down a brick at a time. When the radioactive fluorine atoms come in contact with this partially destroyed bone, they find many places they can fit in—more so than in normal bone. The increased radiation from the tumor area signals the possibility of a bone tumor (see Chapter 17 for more details).

3.2. HOW STRONG ARE YOUR BONES?

If a mechanical engineer were confronted with the problem of designing the skeleton, he would, of course, need to examine the functions of the different bones since their functions would determine their shape, their internal construction, and the type of material to be used. We have discussed a number of the more obvious functions of bones in the body. Now let us look at how bones have developed to meet our needs.

If you sort all of the approximately 200 bones of the body into various piles according to their shapes, you might come up with five piles: a small pile of flat, plate-like bones such as the shoulder blade (scapula) and some of the bones of the skull; a second pile of long hollow bones such as those found in the arms, legs, and fingers; a third pile of more or less cylindrical bones from the spine (vertebrae); a fourth pile of irregular bones such as from the wrist and ankle; and a fifth pile of bones such as the ribs that do not belong in any of the other piles.

If you were to cut some of the bones apart you would find that they are composed of one or a combination of two quite different types of bone: solid, or *compact,* bone and spongy, or cancellous, bone made up of thin thread-like trabeculae—*trabecular* bone. Figure 3.4*a* shows these two types of bone in an adult femur cut along its long axis. Trabecular bone is predominately found in the ends of the long bones, while most of the compact bone is in the central shaft. Figure 3.4*b* shows a cross-section of a normal vertebra; note that it is almost entirely composed of trabecular bone with the exception of thin plates of compact bone on the surfaces. Trabecular bone is considerably weaker than compact bone due to the reduced amount of bone in a given volume. Osteoporotic bone (Fig. 3.4*c*) is even weaker. On a microscopic level the bone tissue in a trabecula is the same as that in compact bone.

A study of the construction of the femur illustrates how well it is designed for its job. Stress (force per unit area) in a bone can be analyzed the same way as stress in a beam in a building. Figure 3.5*a* shows a horizontal beam supported at the ends with a downward force in the middle. The stresses inside the beam (shown by arrows) are pulling it apart at the bottom (tension) and pushing it together at the top (compression). There is relatively little stress of either type in the center of the beam. For this reason it is common to use an **I** beam, which has a thick top and bottom joined with a thin web, as a support beam in a building (Fig. 3.5*b*). When the force may come from any direction, a hollow cylinder is used to get the maximum strength with a minimum amount of material (Fig. 3.5*c*). It is almost as strong as a solid cylinder of the same diameter. Since the forces on the femur may come from any direction, the hollow cylinder structure of the bone is well suited for support.

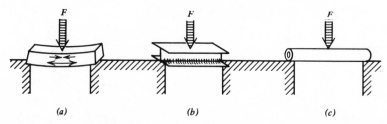

Figure 3.5. **Various types of beams subjected to a force *F*. (*a*) In a simple rectangular beam the greatest stresses are near the top and bottom. There is little stress in the middle of the beam. (*b*) Because the stress in the middle is small, a beam that has less material there—an I beam—can be used. (*c*) A tubular beam can be thought of as a rotated I beam with the center web removed. It is used when the force may come from any direction.**

If you push on one end of a hollow cylinder such as a soda straw, it will tend to buckle near the middle rather than at either end. Extra thickness at the middle would strengthen it. The compact bone of the shaft of the femur is thickest in the center and thinnest at the ends (Fig. 3.4*a*); note again the high quality of the design.

The trabecular patterns at the ends of the femur are also optimized for the forces to which the bone is subjected. Figure 3.6*a* shows schematically the lines of tension and compression in the head and neck of the femur due to the weight on the head. Figure 3.6*b* shows a cross-section of this part of the femur; notice that the trabeculae lie along the lines of force shown in Fig. 3.6*a*. Similarly, in the lower (distal) end of the femur the forces are nearly vertical, as are the trabeculae. There is some cross-banding to reinforce the trabeculae.

What are the advantages of trabecular bone over compact bone? There are at least two. Where a bone is subjected primarily to compressive forces, such as at the ends of the bones and in the spine, trabecular bone gives the strength necessary with less material than compact bone. Also, because the trabeculae are relatively flexible, trabecular bone can absorb more energy when large forces are involved such as in walking, running, and jumping. On the other hand, trabecular bone cannot withstand very well the bending stresses that occur mostly in the central portions of long bones.

Now let us consider some of the mechanical properties of bone, a composite material analogous to fiberglass. As described in Section 3.1, bone is composed of small hard bone mineral crystals attached to a soft flexible collagen matrix. These components have vastly different mechanical properties that also differ from those of bone. The exact nature of the interplay of these two components in producing the remarkable mechani-

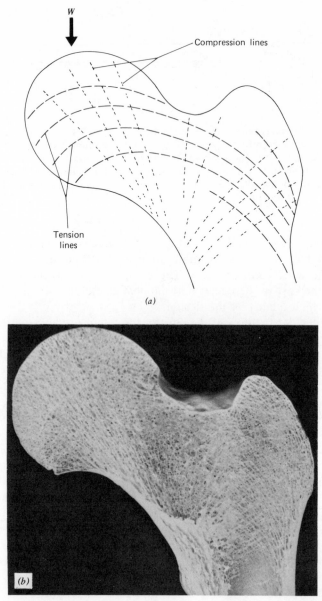

W

Compression lines

Tension
lines

(a)

(b)

Figure 3.6. The head and neck of the femur. (a) The lines of compression and tension due to the weight W of the body. (b) A cross-section showing the normal trabecular patterns. Note that they follow the compression and tension lines.

cal properties of bone is unknown. Nevertheless, the combination provides a material that is as strong as granite in compression and 25 times stronger than granite under tension.

We can make some standard physics and engineering measurements on a piece of compact bone, such as determining its density (or specific gravity); how much it lengthens or compresses under a given force (Young's modulus of elasticity); and how much force is needed to break it by compression, tension, and twisting. We may also determine how its strength depends on the time over which the force is applied and how much elastic energy is stored in it just before it breaks.

The density of compact bone is surprisingly constant throughout life at about 1.9 g/cm^3 (or 1.9 times as dense as water). In old age the bone becomes more porous and disappears from the inside, or *endosteal,* surface. The density of the remaining compact bone is still about 1.9 g/cm^3; it is reduced in strength because it is thinner, not because it is less dense. The physical quantity *bone density* is often confused with *bone mass.* An x-ray of a bone gives an idea of the mass of the bone, not of its density. The confusion is partly due to the use of *density* in connection with the optical density of an x-ray image. In Section 3.4 we discuss instrumentation for measuring bone mass and bone density in patients.

All materials change in length when placed under tension or compression. When a sample of fresh bone is placed in a special instrument for measuring the elongation under tension, a curve similar to that in Fig. 3.7 is obtained. The strain $\Delta L/L$ increases linearly at first, indicating that it is proportional to the stress (F/A)—Hooke's law. As the force increases the length increases more rapidly, and the bone breaks at a stress of about 120 N/mm^2 (~17,000 lb/in.^{2}). The ratio of stress to strain in the initial linear portion is Young's modulus Y. That is,

$$Y = \frac{LF}{A \Delta L} \tag{3.1}$$

Young's moduli for bone and a few common structural materials are given in Table 3.2. It is usually of more interest to calculate the change in length ΔL for a given force F. Equation 3.1 can be rewritten as

$$\Delta L = \frac{LF}{AY} \tag{3.2}$$

Equations 3.1 and 3.2 are valid for both tension and compression (see Example 3.1).

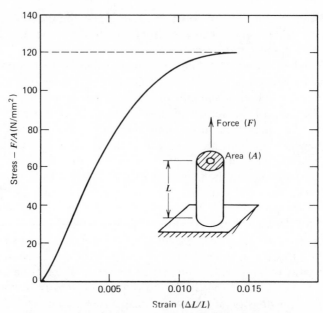

Figure 3.7. When a piece of bone is placed under increasing tension, its stress $\Delta L/L$ increases linearly at first (Hooke's law) and then more rapidly just before it breaks in two at about 120 N/mm² ($\sim$17,000 lb/in.²).

Table 3.2. Strengths of Bone and Other Common Materials

Material	Compressive Breaking Stress (N/mm²)	Tensile Breaking Stress (N/mm²)	Young's Modulus of Elasticity ($\times 10^2$ N/mm²)
Hard steel	552	827	2070
Rubber	—	2.1	0.010
Granite	145	4.8	517
Concrete	21	2.1	165
Oak	59	117	110
Porcelain	552	55	—
Compact bone	170	120	179
Trabecular bone	2.2	—	0.76

Example 3.1

Assume a leg has a 1.2 m shaft of bone with an average cross-sectional area of 3 cm² (3×10^{-4} m²). What is the amount of shortening when all of the body weight of 700 N is supported on this leg?

$$\Delta L = \frac{LF}{AY} = \frac{(1.2 \text{ m}) (7 \times 10^2 \text{ N})}{(3 \times 10^{-4} \text{ m}^2)(1.8 \times 10^{10} \text{ N/m}^2)}$$
$$= 1.5 \times 10^{-4} \text{ m} = 0.15 \text{ mm}$$

The ability of the bones to support the body's weight without breaking is crucial to man's well-being. Of course, they support not only weight but also other forces. In bending over to pick up a heavy object we may develop large forces in the lower spine (see Chapter 2, p. 22). This helps explain why crushed vertebrae of the lower (lumbar) spine are common (Fig. 3.8). Large forces are also produced in such activities as running and jumping. In running, the force on the hip bone when the heel strikes the ground may be four times the body's weight. Even in normal walking the forces on the hip are about twice the body's weight.

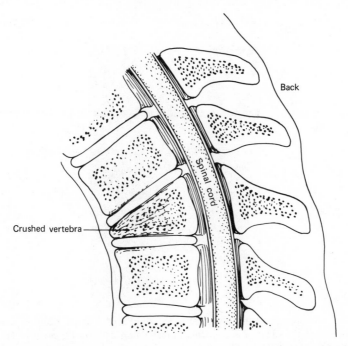

Figure 3.8.   Diagram of a crushed lumbar vertebra. The resulting curvature of the spine produces a hunchback appearance.

What is the built-in safety factor in the bones that support the body's weight? Engineers like to overdesign a support structure so that it can withstand forces of about 10 times the maximum expected forces. Does the femur meet this requirement? Healthy compact bone is able to withstand a compressive stress of about 170 N/mm² (~25,000 lb/in.²) before it fractures (Table 3.2). The midshaft of the femur has a cross-sectional area of about 3.3 cm² (0.5 in.²); it would support a force of about 5.7 × 10⁴ N (12,000 lb, or 6 tons)! The cross-sectional area of the shin bone (tibia) is not as great, but the safety margin is satisfactory for most activities except downhill skiing.

The bones do not normally break due to compression; they usually break due to shear (Fig. 3.9*a* and *b*) or under tension (Fig. 3.9*c*). A common cause of shear is catching the foot and then twisting the leg while falling. A shear fracture often results in a spiral break (Fig. 3.9*b*) in which the bone is apt to puncture the skin. This type of fracture (compound) is more apt to become infected than a fracture in which the bone is not exposed (simple).

The bones are not as strong under tension as they are in compression; a tension stress of about 120 N/mm² (~17,000 lb/in.²) will cause a bone to

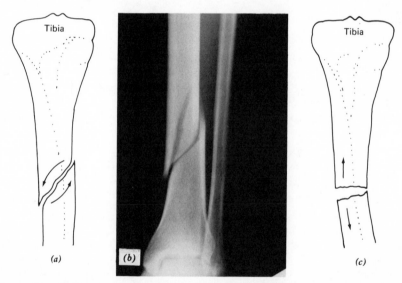

Figure 3.9. **Fractures of the tibia. (*a*) A schematic of a spiral fracture caused by shear (twisting), (*b*) an x-ray of a spiral fracture caused by shear, and (*c*) a schematic of a tension fracture in the tibia.**

break (Fig. 3.7). However, bone is stronger under tension than many common materials (Table 3.2).

Let us consider the forces exerted on a bone during a fall. From Newton's second law, the force exerted during a collision or a fall is equal to the rate of change of momentum, which is simply the momentum of the body divided by the duration of impact. Therefore, the shorter the duration of impact, the greater the force. To reduce the force and thereby reduce the likelihood of fracture, it is necessary to increase the impact time. In both falling down and jumping from an elevation, the impact time can be increased significantly by simply rolling with the fall or jump, thereby spreading the change in momentum of the body over a longer time. A good example of rolling with the impact is the manner in which a parachutist is trained to land; his ankles and knees bend upon impact and his body turns to one side so that he falls on his leg, then on his hip, and then on the side of his chest. If he tried a stiff-legged landing, the force generated would be about 1.42×10^5 N (32,000 lb), which means that each tibia, which is about 3.3 cm^2 in area at the ankle, would bear a stress of about 215 N/mm^2 (31,000 lb/in.^{2}). This value exceeds the maximum compression strength of bone by about 30%. Bone, however, can withstand a large force for a short period without breaking, while the same force over a long period will fracture it. That is, the short-term force developed when you fall or jump, while possibly exceeding the maximum compressive strength of bone, is not as dangerous as the same force applied over a longer period of time. This property is called *viscoelasticity*.

When a bone is fractured, the body can repair it rapidly if the fracture region is immobilized. Even in an elderly woman with osteoporosis the healing process is effective. However, the long period of bed confinement necessary for a fractured hip to heal is very debilitating, and it is important to get the patient on her feet as soon as possible. Metal prosthetic hip joints, pins, nails, and so forth, are often used to repair such damaged bones (Fig. 3.10).

While the details of the growth and repair of bone are not well understood, there is good evidence that local electrical fields may play a role. When bone is bent it generates an electrical charge on its surfaces. It has been suggested that this phenomenon (piezoelectricity) may be the physical stimulus for bone growth and repair. Experiments with animal bone fractures have shown that bone heals faster if an electrical potential is applied across the break. It is too early to tell if this technique can be used successfully on man.

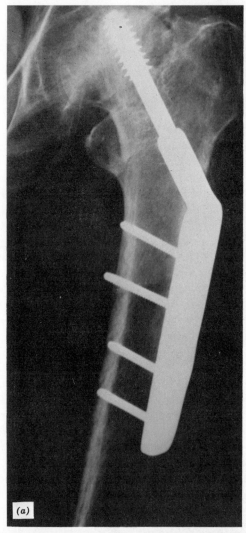

Figure 3.10. Hip prostheses. (a) A weak hip joint can be reinforced by a metal support fastened to the femur. In this instance the femoral neck has been made shorter to reduce the stress. (b) The entire hip joint can be replaced with man-made material. (c) An x-ray of a double hip replacement using prostheses similar to the prosthesis shown in b.

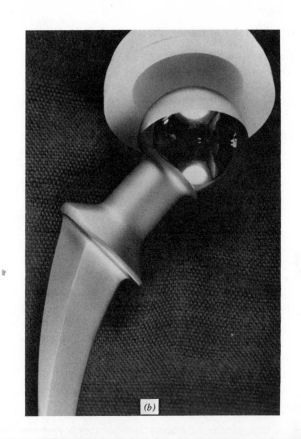

(b)

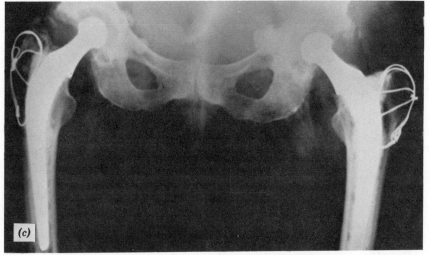

(c)

3.3. LUBRICATION OF BONE JOINTS

Those of us who do not suffer from arthritis take our well-functioning bone joints for granted. Many people are not as fortunate. An analysis of 1000 autopsy reports revealed that over two-thirds of the cadavers had a joint problem in the knee and that about one-third had a similar problem in the hip. There are two major diseases that affect the joints—rheumatoid arthritis, which results in overproduction of synovial fluid in the joint and commonly causes swollen joints, and osteoarthrosis, a disease of the joint itself.

The lubrication of bone joints is not understood in detail, but the essential features are agreed upon. The main components of a joint are shown in Fig. 3.11. The synovial membrane encases the joint and retains the lubricating synovial fluid. The surfaces of the joint are articular cartilage, a smooth, somewhat rubbery material that is attached to the solid bone. A disease that involves the synovial fluid, such as rheumatoid arthritis, quickly affects the joint itself.

The surface of the articular cartilage is not as smooth as that of a good man-made bearing. It has been suggested that its roughness plays a useful role in joint lubrication by trapping some of the synovial fluid. It has also been suggested that because of the porous nature of the articular cartilage, other lubricating material is squeezed into the joint when it is under its greatest stress—when it needs lubrication the most. One theory is that pressure causes lubricating "threads" to squeeze out of the cartilage into the joint; one end of each lubricating thread remains in the cartilage, and

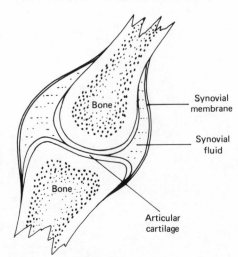

Figure 3.11. The main components of a joint.

as the pressure is reduced the threads pull back into their holes, somewhat like nightcrawler worms when you try to catch them. This *boosted lubrication* is a technique engineers have not yet been able to adapt to industry.

The lubricating properties of a fluid depend on its viscosity; thin oil is less viscous and a better lubricant than thick oil. The viscosity of synovial fluid decreases under the large shear stresses found in the joint. The good lubricating properties of synovial fluid are thought to be due to the presence of hyaluronic acid and mucopolysaccharides (molecular weight, ~500,000) that deform under load.

The coefficient of friction of bone joints is difficult to measure under the usual laboratory conditions. Little, Freeman, and Swanson described the arrangement shown in Fig. 3.12 in the book *Lubrication and Wear in Joints,* edited by Wright (see the bibliography at the end of this chapter). A normal hip joint from a fresh cadaver was mounted upside down with heavy weights pressing the head of the femur into its socket. The weight on the joint could be varied to study the effects of different loads. The whole unit acted like a pendulum with the joint serving as the pivot. From the rate of decrease of the amplitude with time, the coefficient of friction was calculated. The coefficient of friction was found to be independent of the load from 89 to 890 N (20 to 200 lb) and independent of the magnitude of the oscillations. It was concluded that fat in the cartilage helps to reduce the coefficient of friction. For all healthy joints studied, the

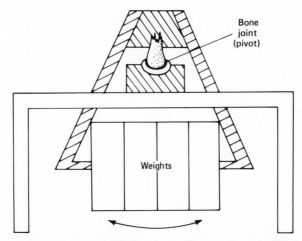

Figure 3.12. Arrangement for determining the coefficient of friction of a joint. The joint is used as the pivot in a pendulum and the decrease in amplitude of the oscillations with time is measured.

coefficient of friction was found to be less than 0.01, much less than that of a steel blade on ice—0.03. (A coefficient of friction of 0.01 means that if there is a 100 lb force on a joint, only 1 lb of force is needed to move it.) When the synovial fluid was removed, the coefficient of friction increased considerably.

3.4.   MEASUREMENT OF BONE MINERAL IN THE BODY

Bone is one of the most difficult organs to study. With the exception of the teeth, the bones are relatively inaccessible. In this section we describe several physical systems for studying the bones *in vivo* (in the living body). There are many other physical techniques for studying bone, but most are used on excised bone samples (*in vitro* studies).

Bone disease is one of the most common problems of the elderly. For example, each year about 150,000 women in the United States break a hip. Most of these women are elderly and have osteoporosis. A few years ago, osteoporosis was difficult to detect until a patient appeared with a broken hip or a crushed vertebra. At that time it was too late to use preventive therapy.

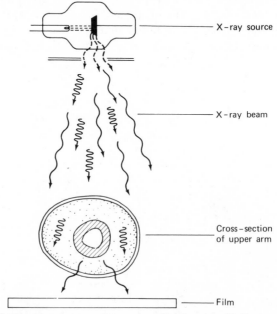

Figure 3.13. Conventional x-rays are not useful for quantitative measurement of bone mineral because the beam is heterogeneous, the scatter in the image is unknown, and film is not a reproducible detector.

The strength of bone depends to a large extent on the mass of bone mineral present, and the most striking feature in osteoporosis is the lower than normal bone mineral mass. Thus a simple technique to measure bone mineral mass *in vivo* with good accuracy and precision (reproducibility) was sought. It was hoped that such a technique could be used to diagnose osteoporosis before a fracture occurred and also to evaluate various types of therapy for osteoporosis. Since bone mineral mass decreases very slowly, 1 to 2% per year, a very precise technique was needed to show changes.

The idea of using an x-ray image to measure the amount of bone mineral present is an old one; it was first tried in 1901! The major problems of using an ordinary x-ray (Fig. 3.13) are (1) the usual x-ray beam has many different energies, and the absorption of x-rays by calcium varies rapidly with energy in this range of energies (see Chapter 16); (2) the relatively large beam contains much scattered radiation when it reaches the film;

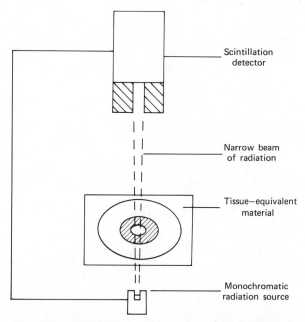

Scintillation detector

Narrow beam of radiation

Tissue—equivalent material

Monochromatic radiation source

Figure 3.14. The basic components used in photon absorptiometry. A radioisotope that emits essentially only one energy, such as iodine 125 (27 keV) or americium 241 (60 keV), serves as the radiation source; the limb is embedded in a uniform thickness of tissue-equivalent material; and the transmitted fraction of the narrow beam is detected by a scintillation detector.

and (3) the film is a poor detector for making quantitative measurements since it is nonlinear with respect to both the amount and the energy of the x-rays. Developing the film can introduce additional variations.

The net result of these problems is that a large change in bone mineral mass (30 to 50%) must occur between the taking of two x-rays of the same patient before a radiologist can be sure that there has been a change. Each of the problems can be reduced by special methods, but the determination of bone mineral mass by this technique (x-ray film densitometry) has been limited to only a few laboratories in the world.

An improved technique based on the same physical principles was developed by one of the authors (JRC) starting in about 1960. The basic components used in this technique, called *photon absorptiometry,* are shown in Fig. 3.14. The three problems with the x-ray technique were largely eliminated by using (1) a monoenergetic x-ray or gamma ray source, (2) a narrow beam to minimize scatter, and (3) a scintillation detector that detects all photons and permits them to be sorted and counted individually. The determination of the bone mineral mass is

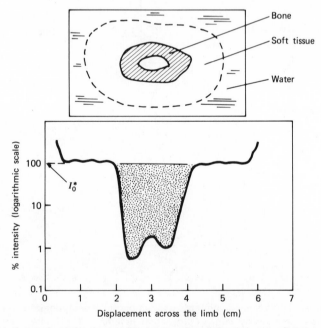

Figure 3.15. A graph of the transmitted intensity of the beam as it traverses the bone during photon absorptiometry. The intensity is plotted on a logarithmic scale. The shaded area is proportional to the bone mineral mass per unit length.

further simplified by immersing the bone to be measured in a uniform thickness of soft tissue (or its x-ray equivalent, e.g., water). Figure 3.15 shows a graph of the logarithm of the transmitted intensity of the beam (log I) as it scans across a bone immersed in a uniform thickness of "tissue." The intensity before the beam enters the bone is called I_o^*. The bone mineral mass (BM) at any point in the beam is proportional to log (I_o^*/I) and is given by BM (g/cm^2) $= k \log (I_o^*/I)$, where k is a constant that can be determined experimentally. This calculation is done electronically for all points in the beam, and the results are integrated to give the bone mineral mass of the slice of bone in grams per centimeter. A modern clinical bone scanner that uses the photon absorption technique is shown in Fig. 3.16. The unit has a reproducibility of 1 to 2% when used by a trained operator.

Another physical technique for measuring bone mineral *in vivo* takes advantage of the fact that nearly all of the calcium in the body is in the bones. This technique is called *in vivo* activation. The whole body is irradiated with energetic neutrons that convert a small amount of the calcium and some other elements into radioactive forms that give off energetic gamma rays, and the emitted gamma rays are then detected and counted. The gamma rays from radioactive calcium can be identified by their unique energy (Fig. 3.17), and the number of them indicates the amount of calcium in the body. The amount of bone mineral is then obtained by multiplying by a constant. Both the source of neutrons and

Figure 3.16.   A modern clinical bone scanner manufactured by Norland Instrument Co., Fort Atkinson, Wis. The arm is held in a rubber bag containing water. The bone mineral mass and bone width appear in digital form on the unit on the left.

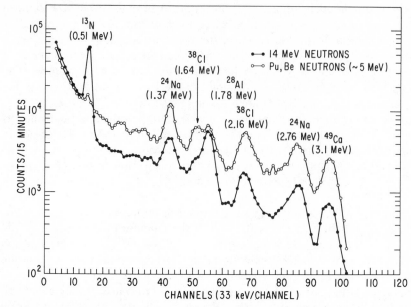

Figure 3.17. A graph of the gamma ray intensities from the body as a function of energy (channels) after whole body irradiation with 14 MeV and about 5 MeV neutrons. The radioactive elements causing the main gamma ray peaks and their energies are given. Note the peak at the right caused by radioactive calcium (^{49}Ca). The area under this peak indicates the amount of calcium (and thus the amount of bone mineral) in the body. (Reprinted from S.H. Cohn, K.K. Shukla, C.S. Dombrowski, and R.G. Fairchild, *Journal of Nuclear Medicine*, Vol. 13, No. 7, with permission of the publisher.)

the special whole-body detecting equipment are expensive and not practical for routine clinical measurements. In addition, because the relatively large radiation exposure is a hazard, it is undesirable to use the technique on healthy subjects to obtain normal data.

BIBLIOGRAPHY

Frost, H. M., *The Laws of Bone Structure,* Thomas, Springfield, Ill., 1964.

Fung, Y. C., N. Perrone, and M. Anliker (Eds.), *Biomechanics: Its Foundations and Objectives,* Symposium on the Foundations and Objectives of Biomechanics, La Jolla, Calif., 1970, Prentice-Hall, Englewood Cliffs, N.J., 1972.

Hall, M. C., *The Architecture of Bone,* Thomas, Springfield, Ill., 1966.

Klopsted, P. E., and P. D. Wilson, *Human Limbs and Their Substitutes,* McGraw-Hill, New York, 1954.

Kraus, H., "On the Mechanical Properties and Behavior of Human Compact Bone" in S. N. Levine (Ed.), *Advances in Biomedical Engineering and Medical Physics,* Vol. 2, Wiley-Interscience, New York, 1968, pp. 169–204.

Wright, V. (Ed.), *Lubrication and Wear in Joints,* Proceedings of a Symposium organized by the Biological Engineering Society and held at The General Infirmary, Leeds, on April 17, 1969, Lippincott, Philadelphia, 1969.

Yamada, H., *Strength of Biological Materials,* edited by F. H. Evans, Williams and Wilkins, Baltimore, 1970.

REVIEW QUESTIONS

1. List six functions of bone in the body.
2. What percentage of normal bone is living cells?
3. What is bone remodeling?
4. At what age does osteoclastic activity begin to exceed osteoblastic activity?
5. What are the major components of bone?
6. What percentage of compact bone is calcium?
7. Bone mineral is believed to be made of what crystalline material?
8. What is the approximate surface area of the bone mineral crystals in the body?
9. What are two advantages of trabecular bone over compact bone?
10. How does bone compare to granite in strength?
11. What is the density of compact bone?
12. Using the information in Fig. 3.7,
    (a) Calculate the maximum tension a bone with a cross-sectional area of 4 cm² could withstand just prior to fracture.
    (b) Determine how much a bone 35 cm long would elongate under this maximum tension.
    (c) Calculate the stress on this bone if a compressive force of 10^4 N were applied to it. How much would this bone shorten?
13. What is the function of synovial fluid?
14. What is the approximate coefficient of friction of a healthy bone joint?
15. Give three problems involved with using an x-ray image to measure bone mineral mass *in vivo*.

CHAPTER 4

Heat and Cold in Medicine

Heat and cold have been used for medical purposes for several thousand years. Galen, an ancient physician, recommended the use of warm water and oil in some treatments, and the application of cold substances on injuries was urged by another early physician, Hippocrates. Throughout the ages, controversy on the therapeutic value of heat and cold has existed. Even today there is much to be learned about these two treatment methods. Much of the progress in this area of medical physics as in many others has been due to the cooperation and collaboration of basic scientists and physicians.

In this chapter we consider elevated and reduced temperatures in medicine. We begin by briefly discussing the physical basis of heat and cold and ways to measure temperature (thermometry). We then discuss mapping the surface temperature of the body (thermography) as a diagnostic tool, heat therapy, and the uses of cold in medicine (cryogenics and cryosurgery).

No medical specialty deals primarily with applications of heat and cold in medicine. Specialists in physical medicine and physical therapy probably use heat and cold the most. Other medical specialists, including family practice physicians, often prescribe heat or cold for therapeutic purposes. Surgeons sometimes use extreme cold (cryosurgery), and radiologists are often involved in interpreting thermographic images.

4.1. PHYSICAL BASIS OF HEAT AND TEMPERATURE

We usually think of temperature from a personal point of view; we know whether we are too hot, too cold, or comfortable. If we want to describe temperature as a physical phenomenon, however, we should try to understand it on a molecular scale. Matter is composed of molecules that are in motion. In a gas or liquid the molecules move about, hitting one another or the walls of the container; even in a solid the molecules have some motion about the sites that they occupy within the crystal structure. The fact that the molecules move means that they have kinetic energy, and this kinetic energy is related to the temperature.

The average kinetic energy of the molecules of an ideal gas can be shown to be directly proportional to the temperature; liquids and solids show a similar temperature dependence. In order to increase the temperature of a gas it is necessary to increase the average kinetic energy of its molecules. This can be done by putting the gas in contact with a flame. The energy transferred from the flame to the gas causing the temperature rise is called *heat*.

If enough heat is added to a solid, it melts, forming a liquid. The liquid may be changed to a gas by adding more heat. Adding still more heat converts the gas to ions.

While adding heat to a substance increases its molecular kinetic energy, thereby increasing its temperature, the reverse is also true; heat can be removed from a substance to lower the temperature. Low temperatures are referred to as the *cryogenic region* (from the ancient Greek word *kryos*, which means icy cold). The ultimate in cold is "absolute zero" (−273.15°C), a temperature that is experimentally unattainable.

4.2. THERMOMETRY AND TEMPERATURE SCALES

Temperature is difficult to measure directly, so we usually measure it indirectly by measuring one of many physical properties that change with temperature. We then relate the physical property to temperature by a suitable calibration.

In the United States the most common temperature scale is the fahrenheit (°F) scale. Water freezes at 32°F and boils at 212°F, and the normal body temperature (rectal) is about 98.6°F. Fahrenheit devised this scale in 1724 so that 100°F would represent the normal body temperature and 0°F would represent the coldest temperature man could then produce (by mixing ice and salt). The United States is the only major country still using this temperature scale. Most scientists in the United States use the celsius (°C) scale (formerly called the centigrade scale), which is in common use throughout most of the world. Water freezes at 0°C and boils at 100°C, and the normal body temperature (rectal) is about 37°C.

Another important temperature scale used for scientific work is the kelvin (°K), or absolute, scale, which has the same degree intervals as the celsius scale; 0°K (absolute zero) is −273.15°C. On the absolute scale water freezes at 273.15°K and boils at 373.15°K, and the normal body temperature (rectal) is about 310°K (Fig. 4.1). This temperature scale is not used in medicine.

You can estimate someone else's body temperature by placing your hand on his forehead and using the temperature sensors in your hand to compare his body temperature with yours. This technique is satisfactory for qualitative purposes, but it is not very scientific. The most common way to measure temperature is with a glass fever thermometer containing mercury or alcohol. The principle behind this thermometer is that an increase in the temperature of different materials usually causes them to expand different amounts. In a fever thermometer, a temperature increase causes the alcohol or mercury to expand more than the glass and thus

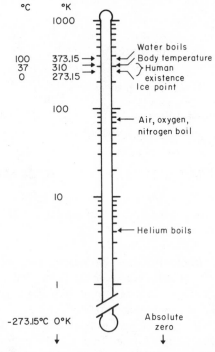

Figure 4.1. Logarithmic scale of temperatures.

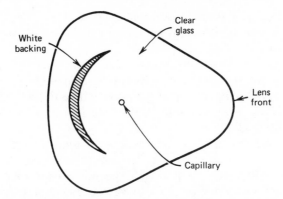

Figure 4.2. Cross-section of the stem of a clinical thermometer. (From 'Thermometry' by Busse, J., in *Medical Physics*, Vol. I by Glasser, O. (Ed.). Copyright © 1944 by Year Book Medical Publishers, Inc., Chicago. Used by permission.)

produces an increase in the level of the liquid. If the liquid expanded the same amount as the glass, the level of the liquid in the stem would remain constant with temperature.

The expansion of the liquid in a thermometer is not large—1 cm³ of mercury increases in volume by only 1.8% in going from 0 to 100°C. In order to show this expansion, thermometers are designed so that the mercury is forced to rise from the bulb in a capillary tube with a very small diameter. The smaller the diameter of the capillary, the greater is the sensitivity of the thermometer; a fever thermometer, which needs to show fractions of degrees, requires a capillary so small—less than 0.1 mm in diameter—that it would be very difficult to read if it were not designed for visibility. Two things increase the visibility of the capillary: the glass case acts as a magnifying glass and an opaque white backing is used (Fig. 4.2).

It is difficult to measure body temperature with a thermometer made for measuring house temperature. Besides being difficult to place under the tongue, a house thermometer would give a low reading because the temperature would fall when the thermometer was removed from the mouth. The capillary of a fever thermometer has a restriction just above the bulb so that after the liquid is forced into the stem by expansion it does not return when the temperature falls. When the thermometer is taken from the mouth it shows the maximum temperature it reached underneath the tongue. In order to return the mercury to the bulb it is necessary to take advantage of some elementary physics involving centrifugal forces. It is more important to know the technique than to understand the physics involved in it. A nurse who understands little physics can reset the thermometer easily with the proper snap of the wrist, while a physicist who has a good grasp of the principles may make many attempts before being successful.

The temperature is usually taken underneath the tongue or in the rectum. Since the thermometer is usually considerably colder than the body it lowers the temperature of the surrounding tissues when it is first inserted. It takes several minutes before the temperature of the tissues rises to the original value.

A number of temperature-sensitive devices other than the glass-liquid thermometer are used in medicine. Two of them are the thermistor and the thermocouple. A thermistor is a special resistor that changes its resistance rapidly with temperature (~5%/°C). Figure 4.3 shows a bridge circuit with a thermistor in one of the legs. Initially the four resistors shown are equal; that is, the bridge is balanced. By symmetry, the voltages at each end of the meter are equal and no current flows through the meter. A temperature change causes the thermistor resistance to change. This unbalances the bridge. The voltages at each end of the meter become unequal, causing a current to flow through the meter, and the resulting meter deflection can be calibrated for temperature. Thermistors are used quite often in medicine because of their sensitivity; with a thermistor it is easy to measure temperature changes of 0.01°C. Because of its small mass, a thermistor has little effect on the temperature of the surrounding tissues and responds rapidly to temperature change. The meter of Fig. 4.3 can be located some distance from the patient, for example, at a nursing station; this permits easy monitoring.

Thermistors are occasionally placed in the nose to monitor the breathing rate of patients by showing the temperature change between inspired cool air and expired warm air. An instrument of this type is called a pneumograph.

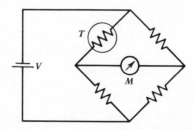

Figure 4.3.   The resistance of a thermistor *T* can be measured with a simple bridge circuit to determine the temperature. The meter *M* can be calibrated directly in degrees celsius or fahrenheit.

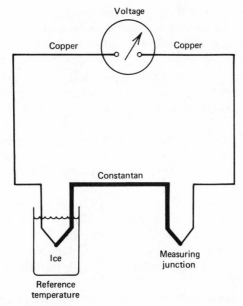

Figure 4.4. Schematic diagram of a thermocouple. The voltage is measured by a potentiometer.

A thermocouple consists of two junctions of two different metals. If the two junctions are at different temperatures, a voltage is produced that depends on the temperature difference. Usually one of the junctions is kept at a reference temperature such as in an ice-water bath (Fig. 4.4). The copper-constantan thermocouple shown in Fig. 4.4 can be used to measure temperatures from -190 to $300°C$. For a $100°C$ temperature difference, the voltage produced is only about 0.004 V (4 mV). Thermocouples can be made small enough to measure the temperature of individual cells.

4.3. THERMOGRAPHY—MAPPING THE BODY'S TEMPERATURE

Measurements of body surface temperature indicate that the surface temperature varies from point to point depending upon external physical factors and internal metabolic and circulatory processes near the skin—blood flow near the skin is the dominant factor. Since variations in these internal processes may be symptomatic of abnormal conditions, many researchers have attempted to accurately measure the surface temperature of the body and relate it to pathologic conditions.

The need for a simple routine method of obtaining a surface temperature map (thermogram) was brought into focus by some studies done in the mid-1950s, when it was found that most breast cancers could be characterized by an elevated skin temperature in the region of the cancer. The surface temperature above a tumor was typically about 1°C higher than that above nearby normal tissue, indicating that a very sensitive temperature-measuring device had to be used. It appeared that if an adequate thermography unit could be constructed, it would be useful for mass screening of women for breast cancer—the most common cancer in women.

One very appealing method of obtaining a thermogram is to measure the radiation emitted from the body. All objects regardless of their temperature emit radiation. If the temperature is sufficiently high (red hot), the radiation is visible. At body temperature the emitted radiation is in the far infrared (IR) region at wavelengths much longer than those observable by the human eye.

The basic equation describing the radiation emitted by a body was given by Max Planck in 1901. For our purposes the Stefan–Boltzmann law for the total radiative power per surface area W is more useful. It is

$$W = e\sigma T^4 \tag{4.1}$$

where T is the absolute temperature; e is the emissivity, which depends upon the emitter material and its temperature; and σ (the Stefan–Boltzmann constant) is 5.7×10^{-12} W/cm² °K⁴. For radiation from the body e is almost 1; thus if we measure W, we can find the temperature T. The power radiated per square centimeter is small, as shown in Example 4.1.

Example 4.1

a. What is the power radiated per square centimeter from skin at a temperature of 306°K ($\sim$33°C)?

$$W = e\sigma T^4 = (5.7 \times 10^{-12})(306)^4 \simeq 0.05 \text{ W/cm}^2$$

b. What is the power radiated from a nude body 1.75 m² (1.75×10^4 cm²) in area?

$$W \simeq (0.05)(1.75 \times 10^4 \text{ cm}^2) = 875 \text{ W}$$

The radiative power received from the surrounding walls at 293°K ($\sim$20°C) would be about 735 W, for a net loss of 140 W. Since normally most of the body is clothed, the loss is considerably smaller than 140 W, but it is still significant (see Chapter 5, p. 99).

Figure 4.5 shows a basic thermographic unit used to measure the radiation emitted from a part of the body. Radiation from a small area of a patient (~5 mm in diameter) is passed by a mirror arrangement through a mechanical chopper to a detector, which is usually cooled to increase its sensitivity. The chopper changes the continuous radiation to an alternating signal so that it can be more easily amplified. The IR transparent filter removes visible light, and the detector converts the IR (or body heat) radiation to an electrical signal that is proportional to the temperature of the surface from which the radiation originated. In order to give a heat picture of the total surface, a mechanical system moves the mirrors so the heat from different body areas can be detected. The position and magnitude of the radiation from each part of the patient are displayed on the cathode ray tube (CRT) of an oscilloscope; the brightness of the image is determined by the temperature, and its position on the screen corresponds to the area of the body being scanned. The CRT displays the different body temperatures as different shades of gray; the hot areas can be shown as either black or white.

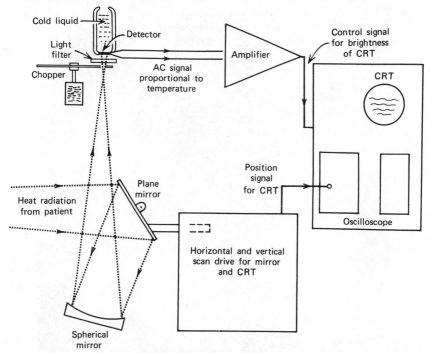

Figure 4.5.   Diagram of a typical thermographic unit for medical use. The plane mirror is mechanically scanned to produce a heat picture of the patient on the CRT of the oscilloscope.

Figure 4.6 shows a commercial instrument used in clinical thermography. This particular instrument can measure temperature differences of 0.2°C and record a thermogram in 2 sec. The image on the CRT can be photographed to obtain a permanent record.

Clothing affects skin temperature and must be removed before thermography. It is necessary to permit the region to be mapped to adapt to room temperature and cool uniformly because this enhances the temperature differences and contrast in thermography. Usually 20 min at 21°C is adequate. Of course, a drafty examination area must be avoided.

Thermography has been most commonly used as an aid in detecting breast cancer. It is customary to compare the heat patterns from the two

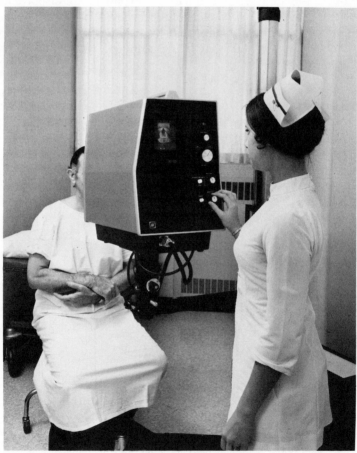

Figure 4.6.   A modern thermography instrument. A temperature map of the patient's face is visible on the monitor. (Courtesy of Spectrotherm, Santa Clara, Calif.)

breasts; a tumor is suggested if one breast is noticeably warmer, since a tumor often increases the blood flow.

Figure 4.7 is a thermogram of a woman's chest. The lower part of the figure is the temperature profile that exists at the horizontal line in the upper part of the scan. The thermogram indicates an elevation of temperature in the right breast that could be due to cancer. Such a thermogram would warrant further diagnostic tests—palpation (feeling), use of low voltage x-rays (mammography with special screen-film systems or xeroradiography), and/or biopsy.

A comment is in order about these additional diagnostic tests. With palpation, it is difficult to detect small tumors (less than 1 cm in diameter). The use of x-rays in the diagnosis of breast cancer has been quite successful, but they present a radiation hazard to the body. A biopsy gives information only about the material excised, and thus cancer tissue near the excised region can be missed.

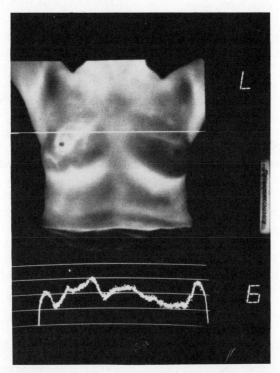

Figure 4.7. Thermogram of a woman's chest. The lighter areas are warmer than the darker ones. The thermogram and the temperature profile below it indicate an elevated temperature of the right breast. (Courtesy of Spectrotherm, Santa Clara, Calif. Taken at the Department of Radiology, Palo Alto Clinic, Palo Alto, Calif.)

The results of using thermography as a screening technique for breast cancer have been disappointing. If 1000 random women of age 45 are studied, about one-third will have abnormal thermograms of the breast, although far fewer than 1% have cancer. Most of the false positive findings are due to different blood flow patterns in the two breasts. Also, on the average only about 40% of women with known breast cancer will show positive thermograms, although one center reported over 70% success. That is, the technique at present has both a high percentage of false negatives (a normal thermogram for a cancer patient) and a high percentage of false positives (an abnormal thermogram for a subject without cancer). It is hoped that improved techniques and instrumentation will increase the diagnostic accuracy of thermography. As a general rule, x-ray studies are much more reliable than thermography for detection of breast cancer—they detect over 80% of known cancers.

Thermography has been used to detect other types of cancer; thermograms of two different patients with enlarged masses in the arm are shown in Fig. 4.8. One thermogram (Fig. 4.8*a*) indicates a distinctly cold area, and the other thermogram (Fig. 4.8*b*) shows a much warmer region. Surgery revealed the first mass to be benign and the second to be malignant.

Thermography has also been used to study the circulation of blood in the head. Differences in temperature between the left and right sides can indicate circulatory problems.

Thermography has had considerable success in reducing leg amputations in diabetics. The blood supply in a diabetic's leg is usually adequate, but if the tissue breaks down and an ulcer is formed, the need for blood in the leg may double. The circulatory problems of the diabetic then become evident; the ulcer does not heal and often becomes infected. With thermography, the presence of a hot spot on the foot can be determined before an ulcer forms. The physician can then use preventive measures such as having the patient wear a special shoe to try to eliminate the hot spot and avoid formation of an ulcer. Preliminary studies resulted in a reduction of about 20% in limb amputations of diabetics in 1975.

4.4. HEAT THERAPY

While it was recognized several thousand years ago that hot baths were therapeutic, it was not until the mid-1800s that the pain-relieving properties of heat were somewhat understood. Two primary therapeutic effects take place in a heated area: there is an increase in metabolism resulting in a relaxation of the capillary system (vasodilation), and there is an increase

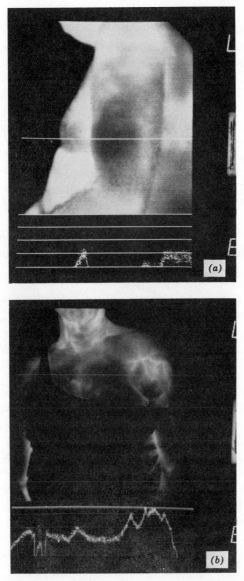

Figure 4.8. Thermograms of two different patients with masses in the arm. (a) A distinctly cold region indicates a benign mass. (b) A warm region indicates the possibility of a malignant tumor. (Courtesy of Spectrotherm, Santa Clara, Calif. Taken at the Department of Radiology, Palo Alto Clinic, Palo Alto, Calif.)

in blood flow as blood moves in to cool the heated area. The relaxation and increased blood flow are beneficial to damaged tissue, although the details of the therapeutic action are not well understood. In this section we briefly consider the physical methods of producing heat in the body. These methods are conductive heating, infrared (IR) radiant heating, radiowave heating (diathermy), and ultrasonic wave heating (ultrasonic diathermy).

The *conductive* method is based on the physical fact that if two objects at different temperatures are placed in contact, heat will transfer by conduction from the warmer object to the cooler one. The total heat transferred will depend upon the area of contact, the temperature difference, the time of contact, and the thermal conductivity of the materials. Hot baths, hot packs, electric heating pads, and occasionally hot paraffin applied to the skin heat the body by conduction. Conductive heat transfer leads to local surface heating since the circulating blood effectively removes heat that penetrates deep into the tissue. Conductive heating is used in treating conditions such as arthritis, neuritis, sprains and strains, contusions, sinusitis, and back pain.

Radiant (IR) *heat* is also used for surface heating of the body. This is the same form of heat we feel from the sun or from an open flame. Man-made sources of radiant heat are glowing wire coils and 250 W incandescent lamps. The IR wavelengths used are between 800 and 40,000 nm (1 nm = 10^{-9} m). The waves penetrate the skin about 3 mm and increase the surface temperature. Excessive exposure causes reddening (erythema) and sometimes swelling (edema). Very prolonged exposure causes browning or hardening of the skin. Radiative heating is generally used for the same conditions as conductive heating, but it is considered to be more effective because the heat penetrates deeper.

When alternating electric current passes through the body, various effects such as heating and electric shock take place. We briefly consider electrical heating here; we discuss it in more detail in Chapter 11. The amount of heat that can be transferred to the body by electrical diathermy increases as the frequency of the current increases. Short-wave diathermy utilizes electromagnetic waves in the radio range (wavelength $\sim$10 m); microwave diathermy uses waves in the radar range (wavelength $\sim$12 cm).

Heat from diathermy penetrates deeper into the body than radiant and conductive heat. It is thus useful for internal heating and has been used in the treatment of inflammation of the skeleton, bursitis, and neuralgia.

Two different methods are used for transferring the electromagnetic energy into the body in short-wave diathermy. In one, the part of the body to be treated is placed between two metal plate-like electrodes energized

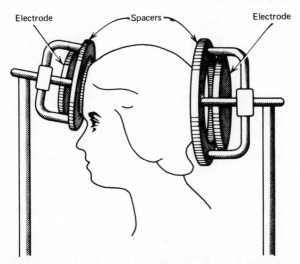

Electrode   Spacers   Electrode

Figure 4.9. Location of capacitor plates for short wave diathermy. (Adapted from Lehmann, J.F.: "Diathermy," *in* Krusen, F.H., Kottke, F.J., and Ellwood, P.M. (Eds.): *Handbook of Physical Medicine and Rehabilitation*, **2nd edition. Philadelphia, W.B. Saunders Company, 1971, p. 286.)**

by the high-frequency voltage (Fig. 4.9). The body tissue between the plates acts like an electrolytic solution. The charged particles are attracted to one plate and then the other depending upon the sign of the alternating voltage on the plates; this results in resistive (joule) heating. Different body materials react differently to the waves, and this effect provides some selectivity in treatments.

The second method of transferring short-wave energy into the body is magnetic induction (Fig. 4.10). In induction diathermy, either a coil is placed around the body region to be treated or a "pancake" coil is placed near that part of the body. The alternating current in the coil results in an alternating magnetic field in the tissues. Consequently alternating (eddy) currents are induced, producing joule heating in the body region being treated.

Short-wave diathermy heats the deep tissues of the body. It has been used in relieving muscle spasms; pain from protruded intervertebral discs, degenerative joint disease, and bursitis; and as a deep heating agent for joints with minimal soft tissue coverage such as the knee, elbow, and ankle.

Microwave diathermy, another form of electromagnetic energy, is usually easier to apply than short-wave diathermy. We are all aware of the tissue heating ability of microwave home ovens. Microwave diathermy

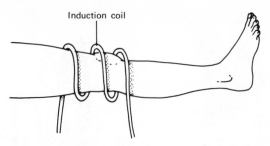

Figure 4.10.   Location of induction coil around knee for short-wave diathermy.

developed out of radar research in the 1940s. The microwaves are produced in a special tube called a *magnetron* and are then emitted from the
applicator (antenna). The antenna is usually designed so that it can be
placed several inches from the region to be treated. The microwaves from
the antenna penetrate deep into the tissues, causing a temperature rise
and deep heating. Microwave diathermy is used in the treatment of
fractures, sprains and strains, bursitis, injuries to tendons, and arthritis.

The frequency used in microwave diathermy is 2450 MHz because this
frequency was the one available after World War II. This is unfortunate
since later research has shown that a frequency closer to 900 MHz would
be more effective in therapy, causing more uniform heating around bony
regions.

Ultrasonic waves are also used for deep heating of body tissue. These
waves are completely different from the electromagnetic waves just discussed. They produce mechanical motion like audible sound waves except the frequency is much higher (usually near 1 MHz—see Chapter 12).
In ultrasonic diathermy, power levels of several watts per square centimeter are usually used and the sound source is directly in contact with the
body. As the ultrasonic waves move through the body the particles in the
tissues move back and forth. The movement is similar to a micromassage
and results in heating of the tissues. Ultrasonic heating has been found
useful in relieving the tightness and scarring that often occur in joint
disease. It greatly aids joints that have limited motion. It is useful for
depositing heat in bones because they absorb ultrasound energy more
effectively than does soft tissue (see Chapter 12). Because ultrasound has
so many diagnostic uses, it is treated in much greater detail in Chapter 12.

Cancer studies in the early 1970s have indicated that heat therapy may
be beneficial in the treatment of cancer when it is combined with radiation
therapy. The tumor is heated by diathermy to about 42°C for 20 to 30 min,
and the radiation treatment is given after the heat treatment. Long-term

studies under controlled conditions are necessary to evaluate this treatment method.

4.5. USE OF COLD IN MEDICINE

Cryogenics is the science and technology of producing and using very low temperatures. The study of low-temperature effects in biology and medicine is called *cryobiology*. It is an active research area. In this section we discuss how cryogenic fluids are produced and stored and how the fluids are used in the long-term storage of biological materials. Cryosurgery is discussed in Section 4.6.

The first successful recorded attempt to cool air for air conditioning took place in 1840. John Gorrie, a physician in Florida, was trying to relieve the suffering of malaria patients by cooling their room. He succeeded in making ice by using the cooling effect of expanding air. (The same cooling effect is produced by aerosol sprays and carbon dioxide fire extinguishers.) Following Gorrie's success a number of achievements were made in liquefying gases. Liquid air ($-196°C$) was produced in 1877; liquid helium ($-269°C$) was produced in 1908.

The storage of cryogenic fluids has always been a problem. Most ordinary liquid-storage containers are unsatisfactory because they absorb a large amount of heat by conduction, convection, and radiation. A significant improvement is the insulated container developed by James Dewar in 1892 and named after him (Fig. 4.11). This container is made of glass or thin stainless steel to minimize conductive losses. It has a vacuum space to essentially eliminate convective losses, and the sides are silvered or polished so that radiation striking the surface is reflected rather than absorbed. The container resembles the familiar Thermos bottle used to store hot and cold drinks. Dewar vessels with capacities of over 100,000 liters have been made.

The problems involving transfer of cryogenic fluids are similar to those of storage, and the transfer line is usually constructed, like the dewar, of two polished concentric metal pipes with a vacuum between the walls to reduce heat transfer to the fluid.

How are cryogenic methods used in medicine? Low temperatures have been used for long-term preservation of blood, sperm, bone marrow, and tissues. Studies of their relationship to the hibernation of animals are under way, and the long-term preservation of man is being considered.

Much interest has been aroused by the idea of using cryogenic methods to cool the body into a state of "suspended animation" so that it can pass

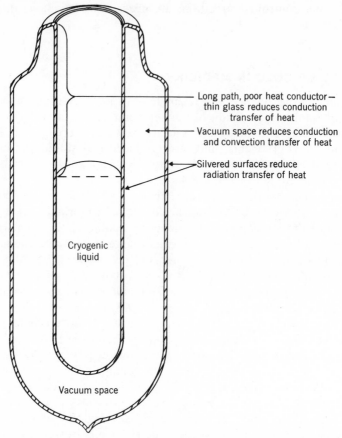

Long path, poor heat conductor — thin glass reduces conduction transfer of heat

Vacuum space reduces conduction and convection transfer of heat

Silvered surfaces reduce radiation transfer of heat

Cryogenic liquid

Vacuum space

Figure 4.11.   Details of a dewar container for cryogenic fluids.

time without aging. This "science" is called *cryonics*. One goal of cryonics is to preserve at a low temperature people with fatal diseases with the hope that in the future they could be revived and their diseases cured.

Some successful work has been done with cooling hamsters down to −5°C (23°F) by freezing 50 to 60% of the water in their bodies and then reviving them. However, the present technology excludes similar cooling of something as complex as man, although sometimes moderate cooling is used in conjunction with surgery. Some simpler human biological systems such as blood, semen, and tissue have successfully been cooled, stored, and revived.

Considerable effort has been expended on problems associated with the preservation and storage of various tissues, but success has so far been limited to simple systems. It has been found that for long-term survival, the tissues should be stored at very low temperatures. Since the biochemical and physical processes that sustain life are temperature dependent, lowering the temperature reduces the rates of the processes. Preservation is much better at the temperature of liquid nitrogen ($-196°C$) than at the temperature of solid carbon dioxide ($-79°C$).

Another important finding involves the freeze-thaw cycle. Survival after freezing is more dependent upon the cooling rate during the freezing cycle than on the warming rate during the thawing cycle. Measurement of the survival of two biomaterials as a function of the cooling rate after freezing and thawing of the materials gives the results shown in Fig. 4.12. The survival curves of different biomaterials as a function of cooling rate have similar shapes, but there is no unique cooling rate that will ensure cell survival for all materials. This puts a severe limitation on preserving biomaterials composed of many different cell types. For many mammalian cells only a few percent survive; thus freezing and thawing offers little hope as a general means of long-term biological storage.

The survival of some cells can be helped by adding a protective agent

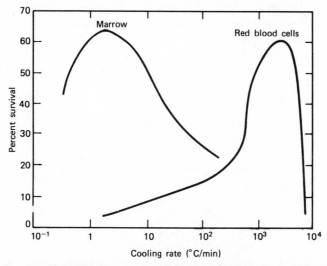

Figure 4.12. Survival behavior as a function of cooling rate for red blood cells and bone marrow. (Adapted from P. Mazur, S.P. Leibo, J. Farrant, E.H.Y. Chu, M.G. Hanna, and L.H. Smith, in G.E.W. Wohlstenholme and M. O'Connor (Eds.), *The Frozen Cell*, A Ciba Foundation Symposium, J. and A. Churchill, London, 1970, p. 82. The red blood cell curve is redrawn from data of G. Rapatz, J.J. Sullivan, and B. Luyet, *Cryobiology*, 5, 1968, pp. 18–25.)

such as glycerol or dimethyl sulfoxide before cooling. The use of additives to enhance survival sometimes presents problems that are more complicated than the original ones involved with the freeze-thaw cycle. For example, the removal of glycerol from blood is sufficiently complicated to limit the process to a few large hospitals.

The conventional noncryogenic method of blood storage involves mixing whole blood with an anticoagulant and storing it at 4°C. About 1% of the red blood cells hemolyze (break) each day, so the blood is not usable after about 21 days. For the more common blood types this does not present a great problem because the blood is usually used rapidly and is easily replaced. For the rare blood types this storage time is insufficient and makes maintaining an adequate supply difficult.

Blood can be stored for a much longer time if it is rapidly frozen. Two techniques are used for this: one uses thin-walled containers (Fig. 4.13); the other is the "blood-sand" method.

The container with thin metal walls is constructed so that the blood volume between the walls is small. After it is filled with blood it is quickly inserted into a liquid nitrogen bath. The frozen blood can be stored indefinitely at the temperature of liquid nitrogen (−196°C).

In the "blood-sand" method, blood is sprayed onto a liquid nitrogen surface and freezes into small droplets. The droplets are about the size of grains of sand—hence the name "blood-sand." The droplets are collected and then stored in special containers, usually at the temperature of liquid nitrogen.

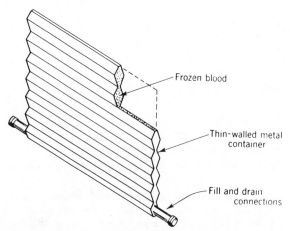

Figure 4.13. Container for whole-blood freezing at cryogenic temperatures. (From *Cryogenic Systems* by R. Barron. Copyright © 1966, McGraw-Hill, New York. Used with permission of McGraw-Hill Book Company.)

Experimental work is under way to form banks for skin, bone, muscle, and organs. These substances are harder to preserve than simple cells such as red blood cells for a number of reasons: (1) the large physical dimensions limit the cooling rate and (2) adding and removing protective agents is difficult. Even so, some work has successfully been carried out with cornea and skin preservation. Organ preservation and re-use is still in the experimental stage.

4.6. CRYOSURGERY

Cryogenic methods are also used to destroy cells; this application is called *cryosurgery*. Cryosurgery has several advantages: (1) there is little bleeding in the destroyed area, (2) the volume of tissue destroyed can be controlled by the temperature of the cryosurgical probe, and (3) there is little pain sensation because low temperatures tend to desensitize the nerves.

One of the first uses of cryosurgery was in the treatment of Parkinson's disease ("shaking palsy"), a disease associated with the basal ganglion of the brain. Parkinson's disease causes uncontrolled tremors in the arms and legs. It is possible to stop the tremors by surgically destroying the part of the thalamus in the brain that controls the transmission of nerve impulses to other parts of the nervous system. A cryosurgical treatment was initially undertaken by Dr. Irving Cooper in New York during the early 1960s. Cooper sought the help of the Union Carbide Company in developing a "cryoknife" to use in cryosurgery. He desired to treat Parkinson's disease by destructively freezing the appropriate region in the thalamus.

The cryosurgical system used by Cooper is shown schematically in Fig. 4.14. The vacuum jacket acts as an insulator for the walls of the variable temperature probe (cannula). In the treatment of Parkinson's disease the tip of the probe is cooled to $-10°C$ and moved into the appropriate regions of the thalamus, causing temporary freezing of these regions. The frozen areas recover if the probe tip is removed in less than 30 sec. The patient must be conscious during the procedure so that the surgeon can observe when the shaking stops; this means that the probe has reached the correct region of the thalamus. This region is then destroyed by freezing for several minutes at temperatures near $-85°C$. After the freezing, the tip is warmed and removed. The destroyed tissue forms a cyst after thawing and does not interfere with normal body functions. The patient almost always shows immediate benefit, and the post-operative recovery period is very short compared to that for a major brain operation. Successful

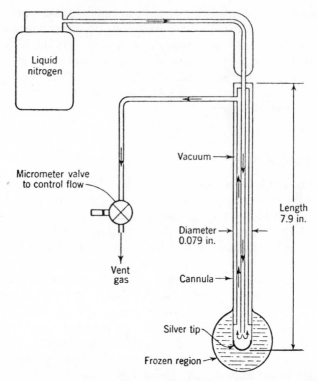

Figure 4.14. Cooper cryosurgery system. The probe-tip temperature is measured by a thermocouple embedded in the tip, and the temperature is recorded on a control console display. The micrometer valve controls the temperature by controlling the flow of gas escaping and thus the flow of liquid nitrogen to the tip. (From *Cryogenic Systems* by R. Barron. Copyright © 1966, McGraw-Hill, New York. Used with permission of McGraw-Hill Book Company.)

results were obtained in more than 90% of the cases in which cryosurgery was used.

Since the early work by Cooper, medication has been developed that can be used to treat most of the patients who have Parkinson's disease. As a result, cryosurgery for Parkinson's disease is now used only for those cases untreatable by medication. However, the methods developed are now used in other types of cryosurgery.

One common use of cryosurgery is in the treatment of tumors and warts. The treatment follows the basic principles already discussed. Cryogenic methods have also been used in several types of eye surgery. We discuss two—the repair of a detached retina and cataract surgery (the removal of a darkened lens).

Occasionally, perhaps as the result of an accident, the retina becomes detached from the wall of the eyeball. This produces a blurred spot in the vision because the light rays are not focused at the correct spot. If a cold tip is applied to the outside of the eyeball in the vicinity of the detachment, a reaction occurs that acts to "weld" the retina to the wall of the eyeball. The technique does not appear to damage the eye.

In cryosurgical extraction of the lens, a cold probe is touched to the front surface of the lens. The probe sticks to the lens, making the lens easy to remove.

4.7. SAFETY WITH CRYOGENICS

Many laboratories and hospitals that use cryogenic fluids have cylindrical containers of gases that are stored at high pressure. Such a cylinder can be dangerous if its valve is accidentally broken off, so these containers should be handled with care and chained in position. When gas is withdrawn, a pressure-reducing regulator should be used. The regulator used for oxygen should not be used for any other gases because of the danger of explosion.

Caution should be exercised when cryogenic liquids or cold gases are used because any contact between these materials and the eyes or the skin results in "freeze burns." Adequate ventilation should be provided in areas in which nitrogen or carbon dioxide is used in order to prevent the oxygen content of the area from being depleted by dilution.

The most care is required when oxygen is used because it greatly enhances combustion. Many materials that do not burn in air will burn in pure oxygen. Thus open flame and smoking must be avoided where oxygen is stored or handled. A good policy is to keep all organic and other flammable substances away from areas used for oxygen storage and to provide adequate ventilation in these areas. If liquid oxygen spills on clothing, the clothing should be removed and allowed to air out for 30 min since clothing is highly flammable and easily ignited when it contains concentrated oxygen.

BIBLIOGRAPHY

Allen, R. J., *Cryogenics,* Lippincott, New York, 1964.
Barnes, R. B., "Thermography," *Ann. N.Y. Acad. Sci.,* **121,** 34–48 (1964).
Barron, R., *Cryogenic Systems,* McGraw-Hill, New York, 1966.
Cravalho, E. G., "Heat Transfer in Biomaterials," in H. E. Stanley (Ed.), *Biomedical Physics and Biomaterial Science,* MIT Press, Cambridge, Mass., 1972, Chapter 11.

Croft, A. J., *Cryogenic Laboratory Equipment,* Plenum, New York, 1970.

Gershon-Cohen, J., "Medical Thermography," *Sci. Amer.*, **216–217,** 94–102 (1967).

Hardy, J. D. (Ed.), *Temperature: Its Measurement and Control in Science and Industry,* Vol. 3, *Biology and Medicine,* Reinhold, New York, 1963.

Huggins, C. E., "Blood Preservation by Freezing," in H. E. Stanley (Ed.), *Biomedical Physics and Biomaterial Science,* MIT Press, Cambridge, Mass., 1972, Chapter 10.

Lehmann, J. F., "Diathermy," in F. H. Krusen, F. J. Kottke, and P. M. Ellwood, Jr. (Eds.), *Handbook of Physical Medicine and Rehabilitation,* 2nd ed., Saunders, Philadelphia, 1971, pp. 273–345.

Meryman, H. T., *Cryobiology,* Academic, New York, 1966.

Popovic, V., and P. Popovic, *Hypothermia in Biology and in Medicine,* Grune & Stratton, New York, 1974.

Raskin, M. M., and M. Viamonte, Jr. (Eds.), *Clinical Thermography,* American College of Radiology, Chicago, 1977.

Thom, H., *Introduction to Shortwave and Microwave Therapy,* 3rd ed., Thomas, Springfield, Ill., 1966.

Went, J. M. van, *Ultrasonic and Ultrashort Waves in Medicine,* Elsevier, Amsterdam, 1954.

REVIEW QUESTIONS

1. What is the normal body temperature on the fahrenheit, celsius, and kelvin scales?

2. What physical quantity changes with temperature in a thermistor? In a thermocouple?

3. Consider a fever thermometer that contains 0.01 cm^3 of mercury. Find the diameter of the capillary if a 1°C change corresponds to a level change of 0.5 cm. Assume the glass does not expand.

4. What techniques are used to increase the visibility of the small capillary of a fever thermometer?

5. How does a thermographic unit measure skin temperature?

6. What are the two primary therapeutic effects of heating the body?

7. List four techniques for heating parts of the body.

8. How does ultrasonic diathermy differ from microwave diathermy?

9. How does a dewar container minimize heat transfer to a cryogenic fluid?

10. (a) What is the optimum cooling rate for preserving red blood cells?
    (b) At the optimum rate, how long would it take to cool red blood cells from 37 to −196°C?

CHAPTER 5

Energy, Work, and Power of the Body

Energy is a basic concept of physics. In the physics of the body energy is of primary importance. All activities of the body, including thinking, involve energy changes. The conversion of energy into work such as lifting a weight or riding a bicycle represents only a small fraction of the total energy conversions of the body. Under resting (basal) conditions about 25% of the body's energy is being used by the skeletal muscles and the heart, 19% is being used by the brain, 10% is being used by the kidneys, and 27% is being used by the liver and spleen.

The body's basic energy (fuel) source is food. The food is generally not in a form suitable for direct energy conversion. It must be chemically changed by the body to make molecules that can combine with oxygen in the body's cells. We do not discuss this complex chemical process (Krebs cycle) here. From a physics viewpoint we can consider the body to be an energy converter that is subject to the law of conservation of energy.

The body uses the food energy to operate its various organs, maintain a constant body temperature, and do external work, for example, lifting. A small percentage (~5%) of the food energy is excreted in the feces and urine; any energy that is left over is stored as body fat. The energy used to operate the organs eventually appears as body heat. Some of this heat is useful in maintaining the body at its normal temperature, but the rest must be disposed of. (Other energy sources such as radiant solar energy and heat energy from our surroundings can help maintain body temperature but are of no use in body function.)

In this chapter we discuss the conservation of energy in the body (first law of thermodynamics), the conversion of energy in the body, the work done by and power of the body, and how the body loses heat.

5.1. CONSERVATION OF ENERGY IN THE BODY

Conservation of energy in the body can be written as a simple equation:

$$
\begin{bmatrix} \text{change in stored energy} \\ \text{in the body (i.e., food} \\ \text{energy, body fat, and} \\ \text{body heat)} \end{bmatrix} = \begin{bmatrix} \text{heat lost} \\ \text{from the body} \end{bmatrix} + \begin{bmatrix} \text{work done} \end{bmatrix}
$$

This equation, which is really the first law of thermodynamics, assumes that no food or drink is taken in and no feces or urine is excreted during the interval of time considered. In this section we discuss this law.

There are continuous energy changes in the body both when it is doing work and when it is not. We can write the first law of thermodynamics as

$$ \Delta U = \Delta Q - \Delta W \tag{5.1} $$

where ΔU is the change in stored energy, ΔQ is the heat lost or gained, and ΔW is the work done by the body in some interval of time.* A body doing no work ($\Delta W = 0$) and at a constant temperature continues to lose heat to its surroundings, and ΔQ is negative. Therefore, ΔU is also negative, indicating a decrease in stored energy. The energy term ΔU is discussed in Section 5.2, the work term ΔW is discussed in Section 5.3, and the heat term ΔQ is considered in Section 5.4.

It is useful to consider the change of ΔU, ΔQ, and ΔW in a short interval of time Δt. Equation 5.1 then becomes

$$ \frac{\Delta U}{\Delta t} = \frac{\Delta Q}{\Delta t} - \frac{\Delta W}{\Delta t} \tag{5.2} $$

where $\Delta U/\Delta t$ is the rate of change of stored energy, $\Delta Q/\Delta t$ is the rate of heat loss or gain, and $\Delta W/\Delta t$ is the rate of doing work, that is, the mechanical power.

Equation 5.2, which is used extensively in this chapter, is merely another form of the first law of thermodynamics. It tells us that energy is conserved in all processes, but it does not tell us whether or not a process can occur. For example, according to the first law if we put heat into the

---

*Conventionally, the first law is written as $\Delta Q = \Delta U + \Delta W$, that is, if heat is added to a gas it can increase the internal energy ΔU and also do work ΔW.

body we could expect the body to produce an equal amount of chemical energy or work. The physical law governing the direction of the energy conversion process is given in the second law of thermodynamics. We refer readers interested in more details on this subject to the book by Kleiber listed in the bibliography.

5.2. ENERGY CHANGES IN THE BODY

Several energy and power units are used in relation to the body. Physiologists usually use *kilocalories* (kcal) for food energy and kilocalories per minute for the rate of heat production. The energy value of food referred to by nutritionists as a Calorie (C) is actually a kilocalorie; thus a diet of 2500 C/day is 2500 kcal/day.

The most widely accepted physics unit for energy is the newton-meter or joule (J); power is given in joules per second or watts (W). The energy unit in the cgs system is the erg, and that in the English system is the foot-pound (ft-lb).

A convenient unit for expressing the rate of energy consumption of the body is the *met*. The met is defined as 50 kcal/m^2 of body surface area per hour. For a normal person 1 met is about equal to the energy consumption under resting conditions. A typical man has about 1.85 m^2 of surface area (a woman has about 1.4 m^2), and thus for a typical man 1 met is about 92 kcal/hr or 107 W.

In this chapter we use kilocalories and joules for energy units and kilocalories per second, minute, or hour, watts, and horsepower (hp) for energy rates (power). These units are summarized as follows:

1 kcal = 4184 J
1 J = 10^7 ergs = 0.737 ft-lb
1 kcal/min = 69.7 W = 0.094 hp
100 W = 1.43 kcal/min
1 hp = 642 kcal/hr = 746 W = 550 ft-lb/sec
1 met = 50 kcal/m^2 hr = 58 W/m^2
1 kcal/hr = 1.162 W

Lavoisier was the first to suggest (in 1784) that food is oxidized. He based his arguments on measurements of an experimental animal that showed that oxygen consumption increased during the process of digestion. He explained this effect as work of digestion. We now know that this explanation is incorrect; the correct explanation is that oxidation occurs in the cells of the body.

In oxidation by combustion heat is released. In the oxidation process within the body heat is released as energy of metabolism. The rate of oxidation is called the *metabolic rate*.

Let us consider the oxidation of glucose, a common form of sugar used for intravenous feeding. The oxidation equation for 1 mole of glucose $(C_6H_{12}O_6)$ is

$$C_6H_{12}O_6 + 6O_2 \rightarrow 6H_2O + 6CO_2 + 686 \text{ kcal}$$

That is, 1 mole of glucose (180 g) combines with 6 moles of O_2 (192 g) to produce 6 moles each of H_2O (108 g) and CO_2 (264 g), releasing 686 kcal of heat energy in the reaction. Using this information we can compute a number of useful quantities for glucose metabolism. (Remember that 1 mole of a gas at normal temperature and pressure has a volume of 22.4 liters.)

$$\text{Kilocalories of energy released per gram of fuel} = \frac{686}{180} = 3.80$$

$$\text{Kilocalories released per liter of } O_2 \text{ used} = \frac{686}{22.4 \times 6} = 5.1$$

$$\text{Liters of } O_2 \text{ used per gram of fuel} = \frac{6 \times 22.4}{180} = 0.75$$

$$\text{Liters of } CO_2 \text{ produced per gram of fuel} = \frac{6 \times 22.4}{180} = 0.75$$

Ratio of moles of CO_2 produced to moles of O_2 used—called the
respiratory quotient $(R) = 1.0$

Similar calculations can be done for fats, proteins, and other carbohydrates. Typical caloric values of these food types and of common fuels are given in Table 5.1. Table 5.1 also lists for the various types of food the

Table 5.1. Typical Energy Relationships for Some Foods and Fuels

Food or Fuel	Energy Released per Liter of O_2 Used (kcal/liter)	Caloric Value (kcal/g)
Carbohydrates	5.3	4.1
Proteins	4.3	4.1
Fats	4.7	9.3
Typical diet	4.8–5.0	—
Gasoline	—	11.4
Coal	—	8.0
Wood (pine)	—	4.5

energy released per liter of oxygen consumed; thus by measuring the oxygen consumed by the body, we can get a good estimate of the energy released.

The caloric values in Table 5.1 for the foods are the maximum that might be expected. Not all of this energy is available to the body because part is lost in incomplete combustion. The "unburned" products are released in feces, urine, and flatus (intestinal gas). What remains is the metabolizable energy. The body is usually quite efficient at extracting energy from food. For example, the energy remaining in normal feces is only about 5% of the total energy contained in the consumed food. When the body is at constant temperature the energy that is extracted from food plus the body fat make up the available stored energy.

When completely at rest, the typical person consumes energy at a rate of about 92 kcal/hr, or 107 W, or about 1 met. This lowest rate of energy consumption, called the *basal metabolic rate* (BMR), is the amount of energy needed to perform minimal body functions (such as breathing and pumping the blood through the arteries) under resting conditions. Clinically an individual's BMR is compared to normal values for a person of the same sex, age, height, and weight. The BMR depends primarily upon thyroid function. A person with an overactive thyroid has a higher BMR than a person with normal thyroid function.

Since the energy used for basal metabolism becomes heat which is primarily dissipated from the skin (see Section 5.4), one might guess that the basal rate is related to the surface area or to the mass of the body. Figure 5.1 shows a plot of BMR (kcal/day) for various animals of widely different weights. The slope of the line indicates that the BMR is proportional to mass$^{3/4}$. Thus as animals get larger their BMR increases faster than their surface area but not as fast as their volume (mass).

The metabolic rate depends to a large extent on the temperature of the body. Chemical processes are very temperature dependent, and a small change in temperature can produce a large change in the rate of chemical reactions. If the body temperature changes by 1°C, there is a change of about 10% in the metabolic rate. For example, if a patient has a temperature of 40°C, or 3°C above normal, the metabolic rate is about 30% greater than normal. Similarly, if the body temperature drops 3°C below normal, the metabolic rate (and oxygen consumption) decreases by about 30%. You can see why hibernating at a low body temperature is advantageous to an animal and why a patient's temperature is sometimes lowered during heart surgery.

Obviously, in order to keep a constant weight an individual must consume just enough food to provide for basal metabolism plus physical activities. Eating too little results in weight loss; continued too long it results in starvation. However, a diet in excess of body needs will cause

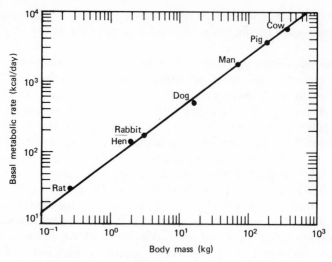

Figure 5.1. Relationship between the basal metabolic rate and the body mass for several different animals.

an increase in weight. Weight loss through dieting and physical exercise is discussed in Example 5.1.

Example 5.1

Suppose you wish to lose 4.54 kg (10 lb) either through physical activity or by dieting.

a. How long would you have to work at an activity of 15 kcal/min to lose 4.54 kg of fat? (Of course, you could not maintain this activity rate very long.)

From Table 5.1 you can expect a maximum of 9.3 kcal/g of fat. If you worked for T minutes, then

$$(T \text{ min}) \left(\frac{15 \text{ kcal}}{\text{min}} \right) = (4.54 \times 10^3 \text{ g}) \left(\frac{9.3 \text{ kcal}}{\text{g}} \right) = 4.2 \times 10^4 \text{ kcal}$$

$$T = 2810 \text{ min} \simeq 47 \text{ hr}$$

Note that a great deal of exercise is needed to lose a few pounds.

b. It is usually much easier to lose weight by reducing your food intake. If you normally use 2500 kcal/day, how long must you diet at 2000 kcal/day to lose 4.54 kg of fat?

$$T = \frac{\text{energy of 4.54 kg fat}}{\text{energy deficit per day}} = \frac{4.2 \times 10^4 \text{ kcal}}{5 \times 10^2 \text{ kcal/day}} \simeq 84 \text{ days}$$

The BMR is sometimes determined from the oxygen consumption when resting. We can also estimate the food energy used in various physical activities by measuring the oxygen consumption. Table 5.2 gives some typical values for various activities.

Oxygen consumption for various organs has been measured, and these values are given in Table 5.3. Note that some of the organs use rather large amounts of energy and that the kidney uses more energy per kilogram than the heart.

5.3. WORK AND POWER

Chemical energy stored in the body is converted into external mechanical work as well as into life-preserving functions. We now discuss external work ΔW, defined as a force F moved through a distance Δx

$$\Delta W = F \, \Delta x$$

Table 5.2. Oxygen Cost of Everyday Activities for a Man with a Surface Area of 1.75 m², Height of 175 cm, and Mass of 76 kg[a]

Activity	O_2 Consumption (liters/min)	Equivalent Heat Production		Energy Consumption (mets—50 kcal/m² hr)
		kcal/min	W	
Sleeping	0.24	1.2	83	0.82
Sitting at rest	0.34	1.7	120	1.15
Standing relaxed	0.36	1.8	125	1.25
Riding in automobile	0.40	2.0	140	1.35
Sitting at lecture (awake)	0.60	3.0	210	2.05
Walking slow (4.8 km/hr)	0.76	3.8	265	2.60
Cycling at 13–17.7 km/hr	1.14	5.7	400	3.90
Playing tennis	1.26	6.3	440	4.30
Swimming breaststroke (1.6 km/hr)	1.36	6.8	475	4.65
Skating at 14.5 km/hr	1.56	7.8	545	5.35
Climbing stairs at 116 steps/min	1.96	9.8	685	6.70
Cycling at 21.3 km/hr	2.00	10.0	700	6.85
Playing basketball	2.28	11.4	800	7.80
Harvard Step Test[b]	3.22	16.1	1120	11.05

[a]Adapted from P. Webb, in J. F. Parker and V. R. West (Eds.), *Bioastronautics Data Book*, National Aeronautics and Space Administration, Washington, D.C., 1973, pp. 859–861.
[b]A test in which the subject steps up and down a 40 cm step 30 times/min for 5 min.

Table 5.3.   Oxygen Use and Metabolic Rate Contribution of the Principal Organs of a Resting, Healthy Man Weighing 65 kg[a]

Organ	Mass (kg)	Average Rate of O_2 Consumption by Experiment (ml/min)	Average Rate of Energy Consumed (kcal/min)	Power per kg (kcal/min/kg)	Percent of BMR
Liver and spleen	—	67	0.33	—	27
Brain	1.40	47	0.23	0.16	19
Skeletal muscle	28.0	45	0.22	7.7×10^{-3}	18
Kidney	0.30	26	0.13	0.42	10
Heart	0.32	17	0.08	0.26	7
Remainder	—	48	0.23	—	19
		250	1.22		100%

[a]Adapted from R. Passmore, in R. Passmore and J.S. Robson (Eds.), *A Companion to Medical Studies*, Vol. I, Blackwell, Osney Mead, England, 1968, p. 4.9.

The force and the motion Δx must be in the same direction. The rate of doing work is the power p; thus for a constant force

$$p = \frac{\Delta W}{\Delta t} = \frac{F \Delta x}{\Delta t} = Fv$$

where $\Delta x / \Delta t$ equals the velocity v.

Obviously, external work is done when a person is climbing a hill or walking up stairs. We can calculate the work done by multiplying the person's weight (mg) by the vertical distance (h) moved. When a man is walking or running at a constant speed on a level surface, most of the forces act in the direction perpendicular to his motion. Thus, the external work done by him appears to be zero. However, his muscles are doing internal work which appears as heat in the muscle and causes a rise in its temperature. This additional heat in the muscle is removed by blood flowing through the muscle, by conduction to the skin, and by sweating. These processes are considered in Section 5.4.

In this section we want to study the human body as a machine for doing external work. This topic lends itself well to experiment. For example, we can measure the external work done and power supplied by a subject riding on an ergometer, a fixed bicycle that can be adjusted to vary the amount of resistance to the turning of the pedals (Fig. 5.2). We can also measure the oxygen consumed during this activity. The total food energy

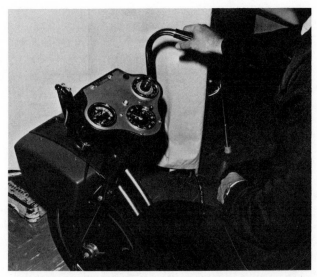

Figure 5.2.   The ergometer, a stationary bicycle with adjustable friction that permits studies of oxygen consumption under various work loads. One of the meters indicates the power produced.

consumed can be calculated since 4.8 to 5.0 kcal are produced for each liter of oxygen consumed.

The efficiency of the human body as a machine can be obtained from the usual definition of the efficiency ϵ:

$$\epsilon = \frac{\text{work done}}{\text{energy consumed}}$$

Efficiency is usually lowest at low power but can increase to 20% for trained individuals in activities such as cycling and rowing. Table 5.4 shows the efficiency of man for several activities along with the efficiency of several mechanical engines. It is difficult to compare different activities (such as swimming and cycling).

Studies have shown that cycling is one of our most efficient activities (see Example 5.2). For a trained cyclist the efficiency approaches 20% with an external power production of 370 W (0.5 hp) and a metabolic rate of 1850 W. If the cyclist is on level ground and moving at a constant speed there is no change in potential or kinetic energy and the power supplied is used primarily to overcome wind resistance and friction of tire flexing. (See *Bicycling Science: Ergonomics and Mechanics,* listed in the bibliography.)

Table 5.4. Mechanical Efficiency of Man and Machines

Task or Machine	Efficiency (%)
Cycling	~20
Swimming (on surface)	< 2
(underwater)	~ 4
Shoveling	~ 3
Steam engine	17
Gasoline engine	38

Example 5.2

Compare the energy required to travel 20 km on a bicycle to that needed by an auto for the same trip. Gasoline has 11.4 kcal/g and a density of 0.68 kg/liter. Assume that the auto can travel 8.5 km on a liter of gasoline.

The auto requires 2.35 liters to travel the 20 km.

$$(2.35 \text{ liters}) (0.68 \text{ kg/liter}) = 1.6 \text{ kg of gasoline}$$
$$(1.6 \times 10^3 \text{ g}) (11.4 \text{ kcal/g}) = 1.8 \times 10^4 \text{ kcal for 20 km}$$

The energy consumption for bicycling at 15 km/hr (~9 mph) (Table 5.2) is 5.7 kcal/min, so (5.7 × 80) or 456 kcal is used in the 80 min needed to travel the 20 km. It thus takes almost 40 times more energy to move by car than by bicycle.

The maximum work capacity of the body is variable. For short periods of time the body can perform at very high power levels, but for long-term efforts it is more limited. Experimentally it has been found that long-term power is proportional to the maximum rate of oxygen consumption in the working muscles. For a healthy man this consumption is typically 50 ml/kg of body weight each minute.

The body supplies instantaneous energy for short-term power needs by splitting energy-rich phosphates and glycogen, leaving an oxygen deficit in the body. This process can only last about a minute and is called the anaerobic (without oxygen) phase of work; long-term activity requires oxygen (aerobic work). Figure 5.3 shows these phases of work for a cyclist.

5.4. HEAT LOSSES FROM THE BODY

Birds and mammals are referred to as *homeothermic* (warm-blooded), while other animals are considered *poikilothermic* (cold-blooded). The

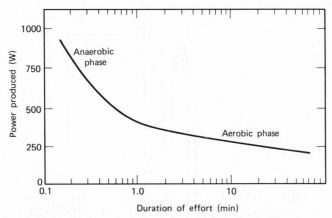

Figure 5.3. Typical power output on a bicycle versus duration of effort for an average healthy adult. Anaerobic work can only be maintained about 1 min.

terms *warm-blooded* and *cold-blooded* are misleading, for a poikilothermic animal such as a frog or a snake will have a higher body temperature on a hot day than a mammal. Birds and mammals both have mechanisms to keep their body temperatures constant despite fluctuations in the environmental temperature. Constant body temperatures permit metabolic processes to proceed at constant rates and these animals to remain active even in cold climates.

Because the body is at a constant temperature it contains stored heat energy that is essentially constant as long as we are alive. However, when metabolic activity ceases at death, the stored heat is given off at a predictable rate until the body cools to the surrounding temperature. The body temperature can thus be used to estimate how long a person has been dead.

Although the normal body (core) temperature is often given as 37°C, or 98.6°F, only a small percentage of people have exactly that temperature. If we measured the temperatures of a large number of healthy people, we would find a distribution of temperatures, with nearly everyone falling within ±0.5°C (~1°F) of the normal temperature. The rectal temperature is typically 0.5°C (~1°F) higher than the oral temperature. The temperature depends upon the time of the day (lower in the morning); the temperature of the environment; and the amount of recent physical activity, the amount of clothing, and the health of the individual. The rectal temperature after hard exercise may be as high as 40°C (104°F).

Figure 5.4 is a schematic diagram of the body's heating and cooling system. The figure does not show food, drink, and wastes (feces, flatus, and urine) or the energy that appears as external work. The heat is

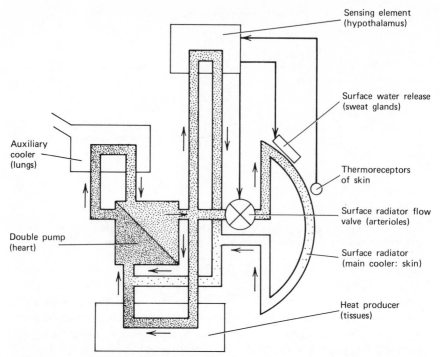

Figure 5.4.   Schematic of the heat loss system of man. The amount of heat in the blood is indicated by the density of the dots. (From R.W. Stacy, D.T. Williams, R.E. Worden, and R.O. McMorris, *Essentials of Biological and Medical Physics*, McGraw-Hill, New York, 1955, p. 158.)

generated in the organs and tissues of the body; most of it is removed by several processes that take place on the skin's surface. The main heat loss mechanisms are radiation, convection, and evaporation (of perspiration). In addition, some cooling of the body takes place in the lungs where the inspired air is usually heated and vaporized water is added to the expired air. Eating hot or cold food may also heat or cool the body.

For the body to hold its temperature close to its normal value it must have a thermostat analogous to a home thermostat that maintains the temperature of the rooms nearly constant. The hypothalamus of the brain contains the body's thermostat. If the core temperature rises, for example, due to heavy exertion, the hypothalamus initiates sweating and vasodilation, which increases the skin temperature. Both of these reactions increase the heat loss to the environment. If the skin temperature drops, the thermoreceptors on the skin inform the hypothalamus and it initiates shivering, which causes an increase in the core temperature.

The rate of heat production of the body for a 2400 kcal/day diet (assuming no change in body weight) is about 1.7 kcal/min or 120 J/sec (120 W). If the body is to maintain a constant temperature it must lose heat at the same rate. The actual amount of heat lost by radiation, convection, evaporation of sweat, and respiration depends on a number of factors: the temperature of the surroundings; the temperature, humidity, and motion of the air; the physical activity of the body; the amount of the body exposed; and the amount of insulation on the body (clothes and fat). We now discuss each of the mechanisms of heat loss for the case of a nude body.

All objects regardless of their temperature emit electromagnetic radiation (see Chapter 4, p. 70). In general, the amount of energy emitted by the body is proportional to the absolute temperature raised to the fourth power. The body also receives radiant energy from surrounding objects. The approximate difference between the energy radiated by the body and the energy absorbed from the surroundings can be calculated from the equation

$$H_r = K_r A_r e (T_s - T_w)$$

where H_r is the rate of energy loss (or gain) due to radiation, A_r is the effective body surface area emitting the radiation, e is the emissivity of the surface, T_s is the skin temperature (°C), and T_w is the temperature of the surrounding walls (°C). K_r is a constant that depends upon various physical parameters and is about 5.0 kcal/m² hr °C. The emissivity e in the infrared region is independent of the color of the skin and is very nearly equal to 1, indicating that the skin at this wavelength is almost a perfect absorber and emitter of radiation. (If we could see the deep infrared emitted by the body we would all be "black.")

Under normal conditions a large fraction of our energy loss is due to radiation even if the temperature of the surrounding walls is not much lower than body temperature. For example, if a nude body has an effective surface area of 1.2 m² and a skin temperature of 34°C, it will lose about 54 kcal/hr to walls maintained at 25°C (77°F). This amounts to about 54% of the body's heat loss. Most of the remaining heat loss is due to convection.

The heat loss due to convection (H_c) is given approximately by the equation

$$H_c = K_c A_c (T_s - T_a)$$

where K_c is a constant that depends upon the movement of the air, A_c is the effective surface area, T_s is the temperature of the skin, and T_a is the temperature of the air. When the body is resting and there is no apparent wind, K_c is about 2.3 kcal/m² hr °C. When the air temperature is 25°C, the skin temperature is 34°C, and the effective surface area is 1.2 m², the nude body loses about 25 kcal/m² hr by convection. This amounts to about 25% of the body's heat loss. When the air is moving, the constant K_c increases according to the equation $K_c = 10.45 - v + 10 \sqrt{v}$, where the wind speed v is in meters per second (Fig. 5.5). This equation is valid for speeds between 2.23 m/sec (5 mph) and 20 m/sec (45 mph).

The equivalent temperature due to moving air is called the wind chill factor and is determined by the actual temperature and wind speed. For example, at an actual temperature of −20°C and a wind speed of 10 m/sec (a stiff breeze), the cooling effect on the body is the same as −40°C on a calm day (Table 5.5). See the book by Mather listed in the bibliography for more details.

The method of heat loss that most of us are familiar with is the evaporation of sweat. Under normal temperature conditions and in the absence of hard work or exercise, this method of cooling is rather unimportant compared to radiative and convective cooling. Under extreme conditions of heat and exercise, a man may sweat more than 1 liter of liquid per hour. Since each gram of water that evaporates carries with it the heat of vaporization of 580 calories, the evaporation of 1 liter carries with it 580 kcal. Of course, the sweat must evaporate from the skin in order to give this cooling effect; sweat that runs off the body provides essentially no

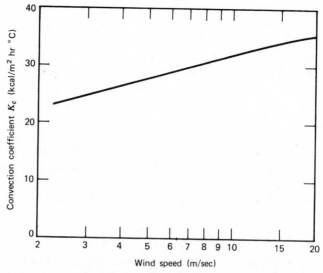

Figure 5.5. Dependence of convection coefficient K_c on wind speed for exposed skin.

Table 5.5. Wind Chill Factor Chart

Wind Speed (m/sec)	*Actual Temperature* (°C)						
	30	20	10	0	−10	−20	−30
	Equivalent Temperature (°C)						
2.23	30	20	10	0	−10	−20	−30
5	29	17	5	−7	−19	−31	−43
10	29	15	1	−13	−27	−40	−54
15	29	14	−1	−16	−30	−45	−60
20	28	13	−2	−17	−32	−48	−63

cooling. The amount evaporated depends upon the air movement and the relative humidity.

There is some heat loss due to perspiration even when the body does not feel sweaty. It amounts to about 7 kcal/hr, or 7% of the body's heat loss. A similar loss of heat is due to the evaporation of moisture in the lungs. When we breathe in air, it becomes saturated with water in the lungs. The additional water in the expired air carries away the same amount of heat as if it were evaporated from the skin. Also, when we inspire cold air, we warm it to body temperature and lose heat. Under typical conditions the total respiratory heat loss is about 14% of the body's heat loss.

Since the radiation of heat from the body and the transfer of heat to the air depend upon the skin temperature, any factors that affect the skin temperature also affect the heat loss. The body has the ability to select the path for blood returning from the hands and feet. In cold weather blood is returned to the heart via internal veins that are in contact with the arteries carrying blood to the extremities. In this way some of the heat from the blood going to the extremities is used to heat the returning blood. This *counter-current* heat exchange lowers the temperature of the extremities and reduces the heat loss to the environment. In the summertime or in a warm environment, the returning venous blood flows near the skin, raising the temperature of the skin and thus increasing the heat loss from the body.

Our discussion of heat loss mechanisms has so far been concerned with heat loss from the nude body—an interesting, but somewhat uncommon case. Including the insulation of clothing in the heat loss equations makes the calculations more difficult. The optimum skin temperature for comfort is about 33°C (92°F). This temperature can be maintained by suitably adjusting the clothing to the activity. Studies with clothing have led to the definition of a unit of clothing, the *clo*, which corresponds to the insulating value of clothing needed to maintain a subject sitting at rest in comfort in a room at 21°C (70°F) with air movement of 0.1 m/sec and air humidity of less than 50%. One clo of insulation is equal to a lightweight business suit. Obviously, 2 clos of clothing would enable a man to withstand a colder temperature than 1 clo. Likewise, a man would need a larger clo value to remain comfortable when he is inactive than when he is active. It is possible to determine the optimum clothing for comfort under various environmental conditions of temperature and air movement and for different physical activities. For example, studies show that an individual in the arctic needs clothing with insulation equal to about 4 clos. (Fox fur has an insulating value of about 6 clos.)

BIBLIOGRAPHY

Altman, P. L., and D. S. Dittmer (Eds.), *Metabolism*, Federation of American Societies for Experimental Biology, Bethesda, Md., 1968.

Benzinger, T. H., "The Human Thermostat," *Sci. Amer.*, **204**, 134–137 (1961).

Kleiber, M., *The Fire of Life: An Introduction to Animal Energetics*, Wiley, New York, 1961.

Mather, J. R., *Climatology: Fundamentals and Applications*, McGraw-Hill, New York, 1974.

Seagrave, R. C., *Biomedical Applications of Heat and Mass Transfer*, Iowa State U. P., Ames, Iowa, 1971.

Webb, P., "Work, Heat and Oxygen Costs," in J. F. Parker and V. R. West (Eds.), *Bioastronautics Data Book*, National Aeronautics and Space Administration, Washington, D.C., 1973, pp. 847–879.

Whitt, F. R., *Bicycling Science: Ergonomics and Mechanics*, edited and enlarged by D. G. Wilson, MIT Press, Cambridge, Mass., 1975.

REVIEW QUESTIONS

1. Under resting (basal) conditions, what percent of the body's energy is being used by the skeletal muscles and the heart?
2. What is a met?
3. For a hypothetical animal that has a mass of 700 kg,
   (a) Use Fig. 5.1 to estimate the basal metabolism rate.
   (b) Assuming 5 kcal/g of food, estimate the minimum amount of food needed each day.
4. By what percent does your metabolic rate increase if you have a fever 2°C above normal?
5. (a) What is the energy required to walk 20 km at 5 km/hr?
   (b) Assuming 5 kcal/g of food, calculate the grams of food needed for the walk.
6. Which organ uses the greatest power per kilogram?

7. Suppose that the elevator is broken in the building in which you work and you have to climb 9 stories—a height of 45 m above ground level. How many extra calories will this external work cost you if your mass is 70 kg and your body works at 15% efficiency?
8. What is the approximate maximum work efficiency of a trained cyclist? How does this compare to the maximum work efficiency of a steam engine?
9. A 70 kg hiker climbed a mountain 1000 m high. He reached the peak in 3 hr.
   (a) Calculate the external work done by the climber.
   (b) Assuming the work was done at a steady rate during the 3 hr period, calculate the power generated during the climb.
   (c) Assuming the average O_2 consumption during the climb was 2 liters/min (corresponding to 9.6 kcal/min), find the efficiency of the hiker's body.
   (d) How much energy appeared as heat in the body?
10. What is the anaerobic phase of work? How long will it last?
11. Are humans homeothermic or poikilothermic?
12. What are the main mechanisms of heat loss from the body?
13. (a) Calculate the convective heat loss per hour for a nude standing in a 5 m/sec wind. Assume $T_s = 33°C$, $T_a = 10°C$, and $A_c = 1.2$ m².
   (b) If the wind speed were 2.23 m/sec, find the still air temperature that would produce the same heat loss (the wind chill equivalent temperature).
14. When an individual is in water, the convective heat loss term is greatly increased. For water immersion, $K_c \simeq 16.5$ kcal/m² hr°C. Assuming the BMR of a resting man is 72 kcal/hr, find the water temperature at which the water heat loss is just balanced by the BMR. Assume $A_c = 1.75$ m² and $T_s = 34°C$.
15. Consider a nude male on a beach in Florida. It is a sunny day so he is receiving radiation from the sun at the rate of 30 kcal/hr. He has an effective body surface area of 0.9 m², $T_s = 32°C$, and the temperature of his surroundings is 30°C.
   (a) Find the net energy gained by radiation per hour.
   (b) If there is a breeze at 4 m/sec, find the energy lost by convection per hour.
   (c) If he loses 10 kcal/hr by respiration and his metabolic rate is 80 kcal/hr, how much heat is lost by evaporation?
16. Under typical conditions, what percent of the body heat loss is due to respiration?
17. When is counter-current heat exchange involved in controlling the heat loss to the environment?
18. What is a clo?

CHAPTER 6

Pressure

Pressure is a very common phenomenon in our lives. The weatherman tells us the atmospheric pressure, the service station attendant checks the pressure in our tires, and the doctor measures our blood pressure as part of a physical examination.

Pressure is defined as the force per unit area in a gas or a liquid. For a solid the quantity force per unit area is referred to as *stress*. You probably know that atmospheric pressure is about 10^5 N/m² (14.7 lb/in.²) and that the pressure in a bicycle tire may be as high as 90 lb/in.² In the metric system pressure is measured in dynes per square centimeter or newtons per square meter; the SI unit for the latter is the pascal (Pa). None of these units is in common use in medicine. The most common method of indicating pressure in medicine is by the height of a column of mercury (Hg). For example, a peak (systolic) blood pressure reading of 120 mm Hg indicates that a column of mercury of this height has a pressure at its base equal to the patient's systolic blood pressure. Atmospheric pressure is about 30 in. Hg or 760 mm Hg. Table 6.1 lists some of the common units used to measure pressure and gives atmospheric pressure in each system.

The pressure P under a column of liquid can be calculated from $P = \rho g h$, where ρ is the density of the liquid, g is the acceleration due to gravity, and h is the height of the column. Since the density of mercury is 13.6 g/cm³, a column of water has to be 13.6 times higher than a given column of mercury in order to produce the same pressure. It is sometimes convenient to indicate pressure differences in the body in terms of the height of a column of water (see Example 6.1).

Table 6.1. Some of the Common Units Used to Measure Pressure

	Atmospheres	N/m²	cm H₂O	mm Hg	lb/in.² (psi)
1 atmosphere	1	1.01×10^5	1033	760	14.7
1 N/m²	0.987×10^{-5}	1	0.0102	0.0075	0.145×10^{-3}
1 cm H₂O	9.68×10^{-4}	98.1	1	0.735	0.014
1 mm Hg	0.00132	133	1.36	1	0.0193
1 lb/in.² (psi)	0.0680	6895	70.3	51.7	1

Example 6.1

What height of water will produce the same pressure as 120 mm Hg?

$$P \text{ (120 mm Hg)} = \rho g h = (13.6 \text{ g/cm}^3)(980 \text{ cm/sec}^2)(12 \text{ cm})$$
$$= 1.6 \times 10^5 \text{ dynes/cm}^2$$

For water

$$1.6 \times 10^5 \text{ dynes/cm}^2 = (1.0 \text{ g/cm}^3)(980 \text{ cm/sec}^2)(h \text{ cm H}_2\text{O})$$
$$h = 163 \text{ cm H}_2\text{O}$$

The height of the water can be obtained by multiplying the height of the mercury by 13.6.

Since we live in a sea of air with a pressure of 1 atm, it is easier to measure pressure relative to atmospheric pressure than to measure true, or absolute, pressure. For example, if the pressure in a bicycle tire is 60 lb/in.², the absolute pressure is 60 + 14.7, or nearly 75 lb/in.² The 60 lb/in.² is the *gauge pressure*. Unless we indicate otherwise, all the pressures used in this chapter are gauge pressures.

There are a number of places in the body where the pressures are lower than atmospheric, or negative. For example, when we breathe in (inspire) the pressure in the lungs must be somewhat lower than atmospheric pressure or the air would not flow in. The lung pressure during inspiration is typically a few centimeters of water negative. When a person drinks through a straw the pressure in his mouth must be negative by an amount equal to the height of his mouth above the level of the liquid he is drinking. Other examples of negative pressure are discussed in Chapter 7 when we consider the physics of the lungs and breathing.

Table 6.2 lists some typical pressures in the body. The heart acts as a pump, producing quite high pressure (~100 to 140 mm Hg) to force the blood through the arteries. The returning venous blood is at quite low pressure and, in fact, needs help to get from the legs to the heart. The failure of this return system in the legs often results in varicose veins. Pressure in the circulatory system is discussed in detail in Chapter 8, and

Table 6.2. Typical Pressures in the Normal Body

	Typical Pressure (mm Hg)
Arterial blood pressure	
Maximum (systole)	100–140
Minimum (diastole)	60–90
Venous blood pressure	3–7
Great veins	<1
Capillary blood pressure	
Arterial end	30
Venous end	10
Middle ear pressure	<1
Eye pressure—aqueous humor	20
Cerebrospinal fluid pressure in brain (lying down)	5–12
Gastrointestinal pressure	10–20
Intrathoracic pressure (between lung and chest wall)	−10

pressure in the lungs is discussed in Chapter 7. In this chapter we discuss some of the other pressure systems of the body and high pressure (hyperbaric) oxygen therapy.

6.1. MEASUREMENT OF PRESSURE IN THE BODY

The classical method of measuring pressure is to determine the height of a column of liquid that produces a pressure equal to the pressure being measured. In Section 8.4 we describe how Rev. Hale measured the blood pressure of a horse by seeing how high a column of blood from an artery would rise in a glass tube. An instrument that measures pressure by this method is called a *manometer*. A common type of manometer is a **U**-shaped tube containing a fluid that is connected to the pressure to be measured (Fig. 6.1). The levels in the arms change until the difference in the levels is equal to the pressure. This type of manometer can measure both positive and negative pressures. The fluid used is usually mercury, but water or other low density fluids can be used when the pressure to be measured is relatively small.

The most common clinical instrument used in measuring pressure is the *sphygmomanometer,* which measures blood pressure. Two types of pressure gauges are used in sphygmomanometers. In a mercury manometer type the pressure is indicated by the height of a column of mercury inside a glass tube. In an aneroid type the pressure changes the shape of a sealed flexible container, which causes a needle to move on a dial.

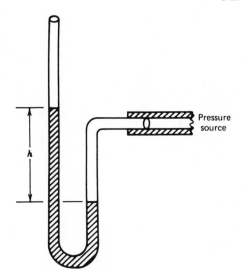

Figure 6.1. A U tube manometer for measuring pressure P. P can be expressed as the height of the fluid, h (e.g., in mm Hg or cm H_2O), or it can be expressed in conventional units of force per unit area by using $P = \rho g h$, where ρ is the density and g is the acceleration due to gravity.

Some parts of the body can act like crude pressure indicators. For example, a person going up or down in an elevator or an airplane is often aware of the change in atmospheric pressure on the ears. When one swallows the pressure in the middle ear equalizes to the outside pressure and the eardrums "pop." It is necessary for the ears to be very sensitive to pressure since pressure changes in an ordinary sound wave are extremely small (see Example 12.1). Another qualitative pressure indicator is the size of the veins on the back of the hand. As a hand is raised slightly above the level of the heart these veins become smaller due to the lower venous blood pressure (see Chapter 8, p. 164).

6.2. PRESSURE INSIDE THE SKULL

The brain contains approximately 150 cm³ of cerebrospinal fluid (CSF) in a series of interconnected openings called ventricles (Fig. 6.2). Cerebrospinal fluid is generated inside the brain and flows through the ventricles into the spinal column and eventually into the circulatory system. One of the ventricles, the aqueduct, is especially narrow. If at birth this opening is blocked for any reason, the CSF is trapped inside the skull and increases the internal pressure. The increased pressure causes the skull to enlarge. This serious condition, called *hydrocephalus* (literally, water-head), is a moderately common problem in infants. However, if the condition is detected soon enough, it can often be corrected by surgically installing a by-pass drainage system for the CSF.

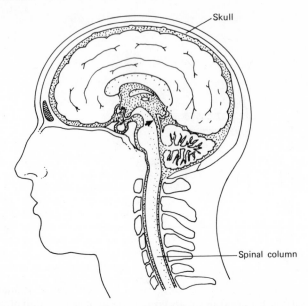

Figure 6.2.   A cross-section of the brain showing the location of the cerebrospinal fluid (shaded area) and the aqueduct (arrow). The fragile brain is supported and cushioned by this fluid.

It is not convenient to measure the CSF pressure directly. One rather crude method of detecting hydrocephalus is to measure the circumference of the skull just above the ears. Normal values for newborn infants are from 32 to 37 cm, and a larger value may indicate hydrocephalus. Another qualitative method of detection, transillumination, makes use of the light-scattering properties of the rather clear CSF inside the skull. Transillumination is discussed in more detail in Section 14.2.

6.3.   EYE PRESSURE

The clear fluids in the eyeball (the aqueous and vitreous humors) that transmit the light to the retina (the light-sensitive part of the eye), are under pressure and maintain the eyeball in a fixed size and shape. The dimensions of the eye are critical to good vision—a change of only 0.1 mm in its diameter has a significant effect on the clarity of vision. If you press on your eyelid with your finger you will notice the resiliency of the eye due to the internal pressure. The pressure in normal eyes ranges from 12 to 23 mm Hg.

The fluid in the front part of the eye, the aqueous humor, is mostly water. The eye continuously produces aqueous humor and a drain system allows the surplus to escape. If a partial blockage of this drain system occurs, the pressure increases and the increased pressure can restrict the blood supply to the retina and thus affect the vision. This condition, called *glaucoma,* produces tunnel vision in moderate cases and blindness in severe cases.

Early physicians estimated the pressure inside the eye by "feel" as they pressed on the eye with their fingertips. Now pressure in the eye is measured with several different instruments, called *tonometers,* that measure the amount of indentation produced by a known force. The tonometers currently used are described in detail in Chapter 15. Tonometers are sometimes calibrated in arbitrary units rather than in millimeters of mercury.

6.4. PRESSURE IN THE DIGESTIVE SYSTEM

The body has an opening through it. This opening, the digestive tract, is rather tortuous; it extends over 6 m from the mouth to the anus. Most of the time it is closed at the lower end and has several other restrictions. Figure 6.3 shows schematically the valves and sphincters (circular muscles) of the digestive tract, which open for the passage of food, drink, and their by-products. The valves are designed to permit unidirectional flow of food. With some effort it is possible to reverse the flow, such as during vomiting (emesis).

The pressure is greater than atmospheric in most of the gastrointestinal

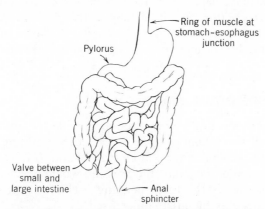

Figure 6.3. The valves and sphincters of the digestive tract.

(GI) system. However, in the esophagus, the pressure is coupled to the pressure between the lungs and chest wall (intrathoracic pressure) and is usually less than atmospheric. The intrathoracic pressure is sometimes determined by measuring the pressure in the esophagus.

During eating the pressure in the stomach increases as the walls of the stomach are stretched. However, since the volume increases with the cube of the radius (R^3) while the tension (stretching force) is proportional to R, the increase in pressure is very slow. A more significant increase in pressure is due to air swallowed during eating. Air trapped in the stomach causes burping or belching. This trapped air is often visible on an x-ray of the chest.

In the gut, gas (flatus) generated by bacterial action increases the pressure. (Flatus is produced in even the most cultured people!) External factors such as belts, girdles, flying, and swimming affect the gut pressure.

One valve, the pylorus, prevents the flow of food back into the stomach from the small intestine. Occasionally a blockage forms in the small or large intestine and pressure builds up between the blockage and the pylorus; if this pressure becomes great enough to restrict blood flow to critical organs, it can cause death. Intubation, the passing of a hollow tube through the nose, stomach, and pylorus, is usually used to relieve the pressure. If intubation does not work it is necessary to relieve the pressure surgically. However, the high pressure greatly increases the risk of infection because the trapped gases expand rapidly when the incision is made. This risk can be reduced if surgery is performed in an operating room in which the external pressure is greater than the pressure in the gut.

The pressure in the digestive system is coupled to that in the lungs through the flexible diaphragm that separates the two organ systems. When it is necessary or desirable to increase the pressure in the gut, such as during defecation, a person takes a deep breath, closes off the lungs at the glottis (vocal folds), and contracts the abdominal muscles.

6.5. PRESSURE IN THE SKELETON

The highest pressures in the body are found in the weight-bearing bone joints. When all the weight is on one leg, such as when walking, the pressure in the knee joint may be more than 10 atm! If it were not for the relatively large area of the joints, the pressure would be even higher (Fig. 6.4). Since pressure is the force per unit area, for a given force the pressure is reduced as the area is increased.

Healthy bone joints are better lubricated than the best man-made bear-

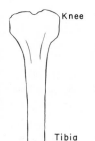

Knee

Tibia

Figure 6.4. The surface area of a bone at the joint is greater than its area either above or below the joint. The larger area at the joint distributes the force, thus reducing the pressure.

ings (see Chapter 3, p. 56). If a conventional lubricant were used in a joint it would be squeezed out and the joint would soon be dry. Fortunately, the system is such that the higher the pressure, the better the lubrication. Joint lubrication is discussed in more detail in Chapter 3.

Bone has adapted in another way to reduce pressure. The finger bones are flat rather than cylindrical on the gripping side, and the force is spread over a larger surface; this reduces the pressure in the tissues over the bones (Fig. 6.5).

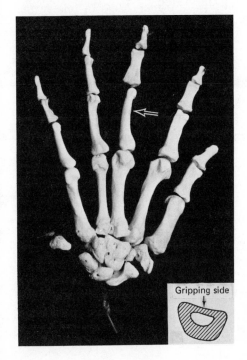

Gripping side

Figure 6.5. The finger bones rotated 90° to show the flat surface (arrow) used for gripping (see cross-section in inset). This flat surface reduces the pressure on the tissues over the bones when we carry something heavy like a suitcase.

6.6. PRESSURE IN THE URINARY BLADDER

One of the most noticeable internal pressures is the pressure in the bladder due to the accumulation of urine. Figure 6.6 shows the typical pressure-volume curve for the bladder, which stretches as the volume increases. One might naively expect the rise in pressure to be proportional to the volume. However, for a given increase in radius R the volume increases as R^3 while the pressure only increases as about R^2. This relationship largely accounts for the relatively low slope of the major portion of the pressure-volume curve in Fig. 6.6. For adults, the typical maximum volume in the bladder before voiding is 500 ml. At some pressure ($\sim$30 cm H_2O) the micturition ("gotta go") reflex occurs. The resulting sizable muscular contraction in the bladder wall produces a momentary pressure of up to 150 cm H_2O. Boys occasionally perform the physics "experiment" of measuring this maximum pressure directly by seeing how high they can urinate on the wall of a building. Normal voiding pressure is fairly low (20 to 40 cm H_2O), but for men who suffer from prostatic obstruction of the urinary passage it may be over 100 cm H_2O.

The pressure in the bladder can be measured by passing a catheter with a pressure sensor into the bladder through the urinary passage (urethra). In direct cystometry the pressure is measured by means of a needle inserted through the wall of the abdomen directly into the bladder (Fig. 6.7). This technique gives information on the function of the exit valves (sphincters) that cannot be obtained with the catheter technique.

The bladder pressure increases during coughing, straining, and sitting up. During pregnancy, the weight of the fetus over the bladder increases

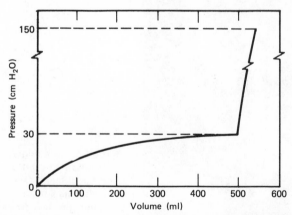

Figure 6.6. The typical pressure-volume relationship in the urinary bladder (cystometrogram).

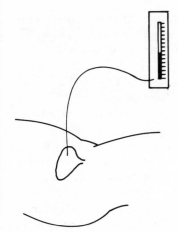

Figure 6.7. In direct cystometry a needle is passed through the wall of the abdomen directly into the bladder.

the bladder pressure and causes frequent urination. A stressful situation may also produce a pressure increase; studying for exams often results in many trips to the toilet due to "nerves."

6.7. PRESSURE EFFECTS WHILE DIVING

Since the body is composed primarily of solids and liquids, which are nearly incompressible, pressure changes do not greatly affect most of it. However, there are gas cavities in the body where sudden pressure changes can produce profound effects. To understand why, we must recall Boyle's law: for a fixed quantity of gas at a fixed temperature the product of the absolute pressure and volume is constant (PV = constant). That is, if the absolute pressure is doubled, the volume is halved. Applications of Boyle's law to scuba diving are given in Example 6.2.

Example 6.2
a. What volume of air at an atmospheric pressure of 1.01×10^5 N/m² is needed to fill a 14.2 liter (0.5 ft³) scuba tank to a pressure of 1.45×10^7 N/m² (2100 lb/in.²)?

$$P_1V_1 = P_2V_2$$
$$(1.01 \times 10^5)\,(V_1) = (1.45 \times 10^7)\,(14.2)$$
$$V_1 = 2 \times 10^3 \text{ liters (72 ft}^3)$$

b. Since at sea level a diver uses about 14.2 liters (0.5 ft³) of air per minute during moderate activity, the tank in (*a*) would last about 144 min. How long would the tank last at a depth of 10 m (33 ft) where the pressure is increased by 1 atmosphere, assuming the same volume use rate?

Since the absolute pressure is twice as great (2 atm), the tank will last only 72 min. (However, no safety-conscious diver would completely empty his tank during a dive for then he would have to surface without air.)

The middle ear is one air cavity that exists within the body (Fig. 13.1). For comfort the pressure in the middle ear should equal the pressure on the outside of the eardrum. This equalization is produced by air flowing through the Eustachian tube, which is usually closed except during swallowing, chewing, and yawning. When diving, many people have difficulty obtaining pressure equalization and feel pressure on their ears. A pressure differential of 120 mm Hg across the eardrum, which can occur in about 1.7 m (5.5 ft) of water, can cause the eardrum to rupture. Rupture can be serious since cold water in the middle ear can affect the vestibular or balance mechanism and cause nausea and dizziness. One method of equalization used by a diver is to raise the pressure in the mouth by holding the nose and trying to blow out; as the pressure equalizes the diver can often "hear" both ears "pop."

A less serious condition is *sinus squeeze*. During a dive the pressure in the sinus cavities in the skull usually equalizes with the surrounding pressure. If a diver has a cold, the sinus cavities may become closed off and not equalize, causing pain. Another pressure effect is pain during and after dives from small volumes of air trapped beneath fillings in the teeth. *Eye squeeze* can occur if goggles are used instead of a facemask; with a facemask the air exhaled from the lungs increases the pressure over the eyes as the descent is made.

If a scuba diver at a depth of 10 m holds his breath and comes to the surface, the air volume will expand by a factor of two and thus cause a serious pressure rise in the lungs. If the lungs are filled to capacity, an ascent of only 1.2 m (4 ft) can cause serious lung damage. All scuba divers learn during training to avoid breath-holding during ascent and to exhale continuously if a rapid ascent is necessary.

The pressure in the lungs at any depth is greater than the pressure in the lungs at sea level. This means that the air in the lungs is more dense underwater and that the partial pressures of all the air components are proportionately higher. The higher partial pressure of oxygen causes more oxygen molecules to be transferred into the blood, and *oxygen poisoning* results if the partial pressure of oxygen gets too high (Fig. 6.8). Usually oxygen poisoning occurs when the partial pressure of oxygen is about 0.8 atm (when the absolute air pressure is about 4 atmospheres), or at a depth of about 30 m (100 ft).

Breathing air at a depth of 30 m is also dangerous because it may result in excess nitrogen in the blood and tissues. This can produce two serious

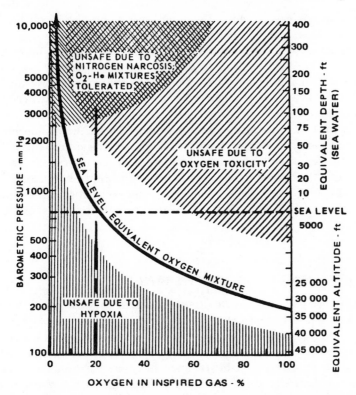

Figure 6.8. At the normal 20% O₂ mixture and at a sea water depth greater than 30 m (100 ft) both oxygen poisoning and nitrogen narcosis can occur. Note that the higher pressures are at the top of the graph and that the pressure decreases as the altitude increases. At an altitude of about 4600 m (~15,000 ft) hypoxia (a shortage of oxygen in the tissues) can occur. (From C.E. Billings in J.F. Parker, Jr., and V.R. West (Eds.), *Bioastronautics Data Book*, 2nd ed., National Aeronautics and Space Administration, Washington, D.C., 1973, p. 2, as adapted from the United States House of Representatives Select Committee on Astronautics and Space Exploration, *Space Handbook: Astronautics and Its Applications*, House Document No. 86, First Session, 86th Congress, Washington, D.C., U.S. Government Printing Office, 1959.)

problems: nitrogen narcosis, which is an intoxication effect (Fig. 6.8), and the *bends,* or decompression sickness, which is an ascent problem. While oxygen is transported primarily by chemical attachment to the red blood cells, nitrogen is dissolved in the blood and tissues. According to Henry's law, the amount of gas that will dissolve in a liquid is proportional to the partial pressure of the gas in contact with the liquid. Thus more nitrogen is dissolved in the blood and from there into the tissues as a diver goes deeper since the pressure of the air and thus the partial pressure of nitrogen are increasing. When the diver ascends, the extra nitrogen in the tissues must be removed via the blood and the lungs. The removal is a slow process, and if the diver ascends too fast bubbles form in the tissues and joints. The bends are quite painful. Stricken divers are usually recompressed in a chamber; the pressure in the chamber is slowly decreased so that the nitrogen can be removed from the tissues via the blood and the lungs.

Other problems can occur during ascent. One of the membranes that separate air and blood in the lung can burst, allowing air to go directly into the bloodstream (air embolism). Air can also become trapped under the skin around the base of the neck or in the middle of the chest. In addition, pneumothorax (lung collapse) can result if air gets between the lungs and the chest wall (see Chapter 7, p. 143). These problems are best treated by a physician.

6.8. HYPERBARIC OXYGEN THERAPY (HOT)

The body normally lives in an atmosphere that is about one-fifth oxygen and four-fifths nitrogen. In some medical situations it is beneficial to increase the proportion of oxygen in order to provide more oxygen to the tissues. Oxygen tents are often used for this purpose. To greatly increase the amount of oxygen, medical engineers have constructed special high pressure (hyperbaric) oxygen chambers. Some are just large enough for a patient, while others are large enough to serve as operating rooms.

Gas gangrene is a disease that killed more than half of its victims before hyperbaric oxygen therapy (HOT) was developed. Since the bacillus that causes gas gangrene cannot survive in the presence of oxygen, almost all gas gangrene patients treated with HOT are cured without the need for amputation—the previous best method of treatment.

In carbon monoxide poisoning the red blood cells cannot carry oxygen to the tissues because the carbon monoxide fastens to the hemoglobin at the places normally used by oxygen. The presence of even a few carbon monoxide molecules on a red blood cell greatly reduces the ability of the cell to transport oxygen. Normally the amount of oxygen dissolved in the blood is about 2% of that carried on the red blood cells. With HOT, the partial pressure of oxygen can be increased by a factor of 15, permitting enough oxygen to be dissolved to fill the body's needs. Many victims of carbon monoxide poisoning are saved with this technique.

Hyperbaric oxygen has been used in conjunction with radiation in the treatment of cancer. (Radiation therapy is discussed in detail in Chapter 18.) The patient was placed inside a transparent plastic tank, and the radiation was beamed through the walls into the tumor. The theory was that more oxygen would make the poorly oxygenated radiation-resistant cells in the center of the tumor more susceptible to radiation damage. This technique, which worked well in the laboratory on cells and mice, did not produce markedly better results than present techniques for humans. The use of oxygen pressures of up to 3 atm required the constant attendance of a physician and a nurse. The patient's eardrums were usually intentionally punctured to aid in the pressure equalization process, and the entire process lasted about 1 hr while conventional treatments take about 10 min.

Like many new developments in medicine, hyperbaric oxygen therapy brought with it new problems. The oxygen atmosphere makes fire a much greater hazard—three astronauts died in the pure oxygen atmosphere on a U.S. spaceship during preliminary tests in 1967. Another problem is the risk of rupture of the tank due to the high pressures used. Such a rupture occurred on at least one occasion, seriously injuring the patient and the physician in attendance. However, physical dangers such as these are usually easier to evaluate and avoid than biological dangers, which are often poorly understood (e.g., air pollution).

BIBLIOGRAPHY

Brummelkamp, W. H., *Hyperbaric Oxygen Therapy in Clostridial Infections, Type Welchii,* Bohn, Haarlem, 1965.

Fundamentals of Hyperbaric Medicine, National Research Council Committee on Hyperbaric Oxygenation, Publication 1298, National Academy of Sciences–National Research Council, Washington, D.C., 1966.

New Science of Skin and Scuba Diving, 4th revised ed., Council for National Cooperation in Aquatics, Associated Press, New York, 1974.

REVIEW QUESTIONS

1. What is negative pressure?
2. Calculate the pressure in millimeters of mercury equal to a pressure of 20 cm H_2O.
3. What causes the bends? How is a victim with the bends treated?
4. Assume you are a shallow-water diver preparing for a 10 m dive into salt water.
   (a) What absolute pressure and gauge pressure will you experience?
   (b) Normally your lungs have an available volume of 6 liters. What will happen to that volume?
   (c) Suppose you cannot equalize the pressure in your middle ear. What will happen during the dive?

5. The venous pressure is typically about 5 mm Hg. Describe a method to measure this pressure.

6. Positive pressure is used in blood transfusions. Suppose a container is placed 1 m above a vein with a venous pressure of 2 mm Hg; if the density of the blood is 1.04 g/cm³, what is the net pressure acting to transfer the blood into the vein?

7. Suppose you are a deep-sea diver preparing for a dive to 30 m.
   (a) What absolute pressure and gauge pressure will you experience?
   (b) What will be your rate of air consumption compared to that at sea level?

8. Negative pressure or suction is often used to drain body cavities. In the drainage arrangement for the gastrointestinal region shown, the negative pressure supplied to the collection bottle is 100 mm Hg and the top end of the tube is 37 cm above the end of the tube in the body. Find the negative pressure at the lower end of the tube.

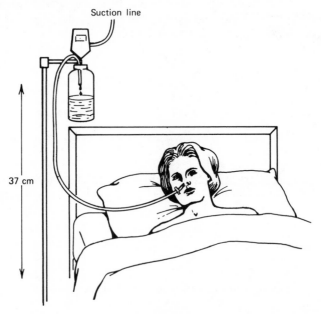

Suction line

37 cm

9. Atmospheric pressure is due to the weight of the air above us. The density of air is 1.3×10^{-3} g/cm³. What is the weight in dynes of 1 cm³ of air? If this weight were spread over 1 cm² what would be the pressure? What fraction of 1 atm would it be?

10. Using the density of air given in Review Question 9, calculate the pressure difference in dynes per square centimeter and in millimeters of mercury between the bottom and the top of a building 30 m tall (8 stories).

CHAPTER 7

The Physics of the Lungs and Breathing

The body is in many ways a machine—a very remarkable machine. It must have a source of energy, a method of converting the energy into electrical and mechanical forms, and a way of disposing of the by-products. In an automobile the source of energy is gasoline; it is combined with air and burned in the cylinders to produce kinetic energy to drive the wheels, and its by-products of noxious gases and heat are disposed of through the exhaust and radiator. In the body the source of energy is food; it is processed in the digestive system and then combined with O_2 in the cells of the body to release energy. Its by-products are disposed of by four routes: (1) the nondigestible components are eliminated as feces (releasing only a small amount of noxious gases), (2) water and other by-products are carried away in the urine, (3) almost 0.5 kg of CO_2 is disposed of via the lungs each day, and (4) heat is dissipated from the body's surface.

The human "machine" really consists of billions of very small "engines"—the living cells of the body. Each of these miniature engines must be provided with fuel, O_2, and a method of getting rid of the by-products. The blood and its vessels (cardiovascular system) serve as the transport system for these engines. The lungs (pulmonary system) serve as the supplier of O_2 and the disposer of the main by-product—CO_2. The blood takes the O_2 to the tissues and removes the CO_2 from the tissues; it must come in close contact with the air in the lungs in order to exchange its load of CO_2 for a fresh load of O_2 (Fig. 7.1). The details of this process are discussed in Section 7.2.

Because of the close cooperation and interactions between the cardiovascular and pulmonary systems, the actions of one system often

119

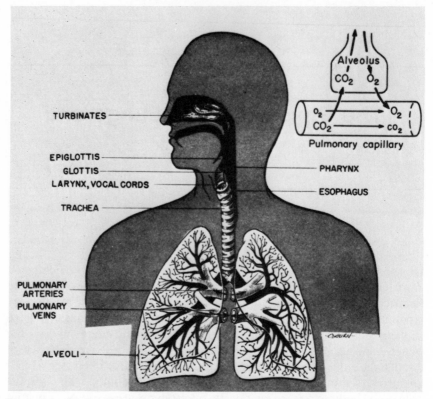

Figure 7.1. The major channels for air and blood in the respiratory system. The diagram illustrates schematically the exchange of O_2 and CO_2 between the air in an alveolus and the blood. (From Guyton, A.C., *Function of the Human Body,* **3rd ed., © W.B. Saunders Company, Philadelphia, 1969, p. 222. Reprinted by permission.)**

affect the other. For example, during breathing the pressure on the major veins in the chest affects the return of blood to the heart. Often a disease of the lungs will produce heart symptoms and vice versa.

The lungs perform other physiologic functions in addition to exchanging O_2 and CO_2. One primary function is keeping the pH (acidity) of the blood constant. The lungs play a secondary role in the heat exchange (see Chapter 5) and fluid balance of the body by warming and moisturizing the air we breathe in (inspire). Our breathing mechanisms provide a controlled flow of air for talking, coughing, sneezing, sighing, sobbing, laughing, sniffing, and yawning. In addition, blocking the air passage generates increased pressure for defecating and vomiting.

An important function of the breathing apparatus is voice production.

Breathing patterns are markedly different during conversation. Since the voice is produced by a controlled outflow of air from the lungs a person inhales rapidly and more deeply before speaking in order to have more time to produce voice sounds. The inhalation time is typically less than 20% of the breath cycle, and the amount inhaled is usually more than twice the usual volume. A singer (especially an opera singer) inhales even more air in a short period of time to further minimize the inhalation part of the cycle. The airway resistance produced by the vocal cords causes a sizable pressure increase in the trachea. Thus the work involved in speaking and singing is considerably greater than the work of normal breathing. However, relatively little of the increased work goes into sound energy. The voice typically has a power of less than 1 mW. Voice production is discussed in Chapter 12.

We breathe about 6 liters of air per minute. (This is also about the volume of blood the heart pumps each minute.) Men breathe about 12 times per minute at rest, women breathe about 20 times per minute, and infants breathe about 60 times per minute. We discuss in Section 7.6 the physical factors that affect the breathing rate.

The air we inspire is about 80% N_2 and 20% O_2. Expired air is about 80% N_2, 16% O_2,* and 4% CO_2. We breathe about 10 kg (22 lb) of air each day. Of this the lungs absorb about 400 liters ($\sim$0.5 kg) of O_2 and release a slightly smaller amount of CO_2. We also saturate the air we breathe with water. When we breathe in dry air, our expired air carries away about 0.5 kg of water each day. (This moisture can be used to clean glasses.) In cold weather some of this moisture condenses and we see our breath.

The air we breathe contains dust, smoke, air-borne bacteria, noxious gases, and so forth, which come into close contact with the blood. The large convoluted surfaces of the lungs with a surface area of about 80 m² have a greater exposure to the environment than any other part of the body including the skin. It is perhaps surprising we don't have more diseases of the lungs. The importance of clean air is obvious.

Each time we breathe, about 10^{22} molecules of air enter our lungs. Remember that 22.4 liters of air contain about 6×10^{23} molecules— Avogadro's number. The total number of molecules in the earth's atmosphere is about 10^{44}. We thus take in $1/10^{22}$ of all the earth's air each time we breathe; in other words, for each molecule we breathe there are 10^{22} more in the earth's atmosphere. The earth's atmosphere is in constant motion, and over a period of centuries there has been thorough mixing of the gases. As a result, each 0.5 liter of air (10^{22} molecules) contains on the

---

*This relatively high percentage of O_2 is the reason that mouth-to-mouth resuscitation is practical and that blowing on a campfire helps get it started.

average one molecule that was present in any 0.5 liter of air centuries ago. An interesting way to think of this is that on the average each of our breaths contains one air molecule that was breathed by Christ in any one of his breaths. Of course, this is true for any famous, infamous, or unknown person who lived many years ago.

7.1. THE AIRWAYS

The principal air passages into the lungs are shown in Fig. 7.2. Air normally enters the body through the nose where it is warmed (if necessary), filtered, and moisturized. The moist surfaces and the hairs in the nose trap dust particles, bugs, and so forth. During heavy exercise, such as jogging, air is breathed in through the mouth and bypasses this filter system. The air then passes through the windpipe (trachea). The trachea divides in two (bifurcates) to furnish air to each lung through the bronchi.

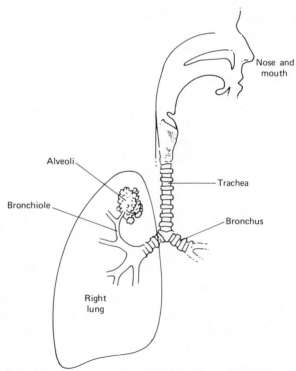

Figure 7.2.   A schematic diagram showing the principal air passages into the lungs.

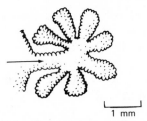

1 mm

Figure 7.3. The structure of the alveoli.

Each bronchus divides and redivides about 15 more times; the resulting terminal bronchioles supply air to millions of small sacs called *alveoli.** The alveoli, which are like small interconnected bubbles (Fig. 7.3), are about 0.2 mm in diameter (a sheet of paper is ~0.1 mm thick) and have walls only 0.4 μm thick. They expand and contract during breathing; they are "where the action is" in the exchange of O_2 and CO_2. Each alveolus is surrounded by blood so that O_2 can diffuse from the alveolus into the red blood cells and CO_2 can diffuse from the blood into the air in the alveolus. At birth the lungs have about 30 million alveoli; by age 8 the number of alveoli has increased to about 300 million. Beyond this age the number stays relatively constant, but the alveoli increase in diameter. The alveoli play such an important role in breathing that we will discuss the physics of the alveoli in more detail in Section 7.5.

In addition to serving as the transport system for the air, the airways remove the dust particles that stick to the moist lining of the various air passages. The body has two mechanisms for clearing the airways of foreign particles. Large chunks are removed by coughing. Small particles are carried upward toward the mouth by millions of small hairs, or cilia. The cilia, which are only about 0.1 mm long, have a waving motion that moves mucus carrying dust and other small particles up the major airways. Each of the cilia vibrates about 1000 times a minute. The mucus moves at about 1 to 2 cm/min (~ 1 mile/week!). You can think of the cilia as an escalator system for the trachea. It takes about 30 min for a particle of dust to be cleared out of the bronchi and trachea into the throat where it is expelled or swallowed.

7.2. HOW THE BLOOD AND LUNGS INTERACT

The primary purposes of breathing are to bring a fresh supply of O_2 to the blood in the lungs and to dispose of the CO_2. In this section we will try to

*Pronounced al-ve'-o-li.

help you understand the physics involved in the exchange of gas between the lungs and the blood.

Blood is pumped from the heart to the lungs under relatively low pressure. The average peak blood pressure in the main pulmonary artery carrying blood to the lungs is only about 20 mm Hg or about 15% of the pressure in the main body circulation. The lungs offer little resistance to the flow of blood. On the average, about one fifth ($\sim$1 liter) of the body's blood supply is in the lungs, but only about 70 ml of that blood is in the capillaries of the lungs getting O_2 at any one time. Since blood is in the pulmonary capillaries for less than 1 sec, the lungs must be well designed for gas exchange; the alveoli of the lungs have extremely thin walls and are surrounded by the blood in the pulmonary capillary system. The surface area between air and blood in the lungs is about 80 m² (about one-half the area of a tennis court). If the 70 ml of blood in the pulmonary capillaries were spread over a surface area of 80 m² the resulting layer of blood would be only about 1 μm thick, less than the thickness of a single red blood cell.

Two general processes are involved in gas exchange in the lungs: (1) getting the blood to the pulmonary capillary bed (perfusion) and (2) getting the air to the alveolar surfaces (ventilation). If either process fails the blood will not be properly oxygenated.

There are three types of ventilation-perfusion areas in the lungs: (1) areas with good ventilation and good perfusion, (2) those with good ventilation and poor perfusion, and (3) those with poor ventilation and good perfusion. In a normal lung the first type accounts for over 90% of the total volume. If the blood flow to part of a lung is blocked by a clot (a pulmonary embolism) that volume will have poor perfusion. If air passages in the lungs are obstructed as in pneumonia, the involved area will have poor ventilation. Many pulmonary diseases cause reductions in perfusion or in ventilation.

The transfer of O_2 and CO_2 into and out of the blood is controlled by the physical law of diffusion. All molecules are continually in motion. In gases and liquids, and to a certain extent even in solids, the molecules do not remain in one location. For example, if you could identify a group of molecules in a room (e.g., from a drop of perfume) in a few minutes you would find that these molecules had moved (diffused) throughout the room. Molecules of a particular type diffuse from a region of higher concentration to a region of lower concentration until the concentration is uniform. In the lungs we are concerned with diffusion in both gas and liquids. In the O_2 and CO_2 exchange in the tissues we are concerned only with diffusion in liquids. The molecules in a gas at room temperature move at about the speed of sound. Each molecule collides about 10^{10} times each second with neighboring molecules, in the process wandering about in a random manner. The most probable distance D a molecule will travel from its origin after N collisions is $D = \lambda \sqrt{N}$, where λ is the mean free path, or the average distance between collisions. In air λ is about 10^{-7} m; in tissue λ is about 10^{-11} m (see Example 7.1).

Example 7.1
What is the typical value of D in air and in tissue for an O_2 molecule after 1 sec if $N = 10^{10}$ in air and $N = 10^{12}$ in tissue?

$$\text{In air } D \simeq 10^{-7} (10^{10})^{1/2} = 10^{-2} \text{ m.}$$

$$\text{In tissue } D \simeq 10^{-11} (10^{12})^{1/2} = 10^{-5} \text{ m.}$$

Diffusion depends on the speed of the molecules; it is more rapid if the molecules are light and it increases with temperature. Since N is proportional to the diffusion time Δt (i.e., $N \propto \Delta t$), we can write that $D \propto \sqrt{\Delta t}$ or $\Delta t \propto D^2$. If $D = 10$ mm after 1 sec, it will take that molecule 100 sec on the average to diffuse 100 mm. In the lungs the distance to be traveled in air is usually a small fraction of a millimeter, and diffusion takes place in a fraction of a second. The diffusion of O_2 and CO_2 in tissue is about 10,000 times slower than it is in air, but the tissue thickness the molecules must diffuse through in the lungs is very small (~ 0.4 μm) and diffusion through the alveolar wall takes place in much less than 1 sec. We discuss diffusion in tissues more in Chapter 8.

To understand the behavior of gases in the lungs it is necessary to review Dalton's law of partial pressures. This law says that if you have a mixture of several gases, each gas makes its own contribution to the total pressure as though it were all alone. Consider a closed liter container of dry air at atmospheric pressure (760 mm Hg). If you removed all the molecules except O_2 from the container the pressure would drop to about 150 mm Hg (i.e., 20% of 760 mm Hg). This is the partial pressure of oxygen pO_2. If only the N_2 molecules were left, the pressure would be about 80% of 760 mm Hg or about 610 mm Hg. Figure 7.4 schematically shows this imaginary experiment. The partial pressure of water vapor in air depends on the humidity. In typical room air the partial pressure of water vapor is 15 to 20 mm Hg; in the lungs at 37°C and 100% relative humidity the partial pressure of water vapor is 47 mm Hg.

Consider what happens in a closed container of blood and O_2. Some O_2 molecules collide with blood and are dissolved. After a while the number of O_2 molecules that are escaping from the blood each second is the same

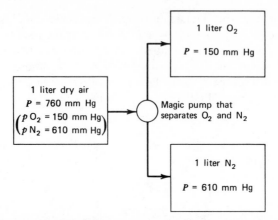

Figure 7.4.   A schematic illustration of Dalton's law of partial pressures. A liter of air at 760 mm Hg pressure can be thought of as a mixture of 1 liter of O₂ at a pressure of 150 mm Hg and 1 liter of N₂ at a pressure of 610 mm Hg.

as the number that are entering it. The blood then has a pO_2 equal to that of the O_2 in contact with it. If the pO_2 in the gas phase is doubled, the amount of O_2 dissolved in the blood will also double. This proportionality is Henry's law of solubility of gases.

The amount of gas dissolved in blood varies greatly from one gas to another. Oxygen is not very soluble in blood or water. At body temperature 1 liter of blood plasma at a pO_2 of 100 mm Hg will hold only about 2.5 cm³ of O_2 at normal temperature and pressure (NTP). At a pCO_2 of 40 mm Hg it will hold about 25 cm³ of CO_2 in solution. If the body had to depend on dissolved O_2 in the plasma to supply O_2 to the cells the heart would have to pump 140 liters of blood per minute at rest instead of the 6 liters/min it actually pumps. As we discuss shortly there is a more efficient method of transporting O_2 and CO_2 that involves the red blood cells.

The different solubilities of O_2 and CO_2 in tissue affect the transport of these gases across the alveolar wall. A molecule of O_2 diffuses faster than a molecule of CO_2 because of its smaller mass. However, because of the greater number of CO_2 molecules in solution, the transport of CO_2 is more efficient than the transport of O_2. If a disease causes the alveolar wall to thicken, the transport of O_2 is hindered more than the transport of CO_2.

The mixture of gases in the alveoli is not the same as the mixture of gases in ordinary air. The lungs are not emptied during expiration. During normal breathing the lungs retain about 30% of their volume at the end of each expiration. This is called the *functional residual capacity* (FRC). At each breath about 500 cm³ of fresh air (pO_2 of 150 mm Hg) mixes with about 2000 cm³ of stale air in the lungs to result in alveolar air with a pO_2

of about 100 mm Hg. The pCO_2 in the alveoli is about 40 mm Hg. Expired air includes about 150 cm^3 of relatively fresh air from the trachea that was not in contact with alveolar surfaces, so expired air has a slightly higher pO_2 and a lower pCO_2 than alveolar air (Table 7.1). The ratio of CO_2 output to O_2 intake is called the *respiratory exchange ratio* or *respiratory quotient R* (see Chapter 5). R is usually slightly less than 1.

Nitrogen from the air does not play any known role in body function. It is dissolved in the blood at its partial pressure. A deep-sea diver breathes air at a much higher pressure underwater than when he is at sea level; the increased partial pressure of N_2 causes more N_2 to be dissolved in his blood and tissues. If the diver surfaces too rapidly some of the N_2 forms bubbles in his joints causing the serious problem of "bends" (see Chapter 6).

During normal breathing the fresh supply of air does not enter the alveoli which are still filled with stale air from previous breaths. Because of its higher concentration, the fresh O_2 rapidly diffuses through the stale air to reach the surface of the alveoli. The O_2 is dissolved in the moist alveolar wall and diffuses through into the capillary blood until the pO_2 in the blood is equal to that in the alveoli. This process takes less than 0.5 sec (Fig. 7.5). Meanwhile the CO_2 in the blood diffuses even more rapidly into the gas in the alveoli until the pCO_2 in the blood is the same as in the alveolar gas.

As mentioned earlier, the blood can carry very little O_2 in solution. Most of the O_2 for the cells is carried in chemical combination with the hemoglobin (Hb) in the red blood cells. A liter of blood can carry about 200 cm^3 of O_2 at NTP by this means while it can carry only 2.5 cm^3 of O_2 in solution. Since most of the O_2 is not in solution, the law of diffusion is altered. The O_2 will combine with or separate from the Hb in a way that depends on the dissociation curve (Fig. 7.6). The Hb leaving the lungs is about 97% saturated with O_2 at a pO_2 of about 100 mm Hg. The pO_2 has to drop by about 50% before the O_2 load of the blood is noticeably reduced.

When the blood reaches the cells and their low pO_2 environment, the O_2

Table 7.1. The Percentages and Partial Pressures of O_2 and CO_2 in Inspired, Alveolar, and Expired Air[a]

	% O_2	pO_2 (mm Hg)	% CO_2	pCO_2 (mm Hg)
Inspired air	20.9	150	0.04	0.3
Alveolar air	14.0	100	5.6	40
Expired air	16.3	116	4.5	32

[a]It is assumed that the inspired air is dry and the expired air is saturated, $pH_2O = 47$ mm Hg.

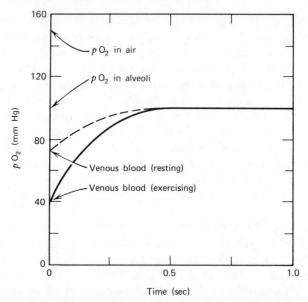

Figure 7.5. The pO_2 of the blood in a pulmonary capillary rises rapidly to the level of the pO_2 in the alveoli as the red blood cells move through the capillary (dashed line). Even during heavy exercise (solid line) the red blood cells are rapidly replenished with O_2.

is dissociated from the Hb and diffuses into the cells. Not all the O_2 leaves the Hb; the amount leaving depends on the pO_2 of the tissues. Under resting conditions the venous blood returns to the heart with about 75% of its load of O_2. The O_2 is retained in the blood because it is not needed in the tissues. During heavy physical labor or exercise the situation in active muscle changes drastically. The pO_2 in the working muscles drops rapidly causing more O_2 to be dissociated from the Hb and to diffuse into the muscles. In addition the body can increase the blood flow to working muscles by a factor of 3. Working muscles can obtain 10 times more O_2 than they consume at rest. For normal people the limiting factor in exercise is not the amount of blood pumped by the heart per minute (cardiac output) or the amount of O_2 supplied to the blood by the lungs, but the speed at which O_2 is transferred to the working muscles.

The dissociation of O_2 from Hb also depends on the pCO_2, the pH (acidity), and the temperature. During exercise the pCO_2, the acidity, and the temperature in working muscles all increase; these increases all shift the curve of Fig. 7.6 to the right and permit the Hb to give up more of its O_2. All these factors thus increase the O_2 to the working muscles. In the

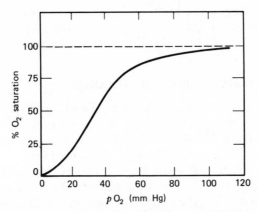

Figure 7.6. The percent O_2 saturation of the blood as a function of pO_2 in the alveoli. At 100% saturation 1 liter of blood can transport 200 ml of O_2 at NTP. This curve is affected by the temperature, the pCO_2, and the pH.

lungs the decrease of pCO_2 due to rapid breathing permits Hb to bind more O_2.

Carbon dioxide is not transported from the tissues by simple diffusion either. (If you are interested in the details see the bibliography at the end of this chapter.) Most of the CO_2 remains in the blood after it has left the lungs ($pCO_2 \simeq 40$ mm Hg). The CO_2 level in the blood is maintained fairly constant by the breathing rate. Excessively rapid breathing (hyperventilation) can lower the pCO_2 in the blood (hypocapnia); this causes mental disturbances and fainting.

In carbon monoxide (CO) poisoning the CO molecules attach very securely to the Hb at places normally used by the O_2. They attach about 250 times more tightly than O_2 and do not easily dissociate into the tissues. In addition to using places normally used to transport O_2, the CO inhibits the release of O_2 from Hb, so even a small amount of CO can seriously reduce the O_2 to the tissues. Cigarette smokers breathe in about 250 cm^3 of CO from each pack, and it is also commonly inhaled by people driving in heavy traffic. Carbon monoxide can cause death by starving the tissues of O_2. Normally the dissolved O_2 in the blood is of no significance, but if a CO victim is placed in a hyperbaric O_2 chamber with an absolute pressure of 3 atm of pure O_2, the pO_2 increases by a factor of 15. The dissolved O_2 in the blood can then supply minimal body needs (see Chapter 6). This therapy cannot be maintained very long because O_2 poisoning can result. Continued use of 1 atm of pure O_2 can cause swelling (edema) of the lung tissues, which reduces O_2 to the blood and ironically results in death from a lack of O_2 (anoxia). Safe levels of pO_2 in "air" are those below 0.5 atm ($\sim$380 mm Hg pO_2) (see Fig. 6.8).

7.3. MEASUREMENT OF LUNG VOLUMES

A relatively simple instrument, the spirometer (Fig. 7.7), is used to measure airflow into and out of the lungs and record it on a graph of volume versus time. Figure 7.8 shows a typical recording for an adult under various breathing conditions. During normal breathing at rest we inhale about 500 cm^3 of air with each breath. This is referred to as the *tidal volume at rest*. At both the beginning and end of a normal breath there is considerable reserve. At the end of a normal inspiration it is possible with some effort to further fill your lungs with air. The additional air taken in is called the *inspiratory reserve volume*. Similarly, at the end of a normal expiration you can force more air out of your lungs. This additional expired air is called the *expiratory reserve volume*. The air remaining in the lungs after a normal expiration is called the *functional residual capacity* (FRC). It is this stale air that mixes with the fresh air of the next breath. During heavy exercise, the tidal volume is considerably larger. You have a fair idea of your lung capacity if you have ever blown up a paper sack or an air mattress. If a person breathes in as deeply as possible (*a* in Fig. 7.8) and then exhales as much as possible (*b* in Fig. 7.8), the volume of air exhaled is his *vital capacity*. However, his lungs will still contain some air—the *residual volume*, which is about 1 liter for an adult. The residual volume can be determined by having the subject breathe in a known volume of an inert gas such as helium and then measuring the fraction of helium in the expired gas. Since the helium and air will mix thoroughly during a single breath, this dilution technique is quite accurate.

A number of clinical tests can be made with the spirometer. The amount of air breathed in 1 min is called the *respiratory minute volume*. The maximum volume of air that can be breathed in 15 sec is called the *maximum voluntary ventilation* and is a useful clinical quantity. The maximum rate of expiration after a maximum inspiration is a useful test for emphysema and other obstructive airway diseases. In some cases the flow rate decreases with excessive expiratory effort. A normal person can expire about 70% of his vital capacity in 0.5 sec, 85% in 1.0 sec, 94% in 2.0 sec, and 97% in 3.0 sec. Normal peak flow rates are 350 to 500 liters/min. The velocity of the expired air can be impressive; if a person coughs or sneezes hard without covering his mouth, the velocity of the air in his trachea can reach Mach 1—the velocity of sound in air! This high velocity

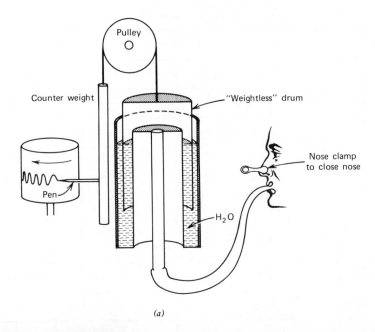

(a)

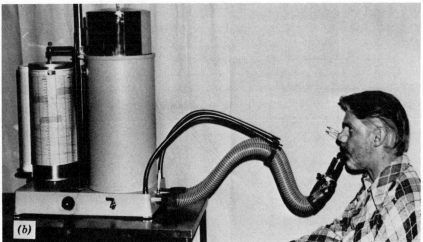

(b)

Figure 7.7. The spirometer is used to measure various quantities of pulmonary function. The airflow in and out of the lungs is recorded on a rotating chart. (a) A cross-section of a spirometer showing how water is used as an air-tight seal to keep air within the counterbalanced drum. (b) One of the authors (JRC) producing the graph shown in Fig. 7.8. The nose clamp forces all air to flow through the mouth.

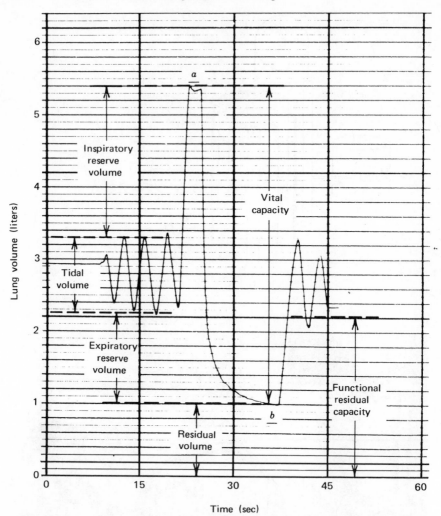

Figure 7.8. A tracing made using the apparatus shown in Fig. 7.7b. It shows the various volumes and capacities of the lungs. Note that during the maximum expiration the outflow is rapid at first; the last 5% takes longer than the first 95%.

can cause partial collapse of the airways because of the Bernoulli effect. In coughing to dislodge a foreign object this partial collapse increases the air velocity and increases the force on the foreign object.

Not all of the air we inspire adds O_2 to the blood. The volume of the trachea and bronchi is called the *anatomic dead space* since air in this space is not exposed to the blood in the pulmonary capillaries. Typically the anatomic dead space is about 150 cm^3. In addition, in some diseases some of the alveolar capillaries are not perfused with blood and the O_2 is not absorbed in these alveoli. This unused volume is called physiologic or alveolar dead space. Air in the dead space does not provide any O_2 to the body. The air in the anatomic dead space after an expiration is taken back into the lungs during the next inspiration. If you increase your dead space by breathing through a long tube, you will recycle more of your own breath. If the tube has a volume equal to your vital capacity you obviously will get no new air and will suffocate.

7.4. PRESSURE-AIRFLOW-VOLUME RELATIONSHIPS OF THE LUNGS

The pressure, airflow, and volume relationships of the lungs during tidal breathing for a normal subject and for a patient with a narrowed airway are shown in Fig. 7.9. The pressure difference needed to cause air to flow

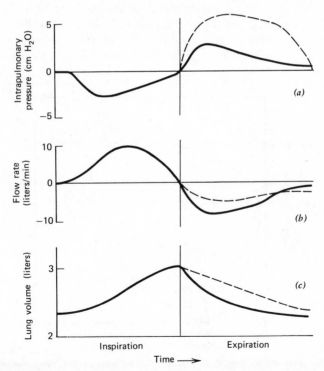

Figure 7.9. Typical pressures (a), flow rates (b), and lung volumes (c) during quiet respiration for a normal individual (solid line) and a patient with a narrowed airway (dashed line). Note the increased pressure and decreased flow rates during expiration due to the narrowed airway.

into or out of the lungs of a healthy individual is quite small. Note that the pressure difference (Fig. 7.9*a*) is only a few centimeters of water for a normal individual. Figure 7.9*b* shows the rate of airflow into and out of the lungs in liters per minute, and Fig. 7.9*c* shows the lung volume during the breathing cycle.

Since the esophagus passes through the chest, it reflects the pressure between the lungs and chest wall (intrapleural or intrathoracic space). It is possible to measure the pressure in the esophagus with a pressure gauge. This pressure is normally negative (~ -10 mm Hg) due to the elasticity of the lungs (see Section 7.6). In Fig. 7.10, the intrathoracic pressure (measured in the esophagus) is plotted versus the tidal lung volume during respiration. Figure 7.11 shows the pressure-volume curves for three different breathing rates—slow, moderate, and fast.

The lungs and chest wall are normally coupled together. The behavior of the system is the result of the combination of the physical characteristics of the two. Figure 7.12 shows curves of volume versus pressure for the chest wall and lungs separately and for the two together. The volume is given as a percentage of the vital capacity. If the chest wall were free of its interaction with the lungs it would have a volume of about two-thirds of the total vital capacity. The lungs by themselves would collapse and have

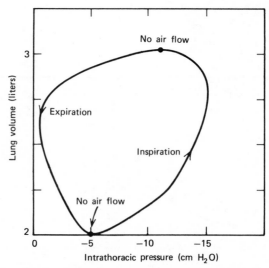

Figure 7.10. The intrathoracic pressure plotted versus the lung volume during respiration for a larger than average tidal volume. (Adapted from Hildebrandt, J., and Young, A.C., in T.C. Ruch and H.D. Patton (Eds.), *Physiology and Biophysics,* **19th ed., © W.B. Saunders Company, Philadelphia, 1965, p. 754.)**

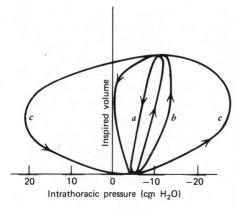

Figure 7.11. The *P-V* curves for three different breathing rates: (a) very slow breathing of about 3 breaths/min: (b) about 40 breaths/min; and (c) maximum breathing rate of about 150 breaths/min. (Adapted from Hildebrandt, J., and Young, A.C., in T.C. Ruch and H.D. Patton (Eds.), *Physiology and Biophysics*, 19th ed., © W.B. Saunders Company, Philadelphia, 1965, p. 754.)

essentially no air volume. Together the lungs and chest wall come to a relaxation volume (FRC) at about 30% of vital capacity.

The combined curve in Fig. 7.12 shows the pressure-volume relationship obtained by filling the lungs to known percentages of the vital capacity. The pressure is measured in the mouth (and lungs) with the nose and mouth closed and the breathing muscles relaxed. For example, at about 60% of vital capacity, the relaxation pressure is 10 cm H_2O. Since the

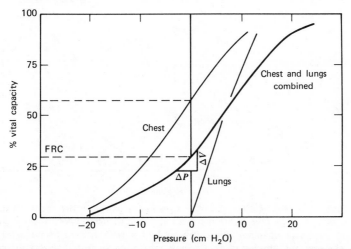

Figure 7.12. The *P-V* curves for the chest alone, the lungs alone, and the chest and lungs combined. The combined curve is the relaxation curve of Fig. 7.13. The slope of the combined curve $\Delta V/\Delta P$ gives the compliance of the lung-chest system. If the vital capacity is 5 liters, $\Delta V/\Delta P \simeq 0.2$ liter/cm H_2O. (Adapted from Hildebrandt, J., and Young, A.C., in T.C. Ruch and H.D. Patton (Eds.), *Physiology and Biophysics*, 19th ed., © W.B. Saunders Company, Philadelphia, 1965, p. 749.)

chest wall is at equilibrium at this volume, this pressure is produced by the elastic properties of the lung. When the same relaxation measurements are made after a forced exhalation the negative pressure values of Fig. 7.12 are obtained.

The relaxation pressure curve is again plotted as a function of the vital capacity in Fig. 7.13. In addition, two other related curves are shown. All of these pressures are measured in the mouth with the nose and mouth closed. Exhaling with the greatest force gives the maximum expiratory effort curve. Inhaling with maximum effort gives the maximum inspiratory effort curve. Forced expiratory effort after a maximum inspiration (100% of vital capacity) compresses the gas according to Boyle's law, PV = constant. The dashed lines a and b show the theoretical curves for the pressure-volume relationship of an ideal gas (PV = constant) at 0% and 100% of vital capacity.

Compliance is an important physical characteristic of the lungs. Compliance is the change in volume produced by a small change in pressure, that is, $\Delta V/\Delta P$ (see Fig. 7.12). Compliance is usually given in liters per centimeter of water. Compliance in normal adults is in the range of 0.18 to 0.27 liter/cm H_2O. It is generally about 25% greater in men over age 60 than in younger men. There is little change in women with age.

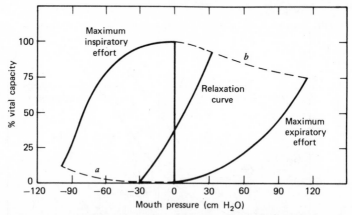

Figure 7.13. *P-V* curves obtained with a pressure gauge in the mouth. The center curve is for the lung-chest system shown in Fig. 7.12. The curve on the right is the maximum pressure obtained when the subject blows as hard as possible on the gauge. The curve on the left is obtained by maximum suction. The dashed curves *a* and *b* are the theoretical curves for Boyle's law *PV* = constant. (Adapted from Hildebrandt, J., and Young, A.C., in T.C. Ruch and H.D. Patton (Eds.), *Physiology and Biophysics*, 19th ed., © W.B. Saunders Company, Philadelphia, 1965, p. 738, after Rahn et al., *Amer. J. Physiol.*, 146, 1946, pp. 161–178.)

A stiff (fibrotic) lung has a small change in volume for a large pressure change and thus it has a low compliance. A flabby lung has a large change in volume for a small change in pressure and has a large compliance. Infants with respiratory distress syndrome (see Section 7.5) have lungs with low compliance. In some diseases, such as emphysema, the compliance increases (see Section 7.9).

During tidal breathing, the *P-V* curve forms a closed loop like those shown in Fig. 7.14. The cycles flow clockwise on the loops. The middle loop represents typical tidal breathing at normal pressure. Loop *b* represents positive pressure breathing where the air supply pressure is about 30 cm H_2O greater than the pressure on the chest wall. Positive pressure breathing is often used therapeutically in resuscitation and in relief of obstructive airway disease. For positive pressure breathing the inspiratory muscles are not used but the expiratory muscles are. Loop *c* in Fig. 7.14 represents negative pressure breathing. This can occur when a per-

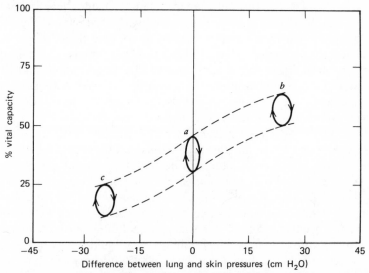

Figure 7.14. *P-V* curves for tidal breathing under three conditions: (*a*) normal breathing where the pressure in the mouth is the same as on the skin, (*b*) positive pressure breathing with a tight-fitting face mask where the breathing muscles must work to expire, and (*c*) snorkel (underwater) breathing where the pressure on the chest is greater than in the mouth and the inspiratory muscles are under continuous tension. (Adapted from Hildebrandt, J., and Young, A.C., in T.C. Ruch and H.D. Patton (Eds.), *Physiology and Biophysics*, 19th ed., © W.B. Saunders Company, Philadelphia, 1965, p. 739, after Rahn et al., *Amer. J. Physiol.*, 146, 1946, pp. 161–178.)

son is underwater and breathing through a tube to the surface (snorkel breathing). In this case the inspiratory muscles never completely relax.

7.5. PHYSICS OF THE ALVEOLI

The alveoli are physically like millions of small interconnected bubbles. They have a natural tendency to get smaller due to the surface tension of a unique fluid lining. This lining, called *surfactant*, is necessary for the lung to function properly. The absence of surfactant in the lungs of some newborn infants, especially prematures, is the cause of the idiopathic* respiratory distress syndrome (RDS), sometimes called hyaline membrane disease. This disease accounts for thousands of infant deaths each year in the United States; it kills more babies than any other disease.

To understand the physics of the alveoli we have to understand the physics of bubbles. The pressure inside a bubble is inversely proportional to the radius and directly proportional to the surface tension γ. The exact relation is $P = 4\,\gamma/R$, a form of Laplace's law. Consider soap bubbles on the ends of a tube with a valve separating them as shown in Fig. 7.15*a*. What happens when the valve is opened to connect them? Because the smaller one has a higher internal pressure it will empty its air into the larger one until the radii of curvature of the large bubble and of the remainder of the small bubble are the same (Fig. 7.15*b*). Although alveoli are not exactly the same as soap bubbles there is a tendency for the smaller alveoli to collapse. The condition that results when a sizable number collapse is called *atelectasis*. The reason most alveoli don't collapse is related to the unique surface tension properties of surfactant.

The surface tension of a fluid can be found by measuring how much force is necessary to pull a loop of wire from a clean liquid surface (Fig. 7.16). The surface tension of a water-air interface is 72 dynes/cm; that of a plasma-air interface is about 40 to 50 dynes/cm; those of detergent solutions in air are from 25 to 45 dynes/cm. A qualitative measure of surface tension is to note how long small bubbles of a liquid survive. The lower the surface tension, the longer they last. In 1955 it was noted that bubbles expressed from the lung were very stable, lasting for hours. It was concluded that they must have a very low surface tension.

The surface tension of the surfactant that lines the alveoli of healthy individuals plays a major role in lung function. The surface tension of the surfactant is not constant. Figure 7.17*b* shows the surface tension of a film of normal lung extract containing surfactant. Note the large decrease of γ

---

*Idiopathic is a useful word—it means of unknown origin.

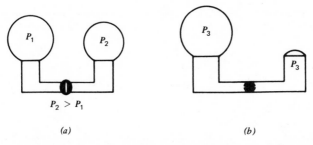

Figure 7.15. The pressure inside a soap bubble depends on its radius. (*a*) When the valve between the two bubbles is closed, the pressure is greater in the smaller bubble ($P = 4\gamma/R$). (*b*) When the valve is opened, the smaller bubble empties into the larger, leaving a spherical cap with the same radius as the new large bubble.

as the area decreases. This characteristic causes the surface tension of the alveoli to decrease as the alveoli decrease in size during expiration. For each alveolus there is a size at which the surface tension decreases sufficiently fast that the pressure starts to drop instead of continuing to increase, and this causes the alveolus to stabilize at about one-fourth its maximum size. Alveoli not covered with surfactant, such as those of infants with RDS, collapse like small bubbles, and quite a large pressure is needed to reopen them. An infant with RDS may not have the energy to breathe with its low compliance lungs.

The *P-V* curves for an excised human lung are shown in Fig. 7.18. If the lung is completely collapsed, a considerable pressure is needed to start its inflation, similar to the extra effort needed to start blowing up a rubber

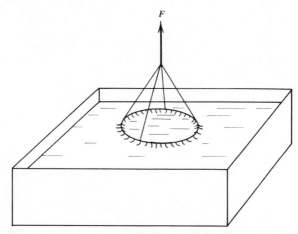

Figure 7.16. The surface tension of a liquid can be measured by determining the force needed to pull a wire loop from a clean liquid surface.

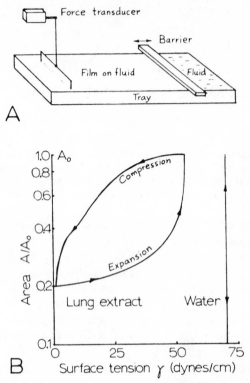

Figure 7.17.   Surface tension as a function of film area. (A) A schematic representation of the apparatus used to measure the surface tension of a film. The tray is filled to the top with fluid, a film of the material to be studied is spread on the surface, a movable barrier is used to compress the film, and a hanging plate balance continuously records the surface tension γ. (B) Graph of surface tension of a lung extract containing surfactant. Note the large decrease in surface tension as the area decreases and the different curve obtained as the area increases. The vertical line at about 70 dynes/cm shows that the surface tension of water is constant with changes in area. (From Hildebrandt, J., and Young, A.C., in T.C. Ruch and H.D. Patton (Eds.), *Physiology and Biophysics*, 19th ed., © W.B. Saunders Company, Philadelphia, 1965, p. 744. Reprinted by permission.)

balloon. From this point the lung inflates rather easily until it is close to its maximum size. The pressure curve during deflation looks quite different. When the pressure has dropped to zero the lung still retains some air. Much less pressure is needed to then reinflate the lung, although reinflation will not follow the deflation curve. A cyclical process in which different curves are followed on the two halves of the cycle is said to show *hysteresis*. The area inside the loop is proportional to the energy lost as heat during the cycle. During normal tidal breathing the hysteresis loop is

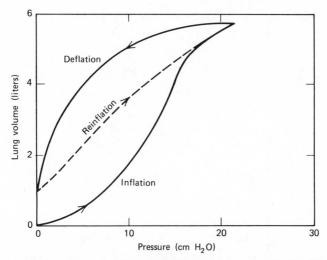

Figure 7.18. Typical *P-V* curve for an excised human lung. More pressure is required for the initial inflation (lower curve) than for reinflation (dashed curve).

quite small, like curve *a* on the normal *P-V* curve in Fig. 7.19. If tidal breathing continues unchanged some of the alveoli collapse, and the hysteresis loop becomes slightly larger and shifts toward higher pressure as shown by curve *b*. A deep breath reopens the alveoli, and the curve shifts back to *a*. We take such a deep breath occasionally without being aware of it (a sigh). During surgery an anesthesiologist will occasionally force a large volume of gas into a patient's lungs to reopen collapsed alveoli. Taping the chest prevents a patient from taking a deep breath and some of his lung space is likely to be lost by collapsed alveoli, or atelectasis.

The *P-V* curve for a lung suffering from a lack of surfactant (RDS) is also shown in Fig. 7.19; the hysteresis loop is shifted to the right and a large pressure must be maintained to keep the lung inflated. Note the lower compliance ($\Delta V/\Delta P$) of this lung. The *P-V* curve of a patient with severe emphysema is also shown in Fig. 7.19; note the increased compliance, the larger residual volume, and the large area inside the hysteresis loop.

7.6. THE BREATHING MECHANISM

Breathing is normally under unconscious control. Although the rate of breathing can be changed at will, a person is unaware of his breathing

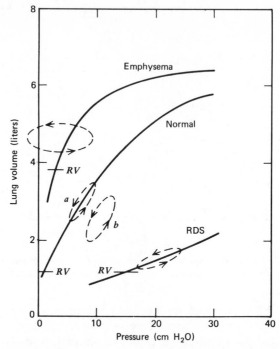

Figure 7.19. *P-V* curves for a normal subject and for patients with emphysema and RDS. The solid lines are static values. The pressure is relative to the intrathoracic pressure. *RV* is the residual volume. Notice the increased *FRC* and *RV* for emphysema. The dashed lines represent the hysteresis loops during tidal breathing. Curve *a* gradually shifts to become curve *b* as atelectasis and surfactant changes occur. A deep breath shifts curve *b* back to curve *a* as the collapsed alveoli are forced open again. (Adapted from Hildebrandt, J., and Young, A.C., in T.C. Ruch and H.D. Patton (Eds.), *Physiology and Biophysics*, 19th ed., © W.B. Saunders Company, Philadelphia, 1965, pp. 740 and 758.)

most of the time unless he is suffering from asthma or emphysema. The physiological control of breathing depends on many factors, but the pH in the respiratory center of the brain exerts primary control.

If a lung were removed from the chest all the air would be squeezed from it and it would collapse to about one-third of its size much as a balloon collapses when air is let out of it. The lung can be thought of as millions of small balloons, all trying to collapse. The lungs don't normally collapse because they are in an airtight container—the chest. As the diaphragm and rib cage move the lungs stay in contact with them. Two forces keep the lungs from collapsing: (1) surface tension between the lungs and the chest wall and (2) air pressure inside the lungs. The surface tension force is similar to that between two pieces of cellophane or saran wrap stuck together. If the lungs overcame this force and pulled away from the chest wall a vacuum would be created since air cannot reach the intrapleural space. Since the air inside the lungs is at atmospheric pressure ($\sim 10^5$ N/m² or $\sim$14.5 lb/in.²) it would push the lungs back in contact with the chest wall. There is normally a negative pressure of 5 to 10 mm Hg in the intrapleural space.

Various muscles are involved in breathing. Intercostal muscles in the chest wall cause the chest to expand when they contract. Other muscles between the neck and chest can also contract to expand the chest. Normally most breathing is done by contracting the diaphragm muscles; these pull the diaphragm down, expanding the lungs. When we inspire, we pull the diaphragm down as shown schematically by the arrow in Fig. 7.20*b*. This produces a slight negative pressure in the lungs and air flows in. When we expire, we relax the diaphragm muscles, the elastic forces in the lungs cause the diaphragm to return to its neutral position, and air flows out of the lungs without any active muscular effort. If the diaphragm muscles are paralyzed the muscles in the chest wall are used for breathing.

If the chest wall is punctured as shown schematically by the open valve in Fig. 7.20*c* the lung collapses, the diaphragm lowers, and the chest wall expands. This condition is known as a pneumothorax (literally, air-chest). Occasionally it is medically desirable to collapse one lung to allow it to

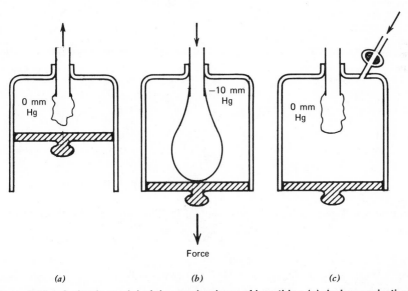

Figure 7.20. A simple model of the mechanisms of breathing (a) during expiration, (b) during inspiration, and (c) during pneumothorax.

"rest." Since each lung is in its own sealed compartment it is possible to collapse one lung only as shown in Fig. 7.21. This is done rather simply by inserting a hollow needle between the ribs (an intercostal puncture) and allowing air to flow into the intrathoracic space. The air trapped in the space is gradually absorbed by the tissues, and the lung expands to normal over a period of a few weeks. Sometimes a lung collapses spontaneously with no known cause. This condition of spontaneous pneumothorax is moderately common in college-age students. As in the medical procedure, the lung returns to normal as the air is absorbed into the surrounding tissues.

Since both the lung and chest wall are elastic, we can represent them with springs (Fig. 7.22). Under normal conditions they are coupled together: the "lung" springs are stretched and the "chest" springs are compressed (Fig. 7.22a). During a pneumothorax, the lungs and chest are independent and the springs representing them go to their relaxed posi-

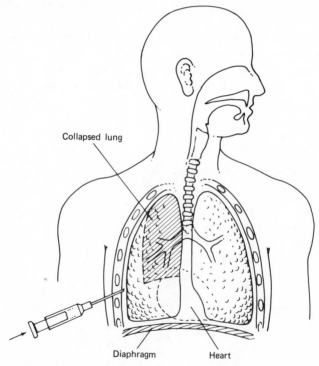

Collapsed lung

Diaphragm　　　Heart

Figure 7.21. **A right pneumothorax is produced by letting in air between the chest wall and the lung. The shaded area shows the outline of the collapsed lung.**

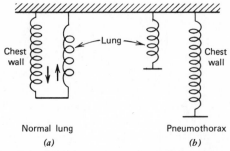

Figure 7.22. (a) A spring model for the lungs. The arrows show the direction of the spring forces. Normally the lung and chest wall are coupled together. (b) During a severe pneumothorax, the springs go to their relaxed positions—the chest enlarges and the lung collapses.

tions as indicated in Fig. 7.22b. The lung collapses, and the chest wall enlarges.

The intrathoracic space is not always at negative pressure. If you close your windpipe and forcefully try to expire, the intrathoracic pressure can become quite high. This is called a Valsalva maneuver. A person does this when he blows up a stiff balloon. Under more physiological conditions he does this just before coughing or sneezing and during the stress of defecation or vomiting. Increasing the pressure in the chest compresses the main vein (vena cava) carrying blood back to the heart and reduces the volume of blood pumped by the heart. The normal negative pressure in the chest helps keep the vena cava open. The blood pressure in the vena cava is only 0.5 cm H_2O near the heart.

7.7. AIRWAY RESISTANCE

We can breathe in more rapidly than we can breathe out. During inspiration the forces on the airways tend to open them further; during expiration the forces tend to close the airways and thus restrict airflow. For a given lung volume, the expiratory flow rate reaches a maximum and remains constant; it might even decrease slightly with increased expiratory force. Patients with obstructive airway disease such as asthma or emphysema find that an increased effort to breathe out decreases the flow rate considerably. These patients unconsciously find some relief by retaining a large amount of air in the lungs, thus keeping their airways as large as possible. They can often inspire at near normal rates so they breathe in rapidly to allow more time for expiration. The pulmonary physics of emphysema is discussed in Section 7.9.

The flow of air in the lungs is analogous to the flow of current in an electrical circuit. "Ohm's law" for air flow looks like Ohm's law for electrical circuits, with voltage replaced by pressure difference ΔP and current replaced by the rate of air flow $\Delta V/\Delta t$ or $\dot{V}$. Airway resistance R_g is the ratio of ΔP to $\dot{V}$. Airway resistance is given in units of pressure per unit flow rate, commonly cm H_2O/(liter/sec). In typical adults $R_g = 3.3$ cm H_2O (liter/sec). R_g depends on the dimensions of the tube and the viscosity of the gas. The situation is complicated by the complexity of the airways. Most of the resistance is in the upper airway passages. The nasal area accounts for about half of R_g, and another 20% is due to the other upper airways. In normal subjects less than 10% of R_g is in the terminal airways. Thus diseases that affect the terminal airways (bronchioles and alveoli) do not appreciably affect the airway resistance until they are far advanced.

The time constant of the lungs is related to the airway resistance R_g and the compliance C. Remember that compliance is $\Delta V/\Delta P$ (Section 7.4). The product R_gC is the time constant for the lung. This is analogous to the time constant RC' for a capacitor C' to discharge through a resistance R in an electrical circuit. The time constant of the lung is complicated since many parts of the lung are interconnected. If one part of the lung has a larger time constant than other parts, it will not get its share of the air and that part of the lung will be poorly ventilated.

7.8. WORK OF BREATHING

The amount of work done in normal breathing accounts for a small fraction of the total energy consumed by the body ($\sim$2% at rest). The primary work of breathing can be thought of as the work done in stretching the springs representing the lung-chest wall-diaphragm system (Fig. 7.23a), which is proportional to the shaded area in Fig. 7.23b; however, this is an oversimplification of the work of breathing. A better model is shown in Fig. 7.24. The resistance of the tissues and the resistance of the gas flow produce heat; these can be represented as a dashpot (R). The springiness of the lung-chest system is represented by the spring C. The inertia I of the mass of the lungs and chest wall must also be overcome; at normal breathing rates, the inertia can be neglected, but at maximum breathing rates (over 100 breaths/min) it is a significant factor. The work of breathing is shown by the total shaded area in Fig. 7.24b; the darker shaded area represents the work against the spring C and the lighter area

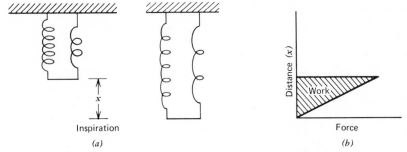

Figure 7.23.   A model of the work of breathing. (a) Position of the springs at ends of breathing cycle. (b) The shaded area represents the work done in stretching the springs a distance x.

represents the work against the resistance. During normal breathing, no work is done during expiration; the muscles relax and the springs "snap back" to expel the air, dissipating the energy in the dashpot R. During strenuous exercise muscles are used to expel air. The work of breathing during heavy exercise may amount to 25% of the body's total energy consumption.

Rapid shallow breathing and slow deep breathing are both less efficient than the normal rate. Most animals adjust their breathing rates at rest to use minimum power. At low breathing rates most of the work is done against the elastic forces of the lung and chest (darker area in Fig. 7.24b);

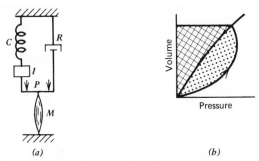

Figure 7.24.   (a) A better model of the work of breathing: C represents the springs of the lung-chest-diaphragm system, R is the resistance to tissue motion and gas flow, I is the inertia of the moving parts, P is the pressure, and M represents the breathing muscles. (b) Work done: the cross-hatched shaded area represents the work against the spring C, and the dotted area represents the work against the resistance R. (Adapted from Hildebrandt, J., and Young, A.C., in T.C. Ruch and H.D. Patton (Eds.), *Physiology and Biophysics*, 19th ed., © W.B. Saunders Company, Philadelphia, 1965, p. 755.)

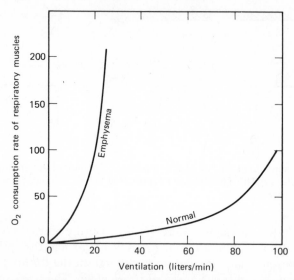

Figure 7.25.   Curves of O₂ consumption rate of the respiratory muscles for a normal subject and for a subject with severe emphysema. (Adapted from E.J.M. Campbell, E.K. Westlake, and R.M. Cherniak, *J. Appl. Physiol.* 11: 303–308, 1957—Fig. 3.)

at fast breathing rates the work against the resistive forces (lighter area in Fig. 7.24*b*) increases.

Another way to determine the work done in breathing is to measure the extra O₂ consumed as the breathing rate is increased under resting conditions. The amount of O₂ consumed is directly related to the calories of food "burned" (see Chapter 5). We assume that the additional O₂ is used in the respiratory muscles. Figure 7.25 shows a typical curve for a normal subject and the curve for a patient with severe emphysema. The latter may use more O₂ in the work of breathing at a faster rate than is provided by his increased ventilation; the amount of O₂ in his general circulation thus falls.

If we compare the energy used in breathing obtained by the O₂ consumption method to the calculated work done using the model shown in Fig. 7.24, we can estimate the efficiency of the breathing mechanism. Because of many uncertainties, the efficiency estimates range from 5 to 10%.

7.9. PHYSICS OF SOME COMMON LUNG DISEASES

Diseases of the lungs account for a large percentage of man's medical problems. It is estimated that 15% of the people in the United States over age 40 have detectable lung disease. Many of these diseases can be understood in terms of physical changes in the lungs. This does not, of course, mean that a physicist can cure them. The physical aspects of some common lung diseases are discussed in this section. The physics of RDS in infants is discussed in Section 7.5.

At rest, only a small fraction of the lungs' capacity is used. Thus a lung disease that reduces the capacity often does not produce noticeable symptoms in its early stages. When the symptoms are noticeable, the disease is fairly well advanced. Many lung function tests force the breathing mechanism to its limits and thus allow detection of changes that are not ordinarily apparent. There are some simple lung tests that should be included in every health check-up.

In emphysema the divisions between the alveoli break down, producing larger lung spaces. This destruction of lung tissue reduces the springiness of the lungs. The lungs become more compliant—a small change in pressure produces a larger than normal change in the volume. While at first glance this would appear to make it easier to breathe, the opposite is true. Much of the work of breathing is done in overcoming the resistance of the airways. In emphysema the airway resistance increases greatly.

Figure 7.26 will help you understand the physics of emphysema. You can think of the elasticity of the tissues in the normal lung as millions of little interconnected springs (Fig. 7.26a). These "springs" tend to collapse the lung and produce the force that pulls on the chest wall. They also

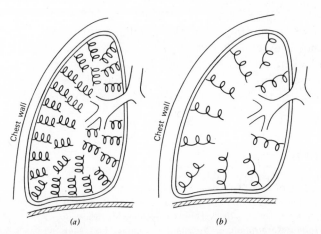

(a)   *(b)*

Figure 7.26. Spring models of (a) a normal lung and (b) a lung with severe emphysema. Note the reduced number and strength of the springs in the model for emphysema. The resulting expansion of the chest wall and narrowing of the major airways causes an increase in airway resistance.

pull on the walls of the airways; this keeps the airways open and helps reduce airway resistance during expiration.

The situation in severe emphysema is shown in Fig. 7.26b. The number of working "springs" has been greatly reduced, and those present are much weaker than normal. This produces two important changes: (1) the lung becomes flabby and expands as the reduced tension allows the chest wall to expand almost to the resting volume of the chest wall without the lung—about 60% of vital capacity (Fig. 7.12); and (2) the tissues do not pull very hard on the airways, permitting the narrowed airways to collapse easily during expiration. This increased airway resistance is the major symptom of severe emphysema. The increased size of the lungs increases the FRC and the residual volume (Fig. 7.19). The chest is overinflated, and the posture is affected: someone with the disease appears barrel-chested. Since a person who has emphysema is unable to blow out a candle, it is simple to test for the disease. Emphysema occurs occasionally in nonsmokers, but the recent large increase in the disease has been primarily among heavy smokers.

In asthma, another common obstructive disease, the basic problem is also expiratory difficulty due to increased airway resistance. Some of this resistance is apparently due to swelling (edema) and mucus in the smaller airways, but much of it is due to contraction of the smooth muscle around the large airways. Lung compliance is essentially normal, but the FRC may be higher than normal since the patient often starts to inspire before completing a normal expiration.

In fibrosis of the lungs the membranes between the alveoli thicken. This has two marked effects: (1) the compliance of the lung decreases, and (2) the diffusion of O_2 into the pulmonary capillaries decreases. The expiratory resistance is essentially normal. A person with the disease will have labored and even painful breathing (dyspnea) or shortness of breath during exercise. Fibrosis of the lungs can occur if the lungs have been irradiated (e.g., in the treatment of cancer), although this is not the only cause.

BIBLIOGRAPHY

Campbell, E. J. M., E. Agostoni, and J. N. Davis, *The Respiratory Muscles, Mechanics and Neural Control,* 2nd ed., Saunders, Philadelphia, 1970.

Clements, J. A., *Surface Tension in the Lungs,* Freeman, San Francisco, 1962.

Duffin, J., *Physics for Anaesthetists,* Thomas, Springfield, Ill., 1976.

Lippold, O., *Human Respiration: A Programmed Course,* Freeman, San Francisco, 1968.

Nunn, J. F., *Applied Respiratory Physiology with Special Reference to Anesthesia,* Appleton-Century-Crofts, New York, 1969.

Ruch, T. C., and H. D. Patton (Eds.), *Physiology and Biophysics,* 19th ed., Saunders, Philadelphia, 1965, Chapters 39 to 42.

REVIEW QUESTIONS

1. List five functions of the breathing mechanism in addition to gas exchange.
2. Calculate the number of O_2 molecules absorbed from a typical breath of 500 cm^3. Assume the O_2 in the air is reduced from 20% to 16% as measured at the mouth.
3. If there are 3×10^8 alveoli in a lung with a functional residual capacity of 2.5 liters, calculate the average volume of an alveolus.
4. Explain qualitatively why the air in the lungs has a pO_2 of only about 100 mm Hg while the pO_2 in ordinary air is about 150 mm Hg.
5. (a) What are the factors that determine how deep you can swim while breathing through a snorkel to the surface?
   (b) Estimate the maximum depth at which you could use a snorkel from Fig. 7.13.
6. If the thickness of the alveolar walls doubled, by what factor would the diffusion time for O_2 to reach the blood change?
7. In Morochocha, Peru, the atmospheric pressure is 447 mm Hg. Find the pO_2 and pN_2 that people in this village inhale.
8. A person's lung volumes were measured and the following results were obtained: vital capacity, 5 liters; residual volume, 1.0 liter; and expiratory reserve volume, 1.5 liters. Find the functional residual capacity.
9. A person inspired maximally and then began breathing from an expandable bag containing 2 liters of 40% helium gas. After a few breaths, the helium concentration in the bag was 10%. What was this person's total lung capacity?
10. The compliance of a normal adult lung is about 0.2 liter/cm H_2O. What is the compliance in m^3/Nm^{-2}?
11. If $R_g = 3$ cm H_2O/(liter/sec), what air flow rate $\dot{V}$ would occur at an expiratory pressure of 100 mm Hg?
12. In a patient with severe emphysema, which of the following are above normal and which are below normal?
    (a) Airway resistance
    (b) Inspiratory reserve volume
    (c) Functional residual capacity
    (d) Vital capacity
    (e) Compliance

CHAPTER 8

Physics of the Cardiovascular System

The cells of the body act like individual engines. In order for them to function they must have (1) fuel from our food to supply energy, (2) O_2 from the air we breathe to combine with the food to release energy, and (3) a way to dispose of the by-products of the combustion (mostly CO_2, H_2O, and heat). Since the body has many billions of cells an elaborate transportation system is needed to deliver the fuel and O_2 to the cells and remove the by-products. The blood performs this important body function. Blood represents about 7% of the body mass or about 4.5 kg (~4.4 liters) in a 64 kg (141 lb) person. The blood, blood vessels, and heart make up the cardiovascular system (CVS). This chapter describes the physical aspects of the CVS.

The blood and its supply of O_2 are so important to the body that the heart is the first major organ to develop in the embryo. Eight weeks after conception the heart is working to circulate blood to the tissues of the fetus. Since the fetus does not have functioning lungs it must obtain its oxygenated blood from its mother via the umbilical cord. The fetal heart has an opening that permits blood to flow from the right atrium to the left atrium, and only about 10% of the blood is circulated to the fetal lungs. After birth the circulation in the infant becomes modified to send much more blood to the lungs. The opening between the right and left atria effectively closes within minutes after birth, although it may take months for the closure to be complete. If closure is not adequate at birth the blood will not be properly oxygenated and the infant will be a "blue baby." A congenital heart defect of this type can now be corrected with surgery.

Several medical specialists are concerned with the CVS. Some physicians who have specialized in internal medicine subspecialize in problems of the blood. They are called *hematologists*,* and they treat patients with blood conditions such as anemia. Other medical specialists called *cardiologists* are primarily concerned with the heart. Cardiologists treat patients who have had heart attacks, and they interpret electrocardiograms. There is also a subspecialty in surgery that deals exclusively with the CVS. In recent years heart surgeons have often made headlines by performing dramatic heart-transplant operations. In other medical fields such as radiology and pediatrics there are also subspecialists who deal primarily with the CVS.

8.1. MAJOR COMPONENTS OF THE CARDIOVASCULAR SYSTEM

The heart is shown in Fig. 8.1. Basically a double pump, it provides the force needed to circulate the blood through the two major circulatory systems: the pulmonary circulation in the lungs and the systemic circulation in the rest of the body (Fig. 8.2). The blood in a normal individual circulates through one system before being pumped by the other section of the heart to the second system.

Let us start with the blood in the left side of the heart and follow its circulation through one complete loop. The blood is pumped by the contraction of the heart muscles from the left ventricle at a pressure of about 125 mm Hg into a system of arteries that subdivide into smaller and smaller arteries (arterioles) and finally into a very fine meshwork of vessels called the capillary bed. During the few seconds it is in the capillary bed the blood supplies O_2 to the cells and picks up CO_2 from the cells. After passing through the capillary bed the blood collects in small veins (venules) that gradually combine into larger and larger veins before it enters the right side of the heart via two main veins—the superior vena cava and the inferior vena cava. The returning blood is momentarily stored in the reservoir (the right atrium), and during a weak contraction (5 to 6 mm Hg) the blood flows into the right ventricle. On the next ventricular contraction this blood is pumped at a pressure of about 25 mm Hg via the pulmonary arteries to the capillary system in the lungs, where it receives more O_2 and where some of the CO_2 diffuses into the air in the lungs to be exhaled. The freshly oxygenated blood then travels via the main veins from the lungs into the left reservoir of the heart (left atrium); during the weak atrial contraction (7 to 8 mm Hg) the blood flows into the left ventricle. On the next ventricular contraction this blood is again pumped

*The combining form *hemo* denotes the blood.

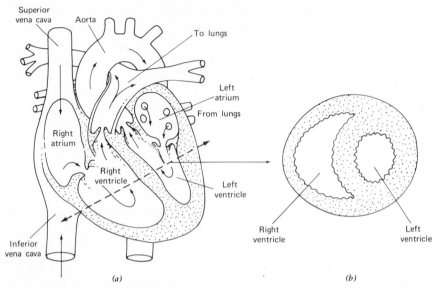

Figure 8.1.   The heart. (a) Note the heavier and stronger muscular walls on the left side where most of the work is done. (b) The cross-section shows the circular shape of the left ventricle; this shape efficiently produces the high pressure needed for the general circulation.

from the left side of the heart into the general circulation. Since a typical adult has about 4.5 liters of blood and each section of the heart pumps about 80 ml on each contraction, about 1 min is needed for the average red blood cell to make one complete cycle of the body.

The heart has a system of valves that, if functioning properly, permit the blood to flow only in the correct direction. If these valves become diseased and do not open or close properly the pumping of the blood becomes inefficient. Modern developments in artificial heart valves have permitted the replacement of natural valves with mechanical valves. It is also possible to implant a heart valve from a cadaver after it has been sterilized by a large amount of ionizing radiation. Thousands of persons in the United States are now living with artificial or implanted heart valves.

The blood volume is not uniformly divided between the pulmonary and systemic circulations. At any one time about 80% of the blood is in the systemic circulation and 20% is in the pulmonary circulation. Of the blood in the systemic circulation, about 15% is in the arteries, 10% is in the capillaries, and 75% is in the veins. In the pulmonary circulation about 7% of the blood is in the pulmonary capillaries and the remaining 93% is

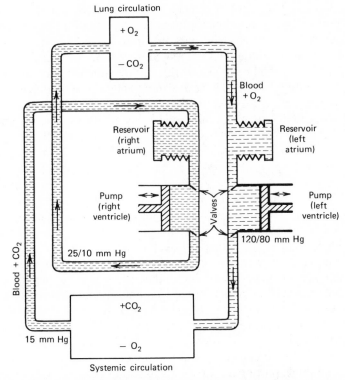

Figure 8.2. The circulatory system can be thought of as a closed loop circulation system with two pumps. One-way valves keep the flow downward through the pumps. The pressures in mm Hg are indicated.

almost equally divided between the pulmonary arteries and pulmonary veins.

While we normally think of blood as bright red, most of the blood in the body is dark red. The venous blood is depleted of the O_2 that makes the blood bright red. The blue tint to the veins in your hands is due to pigmentation in the skin. When you cut yourself venous blood usually flows out, but in a fraction of a second it becomes oxygenated.

To the eye blood appears to be a red liquid slightly thicker than water. When examined by various physical techniques it is found to consist of several different components. The red color is caused by the red blood cells (erythrocytes), flat disks about 7 μm in diameter, which represent about 45% of the volume of the blood. There are about 5×10^6 red blood cells/mm^3 of blood. A nearly clear fluid called blood plasma accounts for the other 55%. The combination of red blood cells and plasma causes blood to have flow properties different from those of a fluid like water.

Besides red blood cells and plasma, there are some important blood components, such as the white blood cells (leukocytes), present in small amounts. White blood cells ($\sim$ 9 to 15 μm in diameter) play an important role in combating disease. There are about 8000 white blood cells/mm^3 of blood. When there is an infection in the body the number of white blood cells (white count) increases. (In one type of blood cancer, leukemia, there is an excessive production of white blood cells.) Different types of white blood cells respond differently to infection, and physicians commonly ask for a *differential count,* that is, a count of the different types of white blood cells.

The blood also contains platelets. Platelets ($\sim$ 1 to 4 μm in diameter) are involved in the clotting function of blood. There are about 3 $\times$ 10^5 platelets/mm^3 of blood.

The blood acts as the transport mechanism for small amounts of hormones that control chemical processes in the body. Certain electrolytes (metal ions) in the blood are crucial to the proper functioning of the body. For example, 100 ml of blood normally contains about 10 mg of calcium. If the amount of calcium in the blood drops below 4 to 8 mg/100 ml the nervous system cannot function normally and death by tetany (muscle spasm) can result.

In the past a blood cell count was usually done by diluting the blood by a known amount, putting a drop on a glass slide under a microscope, and counting the cells. Since this method is very tedious and the accuracy is only about 15%, an easier and more accurate method was sought and developed. The instrument now routinely used in large clinical laboratories for red blood cell counts is the Coulter counter. It was invented by Wallace H. Coulter in the 1950s. The principle of operation is shown in Fig. 8.3. The diluted blood is drawn through a small capillary; the cells essentially go through the capillary one at a time, and as they do they pass between two electrodes that measure the electrical resistance across the capillary. Each red blood cell causes the resistance to change momentarily as it passes. The change in resistance appears as an electrical pulse that is counted in an electronic circuit. Coulter counters have removed much tedium from red blood cell counts (done by medical technologists) and at the same time have improved the accuracy of the measurement.

Unfortunately, Coulter counters cannot distinguish the different types of white blood cells, so differential counts must still be done with a microscope. However, techniques now under development that use a computer to recognize cell shapes may be used in the future for differential counts.

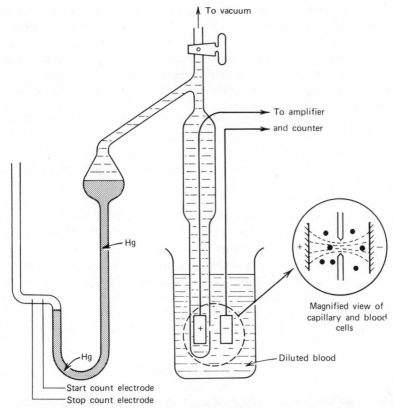

Figure 8.3. A Coulter counter automatically counts blood cells that have been diluted in a conducting solution. The elevated column of Hg produces a reduced pressure and draws the solution through the capillary. As a blood cell passes through the small opening it momentarily increases the resistance between the electrodes. The amplified pulses from the blood cells are counted from the time the Hg touches the start count electrode until the Hg touches the stop count electrode. Thus the blood cells in a fixed volume of solution are counted. The insert is a magnified view of the capillary; the electrical current paths are shown as dashed lines.

8.2. O_2 AND CO_2 EXCHANGE IN THE CAPILLARY SYSTEM

In Chapter 7 we discuss the role of diffusion in the lungs. Oxygen and carbon dioxide also diffuse through tissue. The most probable distance D that a molecule will travel after N collisions with other molecules with an average distance λ between collisions is $D = \lambda \sqrt{N}$. In tissue the density of molecules is about 1000 times greater than in air; therefore, λ is much longer in air than in tissue. A typical value for λ in water, which can serve as a model for tissue, is about 10^{-11} m, and a molecule makes about 10^{12} collisions/sec. Thus after 1 sec in water the most probable diffusion distance is about 10^{-5} m or about a factor of 10^3 less than in air. This very short diffusion distance is the primary reason that the capillaries in tissue must be very close together. In active muscle approximately one-twelfth of the volume is occupied by capillaries. In heart muscle nearly every cell is in contact with a capillary.

If you cut through a piece of active muscle and count the capillaries you will find about 190/mm². The average diameter of the capillaries is about 20μm, although some are only 5 μm in diameter and the red blood cells have to distort to go through. If we assume that the capillaries are evenly distributed at approximately 190/mm², then the total length of the capillaries in each cubic millimeter of muscle is about 190 mm. Since there are about 10^6 mm³ in 1 kg of muscle, there are about 190 km (or over 100 miles) of capillaries! If the capillaries have an average diameter of 20 μm, the surface area of the capillaries in 1 kg of muscle is about 12 m².

Not all capillaries are carrying blood at any one time. In resting muscle only 2 to 5% of the capillaries are functional. The small arteries (arterioles) that supply the capillaries have circular cuffs of muscle (sphincters) that control the flow of blood in the capillary network (Fig. 8.4). When there is a demand for blood the cuffs relax.

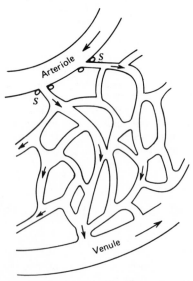

Figure 8.4.   A small section of a capillary bed. A sphincter muscle (S) controls the blood flow into the capillaries.

Starling's law of capillarity describes the flow of fluids into and out of the capillaries. Fluid movement through the capillary wall is the result of two pressures: the hydrostatic pressure P across the capillary wall forcing fluids out of the capillary and the osmotic pressure π bringing fluids in. The capillary pressure varies from about 30 mm Hg where the blood flows in at the arterial end to about 15 mm Hg where the blood leaves the capillary at the venous end. The osmotic pressure is estimated to be about 20 mm Hg into the capillary. Near the arterial end of the capillary there is a net pressure of about 10 mm Hg forcing fluids out of the capillary, and as the hydrostatic pressure drops at the venous end there is a net pressure of about 5 mm Hg favoring reabsorption of the fluids into the capillary. If capillary pressure should rise, for example, due to trauma, more fluids would be forced into the tissues, causing swelling, or edema.

8.3. WORK DONE BY THE HEART

In a typical adult each contraction of the heart muscles forces about 80 ml (about one-third of a cup) of blood through the lungs from the right ventricle and a similar volume to the systemic circulation from the left ventricle. In the process the heart does work.

The pressures in the two pumps of the heart are not the same (Fig. 8.5). In the pulmonary system the pressure is quite low because of the low resistance of the blood vessels in the lungs. The maximum pressure (systole), typically about 25 mm Hg, is about one-fifth of that in the systemic circulation. In order to circulate the blood through the much larger systemic network the left side of the heart must produce pressures that are typically about 120 mm Hg at the peak (systole) of each cardiac cycle. During the resting phase (diastole) of the cardiac cycle the pressure

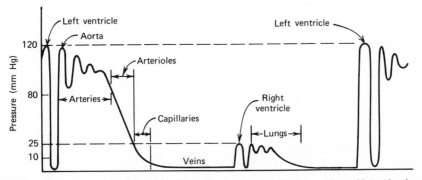

Figure 8.5. The pressure varies throughout the circulatory system. Note the low pressure in the veins and the relatively low pressure in the pulmonary system.

is typically about 80 mm Hg. Note the greater thickness of muscles on the left side of the heart in Fig. 8.1a. The muscle driving the left ventricle is about three times thicker than that of the right ventricle. In addition, the circular shape of the left ventricle is more efficient for producing high pressure than the elliptical shape of the right ventricle (Fig. 8.1b).

For reasons we discuss in Section 8.7, there is little loss of pressure until the blood reaches the arterioles and capillaries. Almost all of the pressure drop occurs across the arterioles and the capillary bed of the circulatory system (Fig. 8.5).

The work W done by a pump working at a constant pressure P is equal to the product of the pressure and the volume pumped ΔV, or $W = P\Delta V$. We can estimate the physical work done by the heart by multiplying its average pressure by the volume of blood that is pumped. Let us assume that the average pressure is 100 mm Hg or about 1.4×10^5 dynes/cm^2. If 80 ml of blood is pumped each second (a pulse rate of 60/min), the work per second is $(80)(1.4 \times 10^5) = 1.1 \times 10^7$ ergs or 1.1 J/sec or a power of 1.1 W. Actually the pumping action takes place in less than one-third of the cardiac cycle and the heart muscle rests for over two-thirds of the cycle. Thus the power during the pumping phase is over three times larger than the average value calculated above.

The heart, like all other engines, is not very efficient. In fact, it is typically less than 10% efficient, and the average power consumption of the heart is 10 W or more. Because of the lower blood pressure in the pulmonary system the power needed there is about one-fifth of that needed by the general circulation. During strenuous work or exercise the blood pressure may rise by 50% and the blood volume pumped per minute may increase by a factor of 5, leading to an increase of 7.5 times in the work done by the heart per minute. We discuss the work of enlarged hearts in Section 8.10.

8.4. BLOOD PRESSURE AND ITS MEASUREMENT

One of the most common clinical measurements is of blood pressure. The first known experimental measurement of blood pressure was made in 1733 by the Rev. Stephen Hales in Great Britain. He bravely connected a 9-ft vertical glass tube to an artery of a horse using the trachea of a goose as a flexible connection and a sharpened goose quill to puncture the artery! He found that the blood rose to an average height of 8 ft above the heart.

During surgery and in intensive care wards, a similar direct measurement of blood pressure is frequently performed. Figure 8.6 shows a

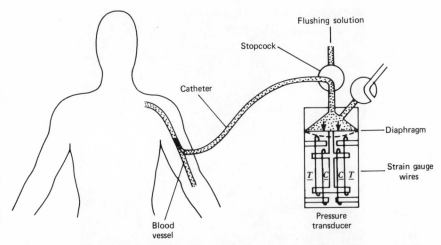

Figure 8.6.   Direct blood pressure measurement. A hollow needle is inserted in the blood vessel, and a catheter (hollow plastic tube) is threaded through the needle. The catheter transmits the blood pressure to the pressure transducer. The blood pressure deflects the diaphragm, causing a change of resistance in the four strain gauge wires. The *T* wires undergo tension and the *C* wires undergo compression.

catheter placed in the arm; during catheterization of the heart, the catheter is advanced into the chambers of the heart. Every few minutes, the stopcock is rotated so that a few milliliters of flushing solution pass through the catheter, thus preventing a clot from forming at the tip. The liquid enters the dome of the pressure transducer, where it pushes down on the metal diaphragm. The resulting bending of the diaphragm moves an armature, around which are wound fine strain gauge wires. The wires are all under tension to begin with, but downward movement of the armature increases the tension T of two wires. The increased tension stretches these wires, makes them narrower, and increases their resistance. The other two wires undergo slight compression C. Compression slackens these wires, makes them fatter, and decreases their resistance. By placing the tension and compression wires in opposite arms of a bridge (similar to Fig. 4.3), a voltage output is obtained that operates a meter or displays pulsatile waveforms on a scope or recorder. Any air bubbles in the system will result in errors in the shape of the recorded pressure waveform.

This direct method of measuring blood pressure is not necessary for routine purposes since reasonably accurate blood pressure measurements can be made by indirect means. The instrument that is commonly used is called a *sphygmomanometer*. It consists of a pressure cuff and gauge on the upper arm and a stethoscope placed over the brachial artery at the

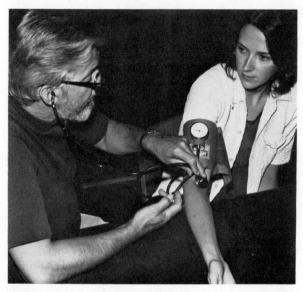

Figure 8.7.   **Blood pressure measurement using a sphygmomanometer. The arterial blood flow to the arm is blocked by an inflated cuff. As the air is gradually released, the stethoscope placed over the brachial artery is used to listen for the Korotkoff sounds. The pressures at which the sounds appear and change are noted on the gauge. (The "doctor" is one of the authors, JRC, and the "patient" is the editorial assistant, Barbara Sandrik.)**

elbow (Fig. 8.7). The pressure cuff is inflated rapidly to a pressure sufficient to stop the flow of blood and the air is gradually released. As the pressure in the cuff drops below the systolic blood pressure the turbulent flow of blood squirting through the artery causes sound vibrations that can be heard in the stethoscope. They are called Korotkoff or K sounds. This onset of K sounds indicates the systolic pressure level. As the pressure falls further, the K sounds become louder and then begin to fade. The point at which the K sounds die out or change indicates the diastolic pressure. For individuals experienced in this technique the reproducibility (precision) of the systolic blood pressure measurement is usually within 2 mm Hg. The reproducibility of the diastolic measurement is not as good (~ 5 mm Hg). The accuracy is dependent on the obesity of the patient and other factors.

The pressure in the circulatory system varies throughout the body. Even in major arteries the pressure varies from one point to another because of gravitational forces. Figure 8.8*a* shows schematically direct measurements of blood pressure made on a standing person; open glass tube manometers are shown connected to arteries in the foot, upper arm,

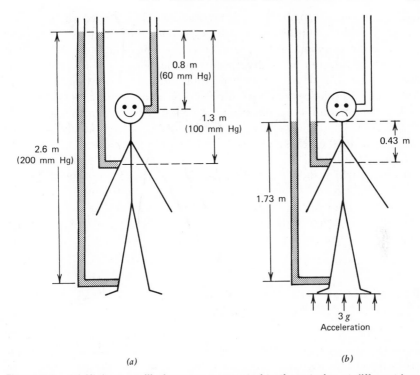

Figure 8.8. (a) If glass capillaries were connected to the arteries at different locations the blood would rise to about the same level. (b) If the body were accelerated
upward at 3 g, the blood would not reach the brain and black-out would result. If the
body were horizontal the blood pressure would be about the same at the three points
instead of differing by a factor of over three as shown here.

and head. In this situation the blood rises to essentially the same level in
all three manometers. The greater pressure P in the foot is due to the
gravitational force ($\rho g h$) produced by the column of blood (of height h)
between the heart and foot added to the pressure at the heart (ρ is the
density of the blood). Similarly, the decreased pressure in the head is due
to the elevation of the head over the heart. Since mercury is about 13
times as dense as blood ($\rho_{Hg} = 13.6$ g/cm^3, $\rho_{blood} = 1.04$ g/cm^3) a column of
mercury would be only one-thirteenth as high as a given column of blood.
That is, if your blood pressure is 120/80 mm Hg (i.e., 120 mm Hg systolic
and 80 mm Hg diastolic) it would be 1560/1040 measured in millimeters of
blood. If the average pressure at your heart is 100 mm Hg, blood in a tube
such as that shown in Fig. 8.8a would rise to an average height of 1300 mm
or 1.3 m above your heart.

If gravity on earth suddenly became three times greater, blood would rise only about 43 cm above the heart and it would not reach the brain of a standing person. This situation can be produced artificially by accelerating the body at 3 g in a vertical direction (Fig. 8.8b). It can also occur in an airplane pulling out of a dive, causing the pilot to black out. These conditions also produce pooling of blood in the legs. Special tight-fitting suits that compress the legs have been designed to reduce this pooling.

A simple method for measuring the venous pressure at the heart is to observe the veins on the back of the hands. When the hands are lower than the heart the veins stand out because of increased venous pressure: As the hands are slowly raised above the level of the heart a point is reached at which the veins collapse; this indicates a pressure of 0 cm of blood. The height of the hand veins above the heart gives the venous pressure at the heart in centimeters of blood. Venous pressure normally averages 8 to 16 cm H_2O (or blood). A pressure in excess of 16 cm H_2O may indicate congestive heart failure (see Section 8.10).

8.5. PRESSURE ACROSS THE BLOOD VESSEL WALL (TRANSMURAL PRESSURE)

As indicated in Fig. 8.5, the greatest pressure drop in the cardiovascular system occurs in the region of the arterioles and capillaries. The capillaries have very thin walls (~ 1 μm) that permit easy diffusion of O_2 and CO_2. In order to understand why they do not burst we must discuss the law of Laplace, which tells us how the tension in the wall of a tube is related to the radius of the tube and the pressure inside the tube.

Consider a long tube of radius R carrying blood at pressure P (Fig. 8.9a). We can calculate the tension T in the wall. The pressure is uniform on the wall, but we can mathematically divide the tube in half as shown in Fig. 8.9b. The force per unit length pushing upward is $2RP$. There is a tension force T per unit length at each edge that holds the top half of the tube to the bottom half. Since the wall is in equilibrium, the force pushing the two halves apart is equal to the tension forces holding them together or $2T = 2RP$ or $T = RP$. For a very small radius the tension is also very small.

Table 8.1 gives some typical pressures and tensions in the blood vessels. For example, the tension in the wall of the aorta is about 156,000 dynes/cm while the tension in a capillary wall is only about 24 dynes/cm. For comparison, a single layer of toilet tissue can withstand a tension of about 50,000 dynes/cm ($\sim$1.3 lb for a standard width sheet) before tearing. (It sometimes seems to be stronger at the perforations.) This tension is about 3000 times greater than that the capillary has to withstand.

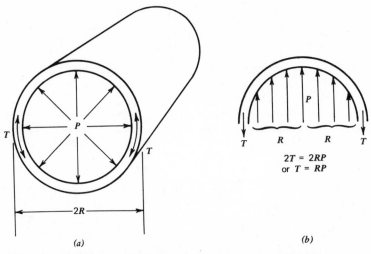

Figure 8.9.   For a long tube of radius R with blood at pressure P (a), we can calculate the tension in the walls (b). The tension is very small for very small vessels, and thus their thin walls do not break.

Table 8.1.   Typical Pressures and Tensions in Blood Vessels

	Mean Pressure		Radius (cm)	Tension (dynes/cm)
	(mm Hg)	(dynes/cm²)		
Aorta	100	1.3×10^5	1.2	156,000
Typical artery	90	1.2×10^5	0.5	60,000
Small capillary	30	4×10^4	6×10^{-4}	24
Small vein	15	2×10^4	2×10^{-2}	400
Vena cava	10	1.3×10^4	1.5	20,000

8.6.   BERNOULLI'S PRINCIPLE APPLIED TO THE CARDIOVASCULAR SYSTEM

You are probably familiar with the Bernoulli principle even though you might not give Bernoulli credit for it. Whenever there is a rapid flow of a fluid such as air or water, the pressure is reduced at the edge of the rapidly moving fluid. For example, the rapid flow of the water in the shower causes a reduced pressure in the vicinity of the shower curtain and it pulls in toward the water. Similarly, when the window in a moving car is first rolled down the reduced pressure caused by the air moving rapidly outside the window causes objects to fly out of the window.

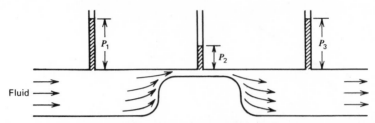

Figure 8.10. As the velocity of the fluid increases in the narrow section of the tube, part of the potential energy (pressure) is converted into kinetic energy so there is a lower pressure P_2 in this section. P_2 is less than P_1 and P_3.

Bernoulli's principle is based on the law of conservation of energy. Pressure in a fluid is a form of potential energy PE since it has the ability to perform useful work. In a moving fluid there is kinetic energy KE due to the motion. This kinetic energy can be expressed as energy per unit volume such as ergs per cubic centimeter. Since 1 erg = 1 dyne cm, 1 erg/cm^3 = 1 (dyne cm)/cm^3 or 1 dyne/cm^2, the unit for pressure in the cgs system. If fluid is flowing through the frictionless tube shown in Fig. 8.10, the velocity increases in the narrow section and the increased kinetic energy KE of the fluid is obtained by a reduction of the potential energy of the pressure in the tube. As the velocity reduces again on the far side of the restriction the kinetic energy is converted back into potential energy and the pressure increases again as indicated on the manometers.

We can calculate the average kinetic energy per unit volume of 1 g ($\sim$ 1 cm^3) of blood as it leaves the heart. Remember that $KE = mv^2/2$. Since the average velocity is about 30 cm/sec, $KE = \frac{1}{2} \times 1 \times 30^2 = 450$ ergs or 450 ergs/cm^3. This kinetic energy is equivalent to a potential energy of 450 dynes/cm^2. Since a pressure of 1 mm Hg corresponds to 1330 dynes/cm^2, this potential energy amounts to less than 0.4 mm Hg. However, during heavy exercise the velocity of the blood being pumped by the heart may be five times its average value during rest, and during the peak of the heartbeat the kinetic energy factor can have a pressure equivalent of 75 mm Hg and can represent 30% of the total work of the heart.

8.7.   HOW FAST DOES YOUR BLOOD FLOW?

As the blood moves away from the heart, the arteries branch and rebranch many times to carry blood to the various tissues. The smallest blood vessels are the capillaries. As discussed in Section 8.2 they are very small ($\sim$ 20 μm in diameter) and there are millions of them. There are so many carrying blood that their total cross-sectional area is equivalent to that of a tube almost 30 cm in diameter! Total cross-sectional areas of the vessels in the circulatory system are shown schematically in Fig. 8.11.

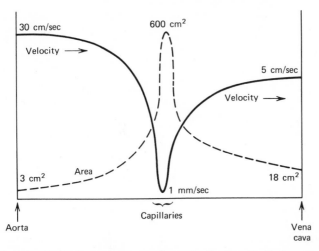

Figure 8.11. The dashed curve shows schematically the change in cross-sectional area of the circulatory system. The velocity of blood flow (solid line) decreases as the total cross-sectional area increases. The total cross-sectional area is obtained by adding the areas of all the vessels at a given distance from the heart. Note that the vena cava returning the blood to the heart has a much larger cross-sectional area than the aorta.

As the blood goes from the aorta into the smaller arteries and arterioles with greater total cross-sectional areas the velocity of the blood decreases much as the velocity of a river decreases at a wide portion. Figure 8.11 also shows schematically the velocity of blood flow in the different portions of the circulatory system. Notice that the blood velocity is related in an inverse way to the total cross-sectional area of the vessels carrying the blood. The velocity equals the flow rate divided by the cross-sectional area. The average velocity in the aorta is about 30 cm/sec; that in a capillary is only about 1 mm/sec. It is in the capillaries that the exchange of O_2 and CO_2 takes place, and this low velocity allows time for diffusion of the gases to occur.

You are undoubtedly aware of the characteristic of a liquid called *viscosity* (η). The syrup you pour on your pancakes pours at a rate different from that of the cream you put in your coffee and the water you pour into a glass. The slipperiness or ease with which a fluid pours is an indication of its viscosity. The cgs unit used to measure viscosity is the *poise*. The SI unit for viscosity is the pascal second (Pas), which equals 10 poises. The viscosity of water is about 10^{-3} Pas at 20°C. The viscosity of thick syrup may be 100 Pas. The viscosity of blood is typically 3×10^{-3} to 4×10^{-3} Pas but depends on the percentage of red blood cells in the blood (the hematocrit). As the hematocrit increases, the viscosity increases

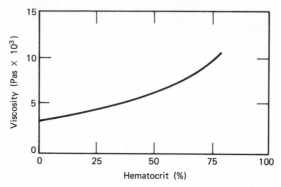

Figure 8.12. As the percent of red blood cells in the blood increases (higher hematocrit) the viscosity increases, decreasing the flow rate.

(Fig. 8.12). Persons with the disease *polycythemia vera* in which there is an overproduction of red blood cells have a high hematocrit and often have circulatory problems. The viscosity of the blood also depends on temperature. As blood gets colder, the viscosity increases and this further reduces the blood supply to cold hands and feet. A change from 37°C to O°C increases the viscosity of blood by a factor of 2.5.

In addition to viscosity, other factors affect the flow of blood in the vessels: the pressure difference from one end to the other, the length of the vessel, and its radius. In order to understand the laws that control the flow of blood in the circulatory system, Poiseuille in the nineteenth century studied the flow of water in tubes of different sizes. The results of his experiments are summarized in Fig. 8.13. Poiseuille's law states that the flow through a given tube depends on the pressure difference from one end to the other $(P_A - P_B)$, the length L of the tube, the radius R of the tube, and the viscosity η of the fluid. If the pressure difference is doubled, the flow rate also doubles. The flow varies inversely with the length and viscosity. If either is doubled, the flow rate is reduced by one-half. Poiseuille's most surprising discovery was related to the dependence of the flow rate on the radius of the tube. As he expected, the flow rate increased as the radius of the tube increased; what was surprising was how rapidly the flow rate increased with small increases in the radius. For example, if the radius is doubled the flow rate increases by 2^4 or a factor of 16. When all of these variables are put together with a constant to keep the units working correctly we get Poiseuille's equation:

$$\text{Flow rate} = (P_A - P_B) \left(\frac{\pi}{8}\right)\left(\frac{1}{\eta}\right)\left(\frac{R^4}{L}\right)$$

In SI units the flow rate will be in cubic meters per second if $P_A - P_B$ is in newtons per square meter, η is in Pas, and R and L are in meters.

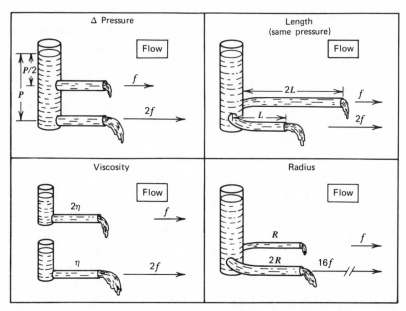

Figure 8.13. Poiseuille's findings. The flow rate through a tube depends on the pressure difference from one end of the tube to the other, the length of the tube, the viscosity of the fluid, and the radius. The radius has the largest influence on flow rate.

Poiseuille's law applies to rigid tubes of constant radius. Since the major arteries have elastic walls and expand slightly at each heartbeat, blood flow in the circulatory system does not obey the law exactly. In addition, the blood's viscosity changes slightly with flow rate; however, this effect is negligible.

Even though the total cross-sectional area of the arterioles is many times greater than that of the aorta, most of the pressure drop occurs across the arterioles because of the great flow resistance produced by the R^4 factor. The next largest drop is across the capillaries (Fig. 8.5).

8.8. BLOOD FLOW—LAMINAR AND TURBULENT

You have probably seen both a slow, smooth, quietly flowing river and a rapid, turbulent, noisy river. The first type of river is analogous to the laminar or streamline flow that is present in most blood vessels; the second is similar to the turbulent flow found at a few places in the circulatory system, for example, where the blood is flowing rapidly past the heart valves.

An important characteristic of laminar flow is that it is silent. If all blood flow were laminar, information could not be obtained from the heart with a stethoscope. The heart sounds heard with a stethoscope are caused by turbulent flow. During a blood pressure measurement, the constriction produced by the pressure cuff on the arm produces turbulent flow and the resulting vibrations can be detected with a stethoscope on the brachial artery.

In laminar flow the blood that is in contact with the walls of the blood vessel is essentially stationary, the layer of blood next to the outside layer is moving slowly, and successive layers move more rapidly just as the water in the middle of a quiet stream moves more rapidly than the water along the banks (Fig. 8.14a). This behavior has an effect on the distribution of red blood cells in the circulatory system.

The red blood cells in an artery are not distributed uniformly; there are more in the center than at the edges (Fig. 8.14b). This produces two effects. When blood enters a small vessel from the side of a main vessel the percentage of red blood cells in that blood (the hematocrit) will be slightly less than in the blood in the main vessel because of the "skim-

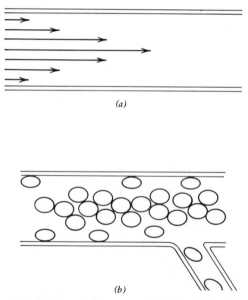

(a)

(b)

Figure 8.14.   **Blood flow in the vessels. (a) In the laminar flow in most of the vessels there is a greater velocity at the center as indicated by the longer arrow. (b) The distribution of red blood cells is not uniform; they are more dense at the center so the blood that flows into small arteries has a smaller percentage of red blood cells than the blood in the main artery.**

ming" effect. The second effect is more important. Because the plasma along the vessel walls is moving more slowly than the red blood cells, the blood in the extremities has a greater percentage of red blood cells than when it left the heart. This causes an increase in the hematocrit in the hands and feet of approximately 10% over the hematocrit of the whole blood. This effect is of some importance when blood volume is measured by the radioisotope dilution technique, which is discussed in Chapter 17.

If you gradually increase the velocity of a fluid flowing in a tube by reducing the radius of the tube, it will reach a critical velocity V_C when laminar flow changes into turbulent flow (Fig. 8.15). The critical velocity will be lower if there are restrictions or obstructions in the tube. Osborne Reynolds studied this property in 1883 and determined that the critical velocity is proportional to the viscosity η of the fluid and is inversely proportional to the density ρ of the fluid and the radius R of the tube, $V_C = K\eta/\rho R$. The constant of proportionality K is called the Reynold's number, and it is approximately equal to 1000 for many fluids, including blood, flowing in long straight tubes of constant diameter. If there are bends or obstructions the Reynold's number becomes much smaller. In the aorta, which has a radius of about 1 cm in adults, the critical velocity $V_C = K\eta/\rho R = (1000)(4 \times 10^{-3} \text{ Pas})/(10^3 \text{ kg/m}^3)(10^{-2} \text{ m}) = 0.4$ m/sec. The velocity in the aorta ranges from 0 to 0.5 m/sec, and thus the flow is turbulent during part of the systole. During heavy exercise the amount of blood pumped by the heart may increase four or five times and the critical velocity will be exceeded for a longer period of time. The heart sounds of a person doing heavy exercise are different from those of a person at rest.

Laminar flow is more efficient than turbulent flow. This is illustrated graphically in Fig. 8.16a. The slope of the curve in the laminar flow region is greater than that in the turbulent flow region. That is, a given increase in pressure causes a greater increase in the laminar flow rate than in the turbulent flow rate. The reduction in efficiency is apparent in the blood

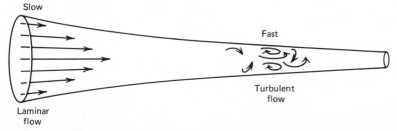

Figure 8.15. **If fluid is flowing in a long tapering tube, the velocity will gradually increase to the point where it exceeds the critical velocity V_C, producing turbulent flow.**

172 *Physics of the Cardiovascular System*

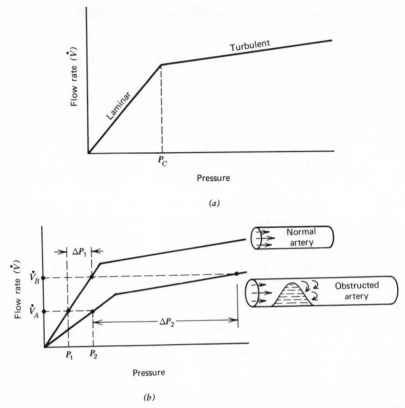

Figure 8.16. (a) When the flow in a tube becomes turbulent (at pressure P_C) the slope of the flow rate versus pressure decreases so that compared to laminar flow a greater increase of pressure is necessary to obtain a given increase in flow rate. (b) In an obstructed artery the pressure needed to produce a given flow rate is greater than in a normal artery of the same size. In addition, if the heart is called upon to increase the flow rate from $\dot{V}_A$ to $\dot{V}_B$ the turbulence produced in the obstructed artery requires a much larger pressure increase (ΔP_2 vs ΔP_1) and thus greater effort from the heart. (Adapted from I.W. Richardson and E.B. Neergaard, *Physics for Medicine and Biology*, Wiley-Interscience, New York, 1972, pp. 46–47.)

flow through an artery with an obstruction (Fig. 8.16*b*). For the flow rate $\dot{V}_A$ a pressure of P_1 is needed for the normal artery and a somewhat higher pressure P_2 is needed for the obstructed artery. If both arteries are required to deliver a new flow rate $\dot{V}_B$, the increase in pressure ΔP_2 (and thus the work) will be much greater for the obstructed artery since the flow will be turbulent.

8.9. HEART SOUNDS

An experienced cardiologist with good hearing can obtain much diagnostic information from the heart sounds. The heart sounds heard with a stethoscope are caused by vibrations originating in the heart and the major vessels. The opening and closing of the heart valves contribute greatly to the heart sounds; turbulent flow occurs at these times and the vibrations produced are often in the audible range. Figure 8.17 shows the sounds heard with a stethoscope from a normal heart. Other sounds may be heard if the heart is not normal. Murmurs may be produced if there is a constriction that causes turbulent flow during part of the cardiac cycle. For example, if the aortic valve is narrow (aortic valve stenosis) blood flow through it during systole will cause a murmur.

The amount and quality of the sound heard depend on the design of the stethoscope as well as on its pressure on the chest, its location, the

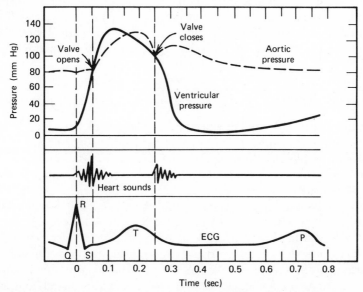

Figure 8.17. The time relationships of the electrocardiogram (ECG), heart sounds (phonocardiogram), and left ventricular and aortic blood pressures. The ventricle begins to contract at time 0, and the aortic valve opens when the ventricular pressure just exceeds the aortic pressure. This contributes to the first heart sound. The closing of the aortic valve contributes to the second heart sound. The first sound is normally longer and louder than the second sound. (Adapted from Scher, A.M., in T.C. Ruch and H.D. Patton (Eds.), *Physiology and Biophysics*, 19th ed., © W.B. Saunders Company, Philadelphia, 1965, p. 557.)

orientation of the body, and the phase of the breathing cycle. (Characteristics of stethoscopes are discussed in Chapter 12.) There are optimum positions for hearing the various heart sounds with a stethoscope. In general, sound is not transmitted well from a liquid to air (see Chapter 12, p. 258), and thus heart sounds are not heard well if the sound must travel through the lung.

The sounds from a normal heart are in the frequency range of 20 to about 200 Hz. This is not the most sensitive range of the human ear (see Chapter 13, p. 297). The sensitivity of the ear is very poor at low frequencies; to be heard, a sound at 20 Hz must be about 10,000 times more intense than a sound at 200 Hz. A normal heart produces some sounds that cannot be heard with a good stethoscope, even under optimum conditions. However, it is possible to electronically amplify these heart sounds to listen to them directly or to record them. Phonocardiography is the graphic recording of heart sounds (Fig. 8.17). The electronic amplifiers used in phonocardiography have a much different response than the human ear so the recordings do not correspond well with what the cardiologist hears. Similarly, an electronically amplified stethoscope distorts the sounds that the physician is accustomed to hearing.

8.10. THE PHYSICS OF SOME CARDIOVASCULAR DISEASES

Heart disease is the number one cause of death in the United States. Because of the many physical aspects of the cardiovascular system, heart diseases often have a physical component. Many of these diseases, for example, increase the work load of the heart or reduce its ability to work at a normal rate.

The work done by the heart is roughly the tension of the heart muscle times how long it acts. Anything that increases the muscle tension or how long it acts will increase the work load of the heart. For example, high blood pressure (hypertension) causes the muscle tension to increase in proportion to the pressure. A fast heart rate (tachycardia) increases the work load since the amount of time the heart muscles spend contracting increases.

The heart disease that causes the most deaths is heart attack. A heart attack is caused by a blockage of one or more arteries to the heart muscle. That portion of the heart muscle without a blood supply dies (the infarct). The blockage does not always immediately affect the electrical signals that control the heart's beating action, and thus a person who has recently had a heart attack may still have a normal electrocardiogram (ECG).

During and after a heart attack the ability of the heart muscle to pump blood to the body is seriously impaired. To reduce the work load of the heart, bed rest and O_2 therapy are prescribed. Giving O_2 increases the O_2 in the blood so that less blood must be pumped to the tissues. This O_2 is probably most beneficial to the heart muscle itself. There are often alternate paths for blood to get to muscles, and these anastomoses in the heart can provide some O_2 to the blocked portion. One of the purposes of a regular exercise program is to stress the cardiovascular system enough to keep these alternate routes open.

Another common heart disease is congestive heart failure. The cause of this disease is not as well understood as the cause of heart attack. It is characterized by an enlargement of the heart and a reduction in the ability of the heart to provide adequate circulation.

For an enlarged heart we can apply the law of Laplace. If the radius of the heart is doubled, the tension of the heart muscle must also be doubled if the same blood pressure is to be maintained. However, since the heart muscle is stretched, it may not be able to produce sufficient force to maintain normal circulation (see Fig. 2.4). The stretched heart muscle is also much less efficient than normal heart muscle; that is, it consumes much more O_2 for the same amount of work.

The medical treatment for congestive heart failure is to reduce the work load of the heart. A dramatic approach is to replace the heart surgically. A number of heart transplants have been successful, but the initial enthusiasm has decreased; relatively few were performed in 1976. Intensive research is under way to develop an implantable mechanical heart or a heart assist device. The major problem has been in developing a material that is compatible with the body over a long period of time. It is estimated that in the United States there are 10,000 patients each year who would be good candidates for implantable artificial hearts. Artificial heart-lung machines, which are used successfully during major surgery, are bulky units that are not practical for routine use.

Patients with a condition in which the heart's electrical signals are inadequate to stimulate heart action have been greatly helped by modern technology. They have received artificial pacemakers to regulate the heartbeat. Pacemakers are discussed in Chapter 10.

Another man-made device that has helped heart patients is the artificial heart valve. Heart valve defects are of two types: the valve either does not open wide enough (stenosis) or it does not close well enough (insufficiency). In stenosis the work of the heart is increased because a large amount of work is done against the obstruction of the narrow opening, and the blood supply to the general circulation is reduced. In insufficiency some of the pumped blood flows back into the heart so that the volume of circulated blood is reduced. Both types of defective valves can be re-

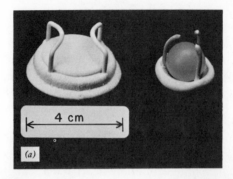

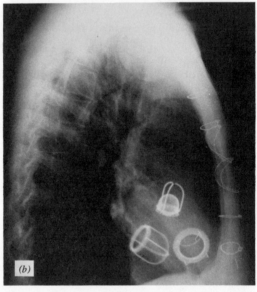

Figure 8.18. (a) Two of the several artificial heart valves used routinely. Such a valve is sutured into the heart to allow blood to flow upward only. (b) An x-ray of a patient showing 3 artificial heart valves in place. At the time the x-ray was taken in 1976 the valves had been in place 6.5 years and the patient was well. (Courtesy of Dr. William Young, University of Wisconsin, Madison.)

placed by artificial valves. Several designs are available (Fig. 8.18). Compatibility between man-made valves and the blood is still a problem. These valves sometimes cause clotting.

Cadaver heart valves are also used as replacements for defective valves. A heart valve is primarily cartilage tissue with relatively few living cells. Before it is transplanted, a cadaver valve is sterilized with radiation from an electron accelerator that is used for treating cancer (see Chapter 18, p. 512).

Many cardiovascular diseases involve the blood vessels. We now discuss the physics of a few of these diseases.

An aneurysm is a weakening in the wall of an artery resulting in an increase in its diameter (Fig. 8.19). The increased diameter increases the

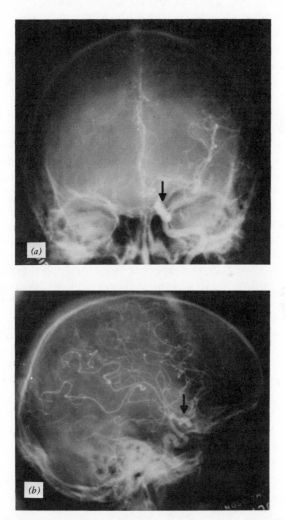

Figure 8.19.   X-rays of the skull from the front (a) and the side (b) showing an aneurysm (arrow). A dye that absorbs x-rays has been injected into the arteries to make them visible.

tension in the wall proportionately. If it were not for the supporting action of the surrounding tissue the wall would blow out the way a bicycle inner tube does under similar conditions. If an aneurysm does rupture it is often fatal—especially if the rupture is in the brain, a type of *cerebrovascular accident* (CVA).

A more common vessel problem is the formation of sclerotic plaques on the walls of an artery. The plaques can cause turbulent flow and produce a noticeable murmur. The narrowing of the artery will cause an increase in the blood velocity in that region with a decrease in wall pressure because of the Bernoulli effect. The plaque may dislodge and travel with the blood until it lodges in a smaller artery. This blockage will shut off the blood supply to the affected part; if it is in the brain, it will produce a stroke, another type of cerebrovascular accident.

A disease that is clinically not as serious as aneurysms and plaques but that often causes embarrassment is varicose veins. Varicose veins can be more than a cosmetic problem since they can develop complications. These enlarged surface veins in the leg result from a failure of the one-way valves in the veins. Consider the blood flow in the lower legs and feet of an erect person. The pressure in a leg vein is approximately 90 mm Hg (or 115 cm of blood) due to the column of blood above it. During walking or other leg exercise, the contraction of the muscles forces the venous blood toward the heart. This action of the muscles on the blood is called the venous pump or muscle pump. At various points along the veins there are one-way flaps or valves that prevent the blood from going back. The action of the muscle pump and the valves results in a venous pressure of about 20 mm Hg during exercise. If these valves become defective and let the blood run back down it will pool in the vein, and the vein will become varicose. Varicose veins may be aggravated by conditions that restrict the return of the blood to the heart such as tight bands at the tops of the legs from girdles. The additional abdominal weight during pregnancy may also restrict venous return. However, some experiments have shown that a pressure cuff above the knee inflated to 90 mm Hg is no hindrance to the action of the muscle pump. The standard treatment for varicose veins is surgical removal of the offending vessels. There are usually adequate parallel veins to carry the blood back to the heart.

8.11. SOME OTHER FUNCTIONS OF BLOOD

Although we have emphasized the role of blood in gas exchange, an equally important function of the blood is to carry the body's liquid wastes to the kidneys. The details of kidney function are not discussed in this book, but by filtration of the blood, the kidneys keep the make-up of the blood very constant despite large fluctuations in our diet. The kidneys are well vascularized in order to filter the blood. Normally 1 to 1.5 liters of blood (one-fifth to one-fourth of the cardiac output) flows through the kidneys each minute. This is far in excess of the amount needed to supply nutrients and oxygen to the kidneys. If a severe blood loss occurs the kidney flow may drop to 0.25 liter/min to permit the blood to be used elsewhere. Although artificial kidneys (dialysis units) are used by thousands of people they fall far short of the kidneys in their ability to regulate body components. Many patients on dialysis units can better be described as surviving than as living.

The blood plays an important role in distributing and dissipating heat in the body (see Chapter 5). The venous blood returning from the limbs can be routed close to the skin to increase heat losses in warm weather. In cold weather it can be routed internally close to the artery carrying blood to the limb; the cool venous blood takes up some of the heat from the warm arterial blood and carries it back to the heart. This counter-current principle keeps down heat losses from the extremities and from the skin in cold weather.

Blood is also involved in the male erection.* On proper (or improper) stimulation arterial blood flows into the penis causing it to enlarge and become rigid due to the fluid pressure. The venous return is slow—the erection helps to block the return of blood to the veins. The blood pressure in the penis during an erection is about the same as the systolic pressure. The same principle keeps plants erect. Their rigidity is due to internal fluid pressure—if you do not water them, the fluid pressure drops and they wilt.

BIBLIOGRAPHY

Brown, B. A., *Hematology: Principles and Procedures,* 2nd ed., Lea and Febiger, Philadelphia, 1976.

Castellanos, A., and L. Lemberg, *Electrophysiology of Pacing and Cardioversion,* Appleton-Century-Crofts, New York, 1969.

Myers, G. H., and V. Parsonnet, *Engineering in the Heart and Blood Vessels,* Wiley-Interscience, New York, 1969.

Ravin, A., *Auscultation of the Heart,* Year Book Medical, Chicago, 1958.

Rushmer, R. F., *Cardiovascular Dynamics,* 3rd ed., Saunders, Philadelphia, 1970.

Samet, P. (Ed.), *Cardiac Pacing,* Grune and Stratton, New York, 1973.

Tavel, M. E., *Clinical Phonocardiography and External Pulse Recording,* 2nd ed., Year Book Medical, Chicago, 1972.

*There are also erectile tissues in females, such as the nipples and clitoris.

REVIEW QUESTIONS

1. Estimate your blood volume from your mass.
2. What causes a "blue baby"?
3. What is the difference between a hematologist and a cardiologist?
4. Estimate the volume of blood your heart pumps to your systemic circulation each day.
5. What percentage of your blood is in your pulmonary circulation?
6. If white blood cells have an average diameter of 12 μm, what percentage of the blood volume is white blood cells?
7. If a platelet has a diameter of about 2 μm, what percentage of the blood volume is platelets?
8. If the average power consumed by the heart is 10 W, what percentage of a 2500 kilocalorie daily diet is used to operate the heart? (4.2 J = 1 calorie)
9. What is a sphygmomanometer?
10. What are Korotkoff sounds?
11. How much does the blood pressure in your brain increase when you change from a standing position to standing on your head?
12. An artery with a 3 mm radius is partially blocked with plaque; in the constricted region the effective radius is 2 mm and the average blood velocity is 50 cm/sec.
    (a) What is the average velocity of the blood in the unconstricted region?
    (b) Would there be turbulent flow in either region?
    (c) For the blood in the constricted region, find the equivalent pressure due to the kinetic energy of the blood.
13. Why can arteries with small diameters have thinner walls than arteries with large diameters carrying blood at the same pressure?
14. What is the approximate velocity of blood in the capillaries?
15. If the radius of an arteriole changed from 50 to 40 μm, how much would the flow rate through it decrease?
16. What is phonocardiography?
17. What is tachycardia?
18. What is an aneurysm?
19. How does the venous blood get from the feet to the heart of a standing person?

CHAPTER 9

Electricity Within the Body

Physical phenomena involving electricity and magnetism have been observed since ancient times. However, only in the last two centuries have scientists begun to understand them. If you were born 200 years ago, you would have had no contact with man-made electricity during your entire life. The extraordinary developments in this area of science have been applied to so many areas that it is now hard to imagine life without electricity.

Electricity plays an important role in medicine. There are two aspects of electricity and magnetism in medicine: electrical and magnetic effects generated inside the body, which are discussed in this chapter, and applications of electricity and magnetism to the surface of the body, which are discussed in Chapter 11. Chapter 10 covers cardiovascular instrumentation.

A number of our modern concepts of electrical activity in the body date back many years. Luigi Galvani made the first contribution in this field in 1786 when he discovered animal electricity in a frog's leg. Since then many years of research have been expended on a wide variety of experiments dealing with electrical effects in and on the body. Basic research in this area is called *neurophysiology*.

The electricity generated inside the body serves for the control and operation of nerves, muscles, and organs. Essentially all functions and activities of the body involve electricity in some way. The forces of muscles are caused by the attraction and repulsion of electrical charges. The action of the brain is basically electrical. All nerve signals to and from the brain involve the flow of electrical currents.

The nervous system plays a fundamental role in nearly every body function. Basically, a central computer (the brain) receives internal and external signals and (usually) makes the proper response. The information is transmitted as electrical signals along various nerves. This efficient communication system can handle many millions of pieces of information at one time with great speed.

In carrying out the special functions of the body, many electrical signals are generated. These signals are the result of the electrochemical action of certain types of cells. By selectively measuring the desired signals (without disturbing the body) we can obtain useful clinical information about particular body functions. In this chapter we discuss some of these electrical signals. The electrical potentials of nerve transmission and the electrical signals seen in the electromyogram (EMG) of the muscle, the electrocardiogram (ECG) of the heart, and the electroencephalogram (EEG) of the brain are the best known. We also discuss some of the less familiar electrical signals of the body, such as those seen in the electroretinogram (ERG) and the electrooculogram (EOG) of the eye, the magnetic signals of the body as shown on the magnetocardiogram (MCG) of the heart and the magnetoencephalogram (MEG) of the brain, and those signals associated with bone growth and biofeedback.

Various medical specialists are involved in the diagnosis and treatment of malfunctions of this internal electrical system. If the problem involves any part of the nervous system it is diagnosed and treated by a *neurologist,* an M.D. who has had three or more years of special training in the study of the nervous system. If the problem requires surgery, it is usually handled by a *neurosurgeon,* an M.D. who specializes in surgery of the nervous system and has had three or more years of training in this area of surgery. Since much of neurosurgery involves the brain, these specialists are sometimes called brain surgeons. *Neuroradiologists* are M.D.s who have taken a three-year residency in diagnostic radiology followed by another year in neuroradiological specialization. Pediatric neurology, a subspecialty of pediatrics, deals with nerve problems in infants and children. Electromyogram tests are usually performed and interpreted by physiatrists, M.D.s who have taken residencies in physical medicine. Cardiologists, specialists in the study and treatment of heart disease, deal with the electrical activity of the heart. Psychiatrists are M.D.s who specialize in the diagnosis, prevention, and treatment of emotional illness and neural disorders. They may use shock therapy and medication. Clinical psychologists are Ph.D.s who have studied behavior and also specialize in the diagnosis and treatment of mental illness; however, they cannot use shock therapy or treat with drugs.

9.1. THE NERVOUS SYSTEM AND THE NEURON

The nervous system can be divided into two parts—the central nervous system and the autonomic nervous system. The central nervous system consists of the brain, the spinal cord, and the peripheral nerves—nerve fibers (neurons) that transmit sensory information to the brain or spinal cord (afferent nerves) and nerve fibers that transmit information from the brain or spinal cord to the appropriate muscles and glands (efferent nerves). The autonomic nervous system controls various internal organs such as the heart, intestines, and glands. The control of the autonomic nervous system is essentially involuntary.

The brain is exceedingly complicated and not well understood. It is the body's most important organ and is given special protection. It is surrounded by three membranes within the protective skull and because it "floats" in the shock-absorbing cerebrospinal fluid (CSF), the 1500 g brain has the effective weight of a 50 g mass. The brain is connected to the spinal cord, which is also surrounded by CSF and is protected by the bone of the spinal column.

The basic structural unit of the nervous system is the *neuron* (Fig. 9.1), a nerve cell specialized for the reception, interpretation, and transmission of electrical messages. There are many types of neurons. Basically, a neuron consists of a cell body that receives electrical messages from other neurons through contacts called *synapses* located on the dendrites or on the cell body. The dendrites are the parts of the neuron specialized for receiving information from stimuli or from other cells. If the stimulus is strong enough, the neuron transmits an electrical signal outward along a fiber called an *axon*. The axon, or nerve fiber, which may be as long as 1 m, carries the electrical signal to muscles, glands, or other neurons.

9.2. ELECTRICAL POTENTIALS OF NERVES

The ability of neurons to receive and transmit electrical signals is fairly well understood. In this section we discuss the electrical behavior of neurons.

Much of the early research on the electrical behavior of nerves was done on the giant nerve fibers of squid. The conveniently large diameter ($\sim$ 1 mm) of these nerve fibers allows electrodes to be readily inserted or attached for measurements.

Across the surface or membrane of every neuron is an electrical potential (voltage) difference due to the presence of more negative ions on the

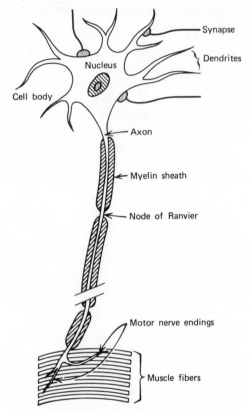

Figure 9.1.   Schematic of a motor neuron.

inside of the membrane than on the outside. The neuron is said to be polarized. The inside of the cell is typically 60 to 90 mV more negative than the outside. This potential difference is called the *resting potential* of the neuron. Figure 9.2 shows schematically the typical concentrations of various ions inside and outside the membrane of an axon. When the neuron is stimulated, a large momentary change in the resting potential occurs at the point of stimulation. This potential change, called the *action potential,* propagates along the axon. The action potential is the major method of transmission of signals within the body. The stimulation may be caused by various physical and chemical stimuli such as heat, cold, light, sound, and odors. If the stimulation is electrical, only about 20 mV across the membrane is needed to initiate the action potential.

We can explain the resting potential by using a model in which a membrane separates a concentrated neutral solution of KCl from one that

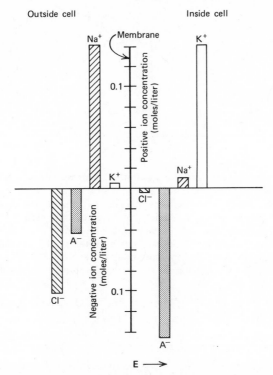

Figure 9.2. Typical concentrations in moles per liter of K$^+$, Na$^+$, Cl$^-$, and large protein ions (A$^-$) inside and outside a cell. The inside of the cell is more negative than the outside by about 60 to 90 mV. The electric field is shown as E.

is less concentrated (Fig. 9.3*a*). The KCl in solution forms K$^+$ ions and Cl$^-$ ions. We assume that the membrane permits K$^+$ ions to pass through it but does not permit the passage of the Cl$^-$ ions. The K$^+$ ions diffuse back and forth across the membrane; however, a net transfer takes place from the high concentration region *H* to the low concentration region *L*. Eventually this movement results in an excess of positive charge in *L* and an excess of negative charge in *H*. These charges form layers on the membrane to produce an electrical force that retards the flow of K$^+$ ions from *H* to *L*. Ultimately a condition of equilibrium exists (Fig. 9.3*b*). Qualitatively, the resting potential of a nerve exists because the membrane is impermeable to the large A$^-$ (protein) ions shown in Fig. 9.2 while it is permeable to the K$^+$, Na$^+$, and Cl$^-$ ions.

Figure 9.4 shows schematically how the axon propagates an action potential. Graphs of the potential measured between point *P* and the outside of the axon are also shown. This axon has a resting potential of

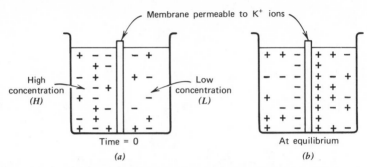

Figure 9.3.   A model of the resting potential. (*a*) **A membrane selectively permeable to K⁺ ions initially separates two KCl solutions of different concentrations; + and − represent the K⁺ and Cl⁻ ions, respectively. The K⁺ ions diffuse from side *H* to side *L*, producing a charge difference (dipole layer) across the membrane and hence a potential.** (*b*) **The dipole layer provides an electrical force that tends to keep K⁺ ions on side *H*. An equilibrium condition is produced when the K⁺ ion movement due to diffusion is balanced by the ion movement due to the electrical force from the dipole layer. The dipole layer is equivalent to a resting potential across the membrane.**

about −80 mV (Fig. 9.4*a*). If the left end of the axon is stimulated, the membrane walls become porous to Na⁺ ions and these ions pass through the membrane, causing it to depolarize. The inside momentarily goes positive to about 50 mV. The reversed potential in the stimulated region causes ion movement, as shown by the arrows in Fig. 9.4*b*, which in turn depolarizes the region to the right (Fig. 9.4*c, d,* and *e*). Meanwhile the point of original stimulation has recovered (repolarized) because K⁺ ions have moved out to restore the resting potential (Fig. 9.4*c, d,* and *e*). The voltage pulse is the action potential. For most neurons and muscle cells, the action potential lasts a few milliseconds; however, the action potential for cardiac muscle may last from 150 to 300 msec (Fig. 9.5).

An axon can transmit in either direction. However, the synapse that connects it to another neuron only permits the action potential to move along the axon away from its own cell body.

Examination of the axons of various neurons with an electron microscope indicates that there are two different types of nerve fibers. The membranes of some axons are covered with a fatty insulating layer called *myelin* that has small uninsulated gaps called nodes of Ranvier every few millimeters (Fig. 9.1); these nerves are referred to as *myelinated nerves.* The axons of other nerves have no myelin sleeve (sheath), and these nerves are called *unmyelinated nerves.* This is a somewhat artificial classification; most human nerves have both types of fibers. Much of the early research on the electrical behavior of nerves was done on the unmyelinated giant nerve fiber of the squid (Fig. 9.4). Myelinated nerves,

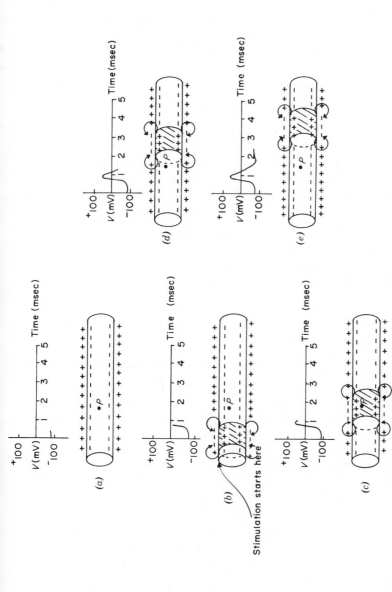

Figure 9.4. The transmission of a nerve impulse along an axon. The graphs show the potential at point P. (a) The axon has a resting potential of about − 80 mV. (b) Stimulation on the left causes Na⁺ ions to move into the cell and depolarize the membrane. (c) The positive current flow on the leading edge, indicated by the arrows, stimulates the regions to the right so that depolarization takes place and the potential change propagates (d and e). Meanwhile K⁺ ions move out of the core of the axon and restore the resting potential (repolarize the membrane). The voltage pulse moving along the nerve is the action potential.

187

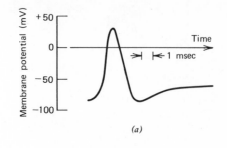

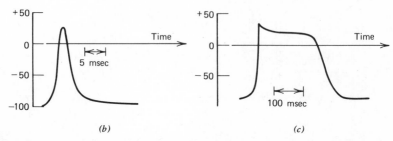

Figure 9.5. **Waveforms of the action potentials from (a) a nerve axon, (b) a skeletal muscle cell, and (c) a cardiac muscle cell. Note the different time scales.**

the most common type in humans, conduct action potentials much faster than unmyelinated nerves.

Myelinated axons conduct differently than the unmyelinated axon shown in Fig. 9.4. The myelin sleeve is a very good insulator, and the myelinated segment of an axon has very low electrical capacitance. The action potential decreases in amplitude as it travels through the myelinated segment just as an electrical signal is attenuated when it passes through a length of cable. The reduced signal then acts like a stimulus at the next node of Ranvier (gap) to restore the action potential to its original size and shape. The conduction in the gap is the same as shown in Fig. 9.4. This process repeats along the axon; the action potential seems to jump from one node to the next, that is, it travels by saltatory conduction.

Two primary factors affect the speed of propagation of the action potential: the resistance within the core of the membrane and the capacitance (or the charge stored) across the membrane. A decrease in either will increase the propagation velocity. The internal resistance of an axon decreases as the diameter increases, so an axon with a large diameter will have a higher velocity of propagation than an axon with a small diameter.

The greater the stored charge on a membrane, the longer it takes to depolarize it, and thus the slower the propagation speed. Because of the low capacitance, the charge stored in a myelinated section of a nerve fiber is very small compared to that on an unmyelinated fiber of the same diameter and length. Hence the conduction speed in the myelinated fiber is many times faster. The unmyelinated squid axons (~1 mm in diameter) have propagation velocities of 20 to 50 m/sec, whereas the myelinated fibers in man (~10 μm in diameter) have propagation velocities of around 100 m/sec. This difference in conduction speed explains why the signal appears to jump from node to node in myelinated nerves.

The advantage of myelinated nerves, as found in man, is that they produce high propagation velocities in axons of small diameter. A large number of nerve fibers can thus be packed into a small bundle to provide for many signal channels. For example, 10,000 myelinated fibers of 10 μm in diameter can be carried in a bundle with a cross-sectional area of 1 to 2 mm^2, whereas 10,000 unmyelinated fibers with the same conduction speed would require a bundle with a cross-sectional area of approximately 100 cm^2, or about 10,000 times larger.

9.3. ELECTRICAL SIGNALS FROM MUSCLES—THE ELECTROMYOGRAM

One means of obtaining diagnostic information about muscles is to measure their electrical activity. In this section we briefly trace the transmission of the action potential from the axon into the muscle, where it causes muscle contraction. The record of the potentials from muscles during movement is called the *electromyogram*, or EMG.

A muscle is made up of many motor units. A motor unit consists of a single branching neuron from the brain stem or spinal cord and the 25 to 2000 muscle fibers (cells) it connects to via motor end plates (Fig. 9.6a). The resting potential across the membrane of a muscle fiber is similar to the resting potential across a nerve fiber. Muscle action is initiated by an action potential that travels along an axon and is transmitted across the motor end plates into the muscle fibers, causing them to contract. The record of the action potential in a single muscle cell is shown schematically in Fig. 9.6b. Such a measurement is made with a very tiny electrode (microelectrode) thrust through the muscle membrane.

Single muscle cells are usually not monitored in an EMG examination because it is difficult to isolate a single fiber. Instead, EMG electrodes usually record the electrical activity from several fibers. Either a surface electrode or a concentric needle electrode is used. A surface electrode attached to the skin measures the electrical signals from many motor units. A concentric needle electrode inserted under the skin measures

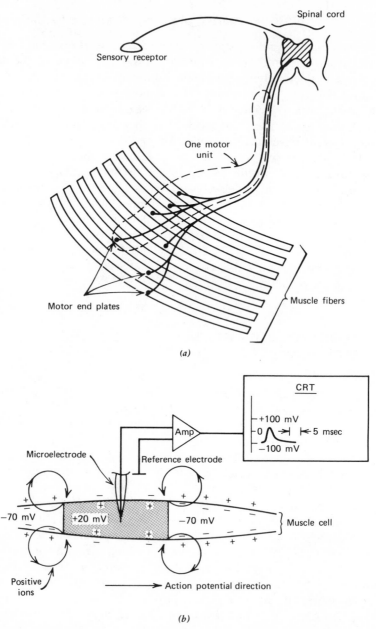

Figure 9.6. (a) Schematic of a neuron originating at the spinal cord and terminating on several muscle cells. The neuron and connecting muscle cells make up a motor unit (dashed line). (b) Instrument arrangement for measuring the action potential in a single muscle cell. The reference electrode is immersed in the fluid surrounding the cell.

single motor unit activity by means of insulated wires connected to its point. Figure 9.7 shows typical EMGs from the two types of electrodes.

A typical arrangement for recording the EMG is shown in Fig. 9.8. The muscle's electrical signals can be displayed directly on one channel of an oscilloscope, and the signals can be integrated and displayed on a second channel. The signals can also be passed through an amplifier and made audible by a loudspeaker. The integrated record (in volt seconds) is a measure of the quantity of electricity associated with the muscle action potentials. Figure 9.9 shows the EMG and its integrated form for different degrees of voluntary muscular contraction. More forceful contractions lead to greater action potential activity. It is easier to evaluate the integrated form of action potential activity because it is a smooth curve. In the

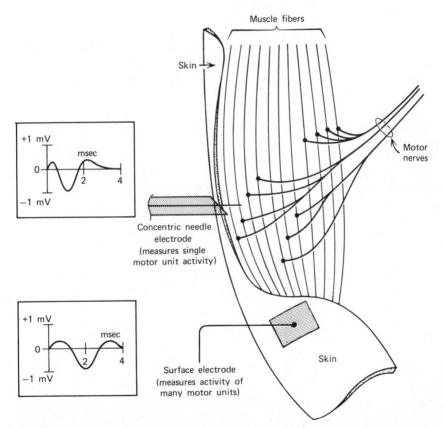

Figure 9.7. Electromyograms obtained with a concentric needle electrode and a surface electrode.

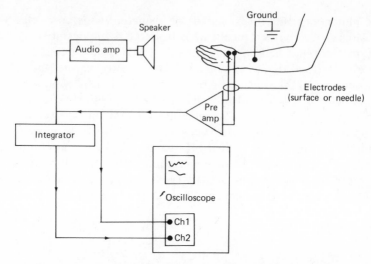

Figure 9.8.   Instrument arrangement for obtaining an EMG.

clinic, the audible EMG and the integrated form are often used to determine the condition of a muscle during contraction.

The EMG can be obtained from muscles or motor units that are stimulated electrically, and this method is often preferred to the voluntary contraction. A voluntary contraction is usually spread over about 100 msec because all the motor units do not fire at the same time; also, each

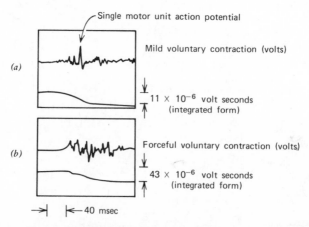

Figure 9.9.   Electromyograms for (a) minimal contraction showing the action potential from a single motor unit and (b) maximal contraction showing the action potentials from many motor units. Note a and b have different scales. (Adapted from P. Strong, *Biophysical Measurements***, Tektronix, Inc., Beaverton, Ore., 1970, p. 183, by permission of Tektronix, Inc. All rights reserved.)**

motor unit may produce several action potentials depending upon the signals sent from the central nervous system. With electrical stimulation, the stimulation time is well defined and all the muscle fibers fire at nearly the same time. A typical stimulating pulse may have an amplitude of 100 V and last 0.1 to 0.5 msec.

An EMG obtained during electrical stimulation of a motor unit is shown in Fig. 9.10. The action potential appears in the EMG after a latency period (the time between stimulation and the beginning of the response). Sometimes the EMGs from symmetrical muscles of the body are compared to each other or to those of normal individuals to determine whether the action potentials and latency periods are similar.

In addition to electrically stimulating the motor units, it is possible to excite the sensory nerves that carry information to the central nervous system. The reflex system can be studied by observing the reflex response at the muscle (Fig. 9.11). At low stimulating levels some of the sensitive sensory nerves are activated but the motor nerves are not and no *M* response is seen (Fig. 9.11*b*). The action potentials of the sensory nerves move to the spinal cord and generate the reflex response that travels along the motor nerves and initiates a delayed *H* response at the muscle. As the stimulus is increased, both the motor nerves and the sensory nerves are

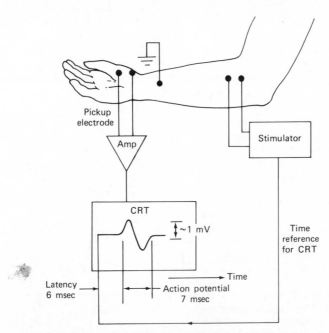

Figure 9.10. Instrument arrangement for obtaining an EMG during electrical stimulation of a motor unit.

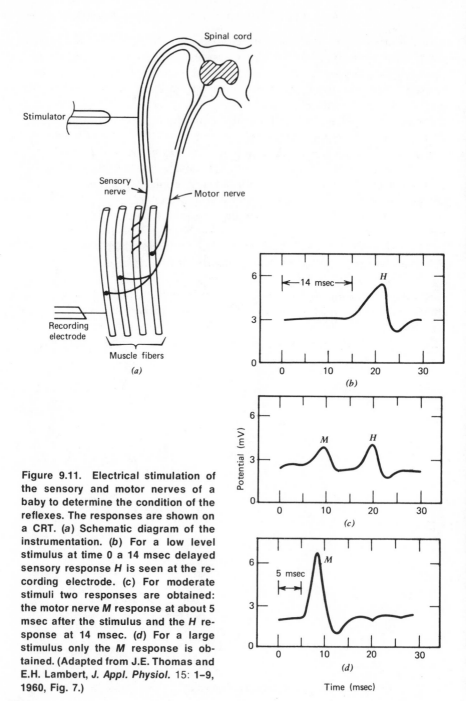

Figure 9.11. Electrical stimulation of the sensory and motor nerves of a baby to determine the condition of the reflexes. The responses are shown on a CRT. (a) Schematic diagram of the instrumentation. (b) For a low level stimulus at time 0 a 14 msec delayed sensory response H is seen at the recording electrode. (c) For moderate stimuli two responses are obtained: the motor nerve M response at about 5 msec after the stimulus and the H response at 14 msec. (d) For a large stimulus only the M response is obtained. (Adapted from J.E. Thomas and E.H. Lambert, *J. Appl. Physiol.* 15: 1–9, 1960, Fig. 7.)

stimulated and both the *M* and the *H* responses are seen (Fig. 9.11c). At large stimulating levels only the *M* response is seen (Fig. 9.11d).

The velocity of the action potential in motor nerves can also be determined. Stimuli are applied at two locations, and the latency period for each response is measured (Fig. 9.12). The difference between the two latency periods is the time required for an action potential to travel the distance between them; the velocity of the action potential is this distance divided by this time.

The conduction velocity for sensory nerves can be measured by stimulating at one site and recording at several locations that are known distances from the point of stimulation (Fig. 9.13). Many times nerve damage results in a decreased conduction velocity. Typical velocities are 40 to 60 m/sec; a velocity below 10 m/sec would indicate a problem.

Electromyograms made during multiple stimulations are used to determine fatigue characteristics of muscles. The major muscles in humans can be restimulated at rates of between 5 and 15 Hz. Normal nerves and muscles show little change during prolonged restimulation as long as the rate of stimulation allows for a relaxation period of about 0.2 sec between

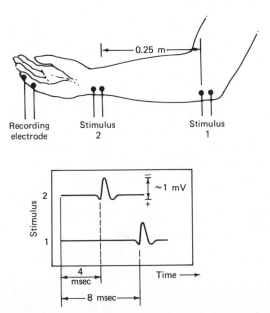

Figure 9.12. Method of measuring the motor nerve conduction velocity. The latency period for the response to stimulus 1 is 4 msec longer than that for the response to stimulus 2 ($\Delta t = 4 \times 10^{-3}$ sec). The difference in distance Δx is 0.25 m; therefore, the nerve conduction velocity $v = \Delta x / \Delta t = 0.25$ m/4 $\times$ 10^{-3} sec = 62.5 m/sec.

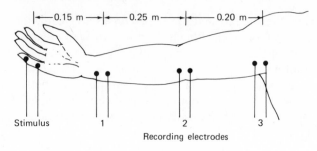

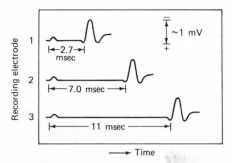

Figure 9.13. The sensory nerve conduction velocity can be determined by stimulating at one location and recording the responses with electrodes placed at known distances. The response traveled the 0.25 m from 1 to 2 in 4.3 msec; the conduction velocity is 0.25 m/4.3 ×10⁻³ sec = 58 m/sec. The conduction velocity from 2 to 3 is 0.20 m/4 × 10⁻³ sec = 50 m/sec.

pulses. A patient with the relatively rare disease *myasthenia gravis* shows muscular weakness when carrying out a repetitive muscular task. The EMG of such a patient shows that in repetitive stimulation the motor nerve to muscle transmission fails.

9.4. ELECTRICAL SIGNALS FROM THE HEART—THE ELECTROCARDIOGRAM

In Chapter 8 we discuss the heart as a double pump. It has four chambers (Fig. 9.14); the two upper chambers, the left and right atria, are synchronized to contract simultaneously, as are the two lower chambers, the left and right ventricles. The right atrium receives venous blood from the body and pumps it to the right ventricle. This ventricle pumps the blood through the lungs, where it is oxygenated. The blood then flows into the left atrium. The contraction of the left atrium moves the blood to the left

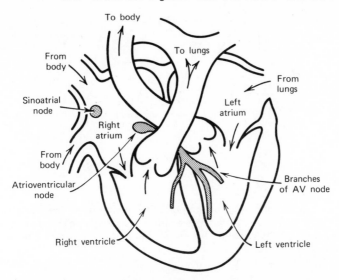

Figure 9.14. The human heart. Note the sinoatrial node, or the pacemaker, and the atrioventricular node, which initiates the contraction of the ventricles.

ventricle, which contracts and pumps it into the general circulation; the blood passes through the capillaries into the venous system and returns to the right atrium.

The rhythmical action of the heart is controlled by an electrical signal initiated by spontaneous stimulation of special muscle cells located in the right atrium. These cells make up the *sinoatrial* (SA) *node,* or the *pacemaker* (Fig. 9.14). The SA node fires at regular intervals about 72 times per minute; however, the rate of firing can be increased or decreased by nerves external to the heart that respond to the blood demands of the body as well as to other stimuli. The electrical signal from the SA node initiates the depolarization of the nerves and muscles of both atria, causing the atria to contract and pump blood into the ventricles. Repolarization of the atria follows. The electrical signal then passes into the *atrioventricular* (AV) *node,* which initiates the depolarization of the right and left ventricles, causing them to contract and force blood into the pulmonary and general circulations. The ventricle nerves and muscles then repolarize and the sequence begins again.

The nerves and muscles of the heart can be regarded as sources of electricity enclosed in an electrical conductor, the torso. Obviously it is not practical to make direct electrical measurements on the heart; diagnostic information is obtained by measuring at various places on the

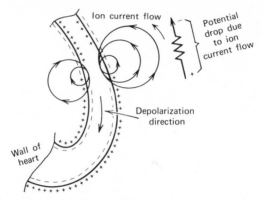

Figure 9.15. Schematic of an action potential moving down the wall of the heart. Some of the ion current, indicated by the circles, passes through the torso, indicated by the resistor. The potential on the chest wall is due to current flow through the resistance of the torso.

surface of the body the electrical potentials generated by the heart. The record of the heart's potentials on the skin is called the *electrocardiogram* (ECG).

The relationship between the pumping action of the heart and the electrical potentials on the skin can be understood by considering the propagation of an action potential in the wall of the heart as shown in Fig. 9.15. The resulting current flow in the torso leads to a potential drop as shown schematically on the resistor. The potential distribution for the entire heart when the ventricles are one-half depolarized is shown by the equipotential lines in Fig. 9.16. Note that the potentials measured on the surface of the body depend upon the location of the electrodes.

The form of the potential lines shown in Fig. 9.16 is nearly the same as that obtained from an electric dipole.* The equipotential lines at other times in the heart's cycle can also be represented by electric dipoles; however, the dipoles for different moments in the cycle would differ in size and orientation. The electric dipole model of the heart was first suggested by A. C. Waller in 1889 and has been modified by many others since.

The electrical (cardiac) potential that we measure on the body's surface is merely the instantaneous projection of the electric dipole vector in a particular direction. As the vector changes with time, so does the pro-

*An electric dipole is produced when equal positive and negative charges are separated from each other. It can be represented by a vector.

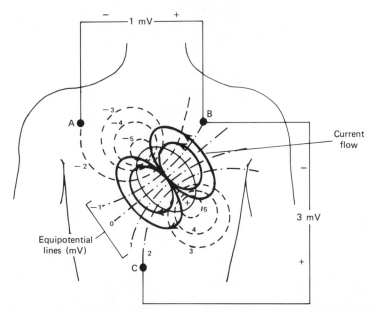

Figure 9.16. **The potential distribution on the chest at the moment when the ventricles are one-half depolarized. Electrodes located at A, B, and C would indicate the potentials at that moment.**

jected potential. Figure 9.17 shows an electric dipole vector along with the three electrocardiographic body planes.

The surface electrodes for obtaining the ECG are most commonly located on the left arm (LA), right arm (RA), and left leg (LL), although the location of the electrodes can vary in different clinical situations; sometimes the hands or positions closer to the heart are used. The measurement of the potential between RA and LA is called Lead I, that between RA and LL is called Lead II, and that between LA and LL is called Lead III (Fig. 9.18). This configuration was pioneered at the turn of the century by Willem Einthoven, a Dutch physiologist, and these three leads are called the standard limb leads. Usually, all three standard limb leads are used in a clinical examination. The potential between any two gives the relative amplitude and direction of the electric dipole vector in the frontal plane (Fig. 9.19).

Three augmented lead configurations, aV_R, aV_L, and aV_F, are also obtained in the frontal plane. For the aV_R lead, one side of the recorder is connected to RA and the other side is connected to the center of two resistors connected to LL and LA (Fig. 9.20). The other two augmented

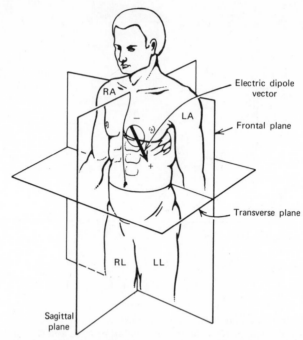

Figure 9.17. Electrocardiographic planes and an electric dipole vector. RA, LA, RL, and LL indicate electrode locations on the right and left arms and legs.

leads are obtained in a similar manner: for the aV_L lead, the recorder is attached to the LA electrode and the resistors are connected to RA and LL; for the aV_F lead, the recorder is attached to the LL electrode and the resistors are connected to RA and LA.

Each ECG tracing maps out a projection of the electric dipole vector, or the electrical activity of the heart, through each part of its cycle. Figure 9.21 shows schematically the Lead II output with the standard symbols for the parts of the pattern. The major electrical events of the normal heart cycle are (1) the atrial depolarization, which produces the P wave; (2) the atrial repolarization, which is rarely seen and is unlabeled; (3) the ventricular depolarization, which produces the QRS complex; and (4) the ventricular repolarization, which produces the T wave (Fig. 9.21).

Figure 9.22 shows the six frontal plane ECGs for a normal subject. Note that in some cases the waveform is positive and in other cases it is negative; the sign of the waveform depends upon the direction of the electric dipole vector and the polarity and position of the electrodes of the measuring instrument.

In a clinical examination, six transverse plane ECGs are usually made

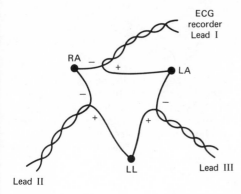

ECG
recorder
Lead I

Figure 9.18. The electrical connections for Leads, I, II, and III. The usual polarities of the recording instrument are indicated for each lead.

in addition to the six frontal plane ECGs. For the transverse plane measurements the negative terminal of the ECG recorder is attached to an *indifferent electrode* at the center point of three resistors connected to RA, LL, and LA (Fig. 9.23*a*), and the other electrode is moved across the chest wall to the six different positions shown in Fig. 9.23. Figure 9.24 shows typical transverse plane ECGs.

ECGs are usually interpreted by cardiologists, who can quickly deter-

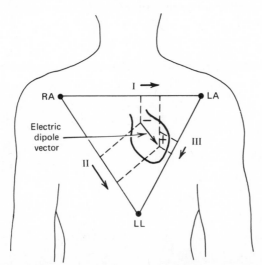

Figure 9.19. Schematic of the electric dipole of the heart projected on the frontal plane. For electrical purposes the three electrodes (RA, LA, and LL) can be thought of as the points of a triangle, the Einthoven triangle. The potential in Lead I at any moment is proportional to the projection of the dipole vector on the line RA-LA; the potentials in Leads II and III are proportional to the projections on the other sides of the triangle.

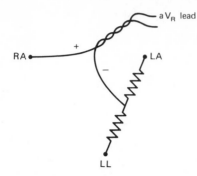

Figure 9.20. An augmented lead is obtained by placing a pair of resistors between two of the electrodes. The center of the resistor pair is used as one of the connections and the remaining electrode is used as the second connection. The arrangement for the aV_R augmented lead is shown.

mine if the patterns are normal and if arrhythmias (rhythm disturbances) exist. However, computers can also be used to analyze ECGs (see Chapter 20). In intensive care areas and during surgery the ECG is usually continuously monitored and displayed on the CRT of an oscilloscope (see Chapter 10, p. 227).

An ECG shows disturbances in the normal electrical activity of the heart. For example, an ECG may signal the presence of an abnormal condition known as heart block. If the normal SA node signal is not conducted into the ventricle, then a pulse from the AV node will control the heartbeat at a frequency of 30 to 50 beats/min, which is much lower

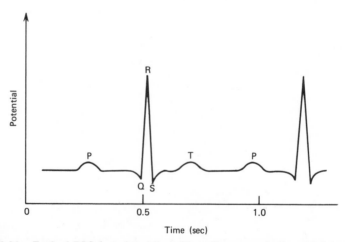

Figure 9.21.   Typical ECG from Lead II position. P represents the atrial depolarization and contraction, the QRS complex indicates the ventricular depolarization, the ventricular contraction occurs between S and T, and T represents the ventricular repolarization.

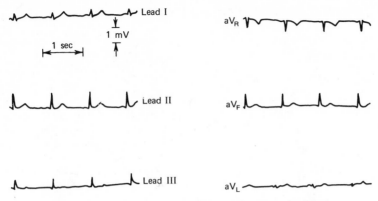

Figure 9.22. Six frontal plane ECGs for a normal subject.

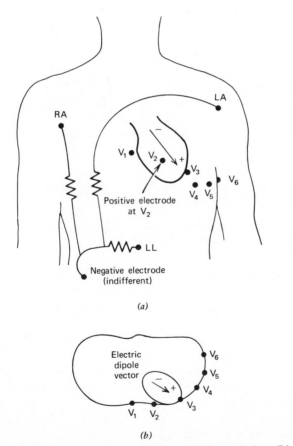

Figure 9.23. Transverse plane ECG positions. (a) Frontal view. (b) Top view.

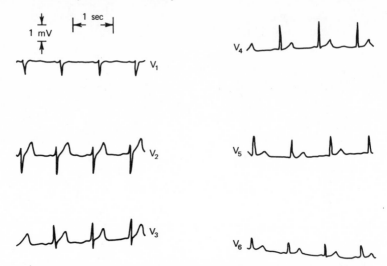

Figure 9.24.   Six transverse plane ECGs for a normal subject.

than normal (70 to 80 beats/min). While a heart block like this could make a patient a semi-invalid, an implanted pacemaker could enable him to live a reasonably normal life (see Chapter 10, p. 231).

9.5.   ELECTRICAL SIGNALS FROM THE BRAIN— THE ELECTROENCEPHALOGRAM

If you place electrodes on the scalp and measure the electrical activity, you will obtain some very weak complex electrical signals. These signals are due primarily to the electrical activity of the neurons in the *cortex* of the brain. They were first observed by Hans Berger in 1929; since then much research has been done on clinical, physiological, and psychological applications of these signals, but a basic understanding is still lacking. One hypothesis is that the potentials are produced through an intermittent synchronization process involving the neurons in the cortex, with different groups of neurons becoming synchronized at different instants of time. According to this hypothesis the signals consist of consecutive short segments of electrical activity from groups of neurons located at various places on the cortex.

The recording of the signals from the brain is called the *electroencephalogram* (EEG). Electrodes for recording the signals are often small discs of chlorided silver. They are attached to the head at locations that depend upon the part of the brain to be studied. Figure 9.25 shows the international standard 10-20 system of electrode location, and Fig. 9.26 shows typical EEGs for several pairs of electrodes. The reference electrode is usually attached to the ear (A_1 or A_2 in Fig. 9.25). In routine exams, 8 to 16 channels are recorded simultaneously. Since asymmetrical activity is often an indication of brain disease, the right side signals are often compared to the left side signals.

The amplitude of the EEG signals is low (about 50 μV), and interference from external electrical signals often causes serious problems in EEG signal processing. Even if the external noise is controlled, the potentials of muscle activity such as eye movement can cause artifacts in the record.

The frequencies of the EEG signals seem to be dependent upon the mental activity of the subject. For example, a relaxed person usually has an EEG signal composed primarily of frequencies from 8 to 13 Hz, or *alpha waves*. When a person is more alert a higher frequency range, the *beta wave* range (above 13 Hz), dominates the EEG signal. The various frequency bands are as follows:

Delta (δ), or slow	0.5 to 3.5 Hz
Theta (θ), or intermediate slow	4 to 7 Hz
Alpha (α)	8 to 13 Hz
Beta (β), or fast	greater than 13 Hz

The EEG is used as an aid in the diagnosis of diseases involving the

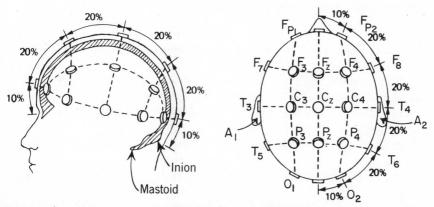

Figure 9.25. International standard 10-20 system of electrode location for EEGs. Lettered electrodes are located at intervals of 10% and 20% of the distances between specific points on the skull. The inion is the bony protuberance at the lower back of the skull and the mastoid is that behind the ear.

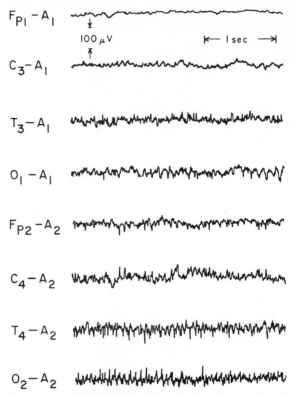

Figure 9.26. Normal EEGs. See Fig. 9.25 for the location of the electrodes. The reference electrode is connected to the ear (A_1 or A_2).

brain. It is most useful in the diagnosis of epilepsy and allows classification of epileptic seizures. The EEG for a severe epileptic attack with loss of consciousness, called a *grand mal* seizure, shows fast high voltage spikes in all leads from the skull (Fig. 9.27*a*). The EEG for a less severe attack, called a *petit mal* seizure, shows up to 3 rounded waves per second followed or preceded by fast spikes (Fig. 9.27*b*).

The EEG aids in confirming brain tumors since electrical activity is reduced in the region of a tumor. Other more quantitative methods for locating brain tumors involve x-ray or nuclear medicine techniques (see Chapters 16 and 17).

The EEG is used as a monitor in surgery when the ECG cannot be used. It is also useful in surgery for indicating the anesthesia level of the patient. During surgery a single channel is usually monitored.

Much research on sleep involves observing the EEG patterns for vari-

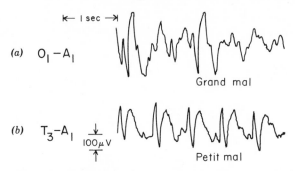

Figure 9.27. Electroencephalograms for two types of epilepsy: (a) grand mal and (b) petit mal.

ous stages of sleep (Fig. 9.28). As a person becomes drowsy, particularly with his eyes closed, the frequencies from 8 to 13 Hz (alpha waves) dominate the EEG. The amplitude increases and the frequency decreases as a person moves from light sleep to deeper sleep. Occasionally an EEG taken during sleep shows a high frequency pattern called *paradoxical sleep* or *rapid eye movement* (REM) *sleep* because the eyes move during this period. Paradoxical sleep appears to be associated with dreaming.

Besides recording the spontaneous activity of the brain, we can measure the signals that result when the brain receives external stimuli such as flashing lights or pulses of sound. Signals of this type are called *evoked responses*. Figure 9.29a shows three EEGs taken during the early stages of sleep with a series of 10 sound pulses (noise) used as an external stimulus. The EEGs show responses to the first few pulses and the last two pulses. The lack of responses in between is called *habituation*.

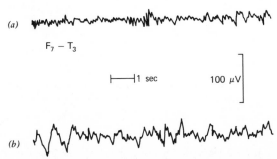

Figure 9.28. Electroencephalograms for two stages of sleep: (a) early sleep and (b) deep (delta wave) sleep. (Courtesy of Dr. Lloyd F. Elfner, Director, Psychoacoustics Laboratory, The Florida State University, Tallahassee.)

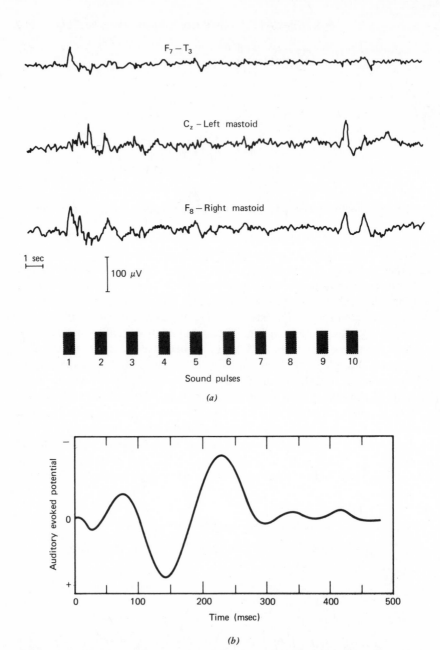

Figure 9.29. (a) An EEG taken during early sleep with noise pulses used as stimuli. (b) Evoked response averaged for 64 sound stimuli. (Courtesy of Dr. Lloyd F. Elfner, Director, and D. Gustafson, Psychoacoustics Laboratory, The Florida State University, Tallahassee.)

Because the evoked response is small, quite often the stimulus is repeated many times and the EEG responses are averaged in a small computer. Random signals such as normal EEG signals tend to average to zero and the evoked response becomes clear. Figure 9.29*b* shows an evoked response averaged for 64 stimuli.

9.6. ELECTRICAL SIGNALS FROM THE EYE—THE ELECTRORETINOGRAM AND THE ELECTROOCULOGRAM

The recording of potential changes produced by the eye when the retina is exposed to a flash of light is called the *electroretinogram* (ERG). One electrode is located in a contact lens that fits over the cornea and the other electrode is attached to the ear or forehead to approximate the potential at the back of the eye (Fig. 9.30).

An ERG signal is more complicated than a nerve axon signal because it is the sum of many effects taking place within the eye. The general form of an ERG is shown in Fig. 9.31. The B wave is the most interesting clinically since it arises in the retina. The B wave is absent in the ERG of a patient with inflammation of the retina that results in pigment changes, or *retinitis pigmentosa*.

The *electrooculogram* (EOG) is the recording of potential changes due to eye movement. For this measurement, a pair of electrodes is attached near the eye (Fig. 9.32*a*). The EOG potential is defined as zero with the eye in the position shown in Fig. 9.32*a* fixed on the reference spot labeled 0°. Figure 9.32*b* shows the EOG potential change for horizontal movement of the eyeball.

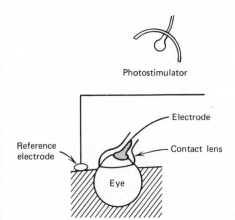

Figure 9.30. The placement of electrodes for obtaining an ERG. The reference electrode is on an ear or the forehead.

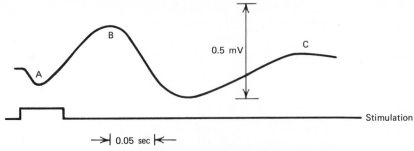

Figure 9.31. Schematic of an ERG. The letters identify portions of a normal ERG.

Electrooculograms provide information on the orientation of the eye, its angular velocity, and its angular acceleration. Some studies have been done to determine the effects of drugs on eye movement and the eye movement involved in sleep and in visual search. Electrooculography is seldom used in the routine practice of medicine.

9.7. MAGNETIC SIGNALS FROM THE HEART AND BRAIN—THE MAGNETOCARDIOGRAM AND THE MAGNETOENCEPHALOGRAM

Since a flow of electrical charge produces a magnetic field, a magnetic field is produced by the current in the heart during depolarization and repolarization. Magnetocardiography measures these very weak magnetic fields around the heart. The recording of the heart's magnetic field is the *magnetocardiogram* (MCG).

The magnetic field around the heart is about 5×10^{-11} tesla (T), or about one-millionth of the earth's magnetic field. (The cgs unit for magnetic fields is the gauss; $1 \text{ T} = 10^4$ gauss.) To measure fields of this size it is necessary to use magnetically shielded rooms and very sensitive magnetic field detectors (magnetometers). One such detector, called a SQUID (Superconducting QUantum Interference Device), operates at about 5°K and can detect both steady (dc) and alternating magnetic fields as small as 10^{-14} T. The SQUID is so sensitive that it can detect the changing magnetic field caused by someone walking past a horseshoe magnet 400 m (0.25 mile) away from it!

Figure 9.33 shows a typical arrangement for obtaining an MCG. The magnetic detector probe in the low-temperature dewar almost touches the subject, and various points on the chest are measured by moving the

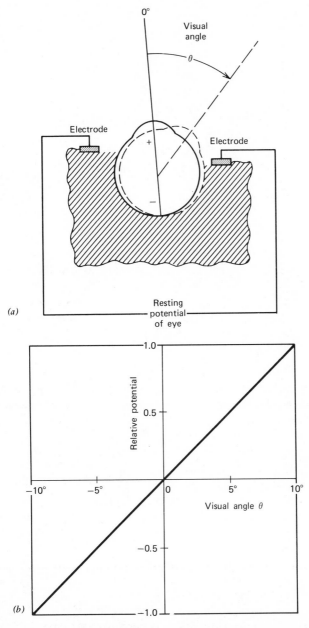

Figure 9.32. (a) For obtaining the EOG an electrode is mounted on each side of the eye. The visual angle is indicated. (b) The change of potential is plotted as a function of visual angle.

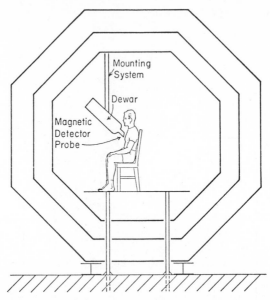

Figure 9.33. Arrangement for obtaining MCGs. The walls of the octagonal room contain five layers of magnetic shielding (only three are shown). All magnetic material such as zippers (pants) are removed from the subject. (From D. Cohen and D. McCaughan, *Amer. J. Cardiol.*, 29, 1972, p. 680.)

dewar. The output of the magnetic detector is recorded at a station outside the shielded room. The total time involved for each MCG is usually less than 1 minute.

The MCG gives information about the heart without the use of electrodes touching the body. Since the MCG and the ECG arise from the same charge movement they have similar features and can be compared. Figure 9.34 shows MCGs taken at different locations on a subject's chest wall along with his ECGs. A close look at the MCGs reveals considerable differences in magnetic fields between location H5 and location H7 due to differences in current flow in the heart.

The MCG provides information not available in an ECG because it measures magnetic fields due to direct currents, which occur in injured muscle and nerve tissue. This information may be useful in diagnosis if, for example, injury currents exist in the heart prior to a heart attack. Further research must be done to determine the usefulness of MCGs.

The SQUID magnetometer has also been used to record the magnetic field surrounding the brain. The recording of this field is called the *magnetoencephalogram* (MEG). During the alpha rhythm, the magnetic field

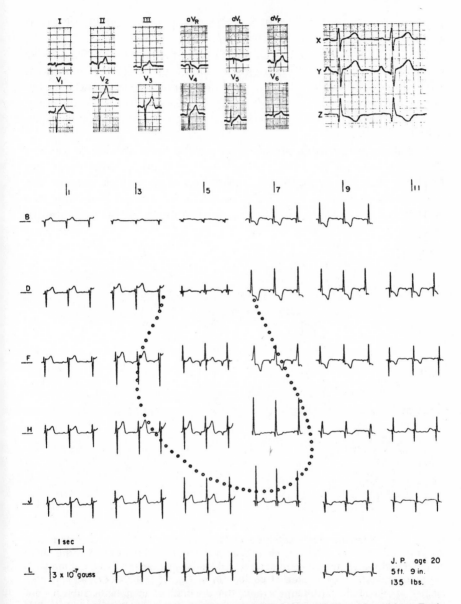

Figure 9.34. MCGs and the standard 12 lead ECGs (above) for a male subject. Each MCG is shown at the chest location where the magnetometer was situated for the recording. The recordings were made 5 cm apart, and H5 is at the tip of the breastbone. The outline of the heart was determined from x-rays. (From D. Cohen and D. McCaughan, *Amer. J. Cardiol.*, 29, 1972, p. 682.)

from the brain is about 1×10^{-13} T. This is almost one-billionth of the earth's magnetic field.

The MEG, like the MCG, can measure the fields resulting from direct currents. It is essentially impossible to obtain this information with the EEG. Figure 9.35 shows both normal and abnormal simultaneously obtained MEGs and EEGs. Note that the MEGs and EEGs are different. The MEG needs further study before its clinical usefulness can be established.

Not all magnetic fields produced within the body are due to ion currents; the body can be easily contaminated with magnetic materials. For example, asbestos workers inhale asbestos fibers, which contain iron oxide particles. The size of the magnetic field from the iron oxide in a worker's lungs can be used to estimate the amount of inhaled asbestos

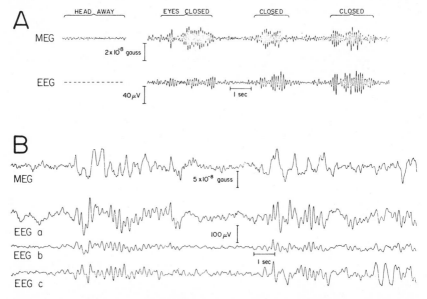

Figure 9.35.   Simultaneously obtained MEGs and EEGs. (A) The alpha rhythm from the brain of a normal subject is clear in both the MEG and the EEG. For the initial period (head away), the magnetometer was taken away from the head. For the MEG the magnetometer was located at position O_1 of Fig. 9.25; the EEG lead was also located at this position. (B) Large events from the brain of an epileptic subject were induced by hyperventilating. The magnetometer was located at the right temple. The three EEG leads were located (a) at the right temple, (b) above the right ear, and (c) at the back of the head. One difference between the MEG and the EEGs is that the 5-Hz waves, present in all three EEGs, are largely missing from the MEG. (From D. Cohen, *Science*, 175 (4022), p. 665. Copyright 1972 by the American Association for the Advancement of Science.)

dust. Typical magnetic fields from asbestos workers' chests are about one-thousandth of the earth's magnetic field (5×10^{-8} T). Figure 9.36 shows scans of persons with magnetic contamination.

9.8. CURRENT RESEARCH INVOLVING ELECTRICITY IN THE BODY

Many electrical phenomena exist in the body, and our understanding of them varies widely. In this section some of the phenomena currently being explored are discussed. As our knowledge about electricity in the body increases we should find more ways to use electricity in the diagnosis and treatment of diseases.

One life process that appears to be electrically controlled is bone growth. Bone contains collagen, which is a piezoelectric material; when a force is applied to collagen a small dc electrical potential is generated. The collagen behaves like an *N* type semiconductor, that is, it conducts current mainly by negative charges. Mineral crystals of the bone (apatite) close to the collagen behave like a *P* type semiconductor, that is, they conduct current by positive charges. At a junction of these two types of semiconductors, current flows easily from the *P* type to the *N* type but not in the other direction. (This is the basic idea of changing an ac signal to a dc signal by rectification.) It is thought that the forces on bones produce potentials by the piezoelectric effect and the *PN* junctions of collagen-apatite produce currents that induce and control bone growth. The currents are proportional to stress (force per unit area), so increased mechanical bone stress results in increased growth.

Another small direct current arises in an injured zone and is called the *injury current*. The electrical potential at a site of injury is higher than that in surrounding areas. This high potential is believed to be associated with limb regeneration in animals like the salamander and with fracture and wound healing in man. Stimulation of fracture sites with a direct current of 1 to 3 nA has been found to promote healing of bone fractures and bone conditions involving poor growth. It also enhances healing of burned areas.

Although the autonomic nervous system is not generally under voluntary control, it can, as previously mentioned, be influenced by external stimuli. One means of influencing the system that has been known for some time is called *biofeedback*. Recently there has been renewed interest in biofeedback. While early research on biofeedback has been promising, many aspects are still not understood. As we learn more about it we may be able to utilize it more in medicine.

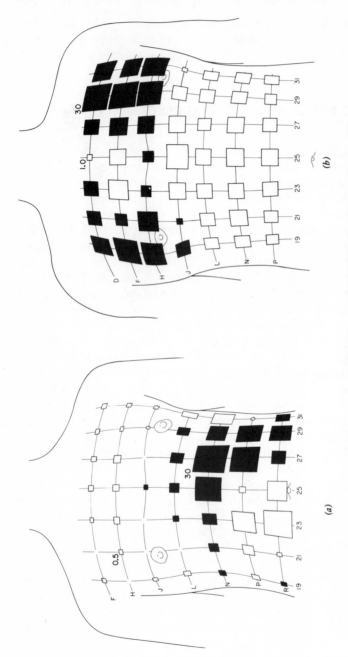

Figure 9.36. The magnetic field distributions from the chests of two subjects with magnetic contamination. The black squares show the fields in one direction while the open squares show the return paths. The area of each square is proportional to the strength of the field; values of the largest and smallest magnetic fields are given in units of 10^{-11} T. These patterns were due to (a) about 100 μg of iron oxide in the stomach of a man who had eaten beans from a can and (b) about 500 μg of iron oxide in the lungs of a welder. (From D. Cohen, *Science*, 180 (4087), p. 747. Copyright 1973 by the American Association for the Advancement of Science.)

The concept of feedback was discussed in Chapter 1. If we want to use feedback to control the output of some device, be it an amplifier or some part of the body, we measure the output to see what is happening and then we feed back this information to the input to affect the output in a desired manner. Negative feedback produces a stable output and is involved with the regulation of many body functions. We see negative feedback in the decrease in the diameter of the iris when a person is subjected to a bright light. The bright light increases the optic nerve signal to the brain, and the brain in turn decreases the diameter of the iris, thus decreasing the optic nerve signal.

In biofeedback the individual is consciously part of the feedback circuit. Sensors that monitor a subject's skin temperature, brain signals, or nerve action provide signals that are amplified and presented to the subject, who then tries through concentration to cause a change in his body to obtain a desired effect.

Through biofeedback body functions that are normally controlled by the autonomic nervous system can be consciously controlled. For example, EEG studies have shown that the alpha rhythm (8 to 13 Hz) indicates a low-arousal or relaxed state of the body, a condition that is often sought in biofeedback studies. If a subject finds that his EEG output changes from alpha rhythm to beta activity (greater than 13 Hz) when a headache is developing, he can, by mental relaxation, persuade his brain and body to return to alpha rhythm, thus forestalling the headache. Muscle relaxation can also be achieved through biofeedback; in this case, the EMG from a tense muscle is the signal presented to the patient. In addition, biofeedback has been used to control high blood pressure, acidity levels in the stomach, and irregular heart activity.

BIBLIOGRAPHY

Cohen, D., "Magnetic Fields Around the Torso: Production by Electrical Activity of the Human Heart," *Science*, **156**, 652–654 (1967).

Cohen, D., J. C. Norman, F. Molokhia, and W. Hood, Jr., "Magnetocardiography of Direct Currents: S-T Segment and Baseline Shifts During Experimental Myocardial Infarction," *Science*, **172**, 1329–1333 (1971).

Cromwell, L., F. J. Weibell, E. A. Pfeiffer, and L. B. Usselman, *Biomedical Instrumentation and Measurements*, Prentice-Hall, Englewood Cliffs, N.J., 1973.

Dubin, D., *Rapid Interpretation of EKG's*, Cover, Tampa, Fla., 1970.

Geddes, L. A., *Electrodes and Measurement of Bioelectric Events*, Wiley-Interscience, New York, 1972.

Goodgold, J., and A. Eberstein, *Electrodiagnosis of Neuromuscular Diseases*, Williams and Wilkins, Baltimore, 1972.

Katz, B., *Nerve, Muscle, and Synapse*, McGraw-Hill, New York, 1966.

Láhoda, F., A. Ross, and W. Issel, *EMG Primer,* Springer-Verlag, Berlin, 1974.
Plonsey, R., *Bioelectric Phenomena,* McGraw-Hill, New York, 1969.
Smorto, M., and J. V. Basmajian, *Clinical Electroneurography: An Introduction to Nerve Connection Tests,* Williams and Wilkins, Baltimore, 1972.
Stevens, C. F., *Neurophysiology: A Primer,* Wiley, New York, 1966.
Strong, P., *Biophysical Measurements,* Tektronix, Beaverton, Ore., 1970.

REVIEW QUESTIONS

1. List five electrical signals from the body that are sometimes recorded.
2. What is a physiatrist?
3. What is the advantage of myelinated nerves over unmyelinated nerves?
4. What is the typical resting potential of a cell?
5. What is the typical conduction velocity of the action potential in a nerve axon?
6. What important role is performed by the SA node of the heart?
7. Give the locations of the electrodes for the standard ECG limb leads.
8. Sketch an augmented ECG lead.
9. What electrical phenomenon in the heart produces the QRS complex of the ECG?
10. What are alpha waves on an EEG? When would a person normally have alpha waves in his EEG?
11. What is an evoked EEG response?
12. What is REM sleep?
13. What is the difference between an ERG and an EOG?
14. What is the difference between an MCG and an MEG?
15. What is biofeedback?

CHAPTER 10

Cardiovascular Instrumentation

John G. Webster,
Department of Electrical and Computer Engineering
University of Wisconsin

People have been dying of heart attacks for years. If you had a heart attack in 1960, you were put to bed and your physician hoped for the best. Now patients who have heart attacks are monitored—their electrocardiogram (ECG) is continuously displayed on an oscilloscope—because the monitor is the best instrument for rapidly detecting life-threatening rhythm disturbances (arrhythmias). Therapy both for preventing these arrhythmias and for curing problems that do occur is also now available.

In this chapter we discuss briefly the origin of the ECG, how the ECG changes under life-threatening conditions, how electrodes must be attached to the body in order to record it, and the problems caused by those electrodes. We also discuss amplifiers, which must be used to boost the ECG signal so that it can be processed, and heart rate meters, which are attached to the amplifiers and have alarms to alert the nurse when heart action stops. Finally we discuss therapeutic devices—defibrillators, which can initiate the heart's normal rhythm, and pacemakers, which can keep it going. These instruments and devices have been responsible for a dramatic increase in the percentage of heart attack patients who recover to lead normal or near-normal lives.

10.1. BIOPOTENTIALS OF THE HEART

As described in Chapter 9, movements of ions into muscle fibers (cells) of the heart cause action potentials, which produce the contraction. Ion

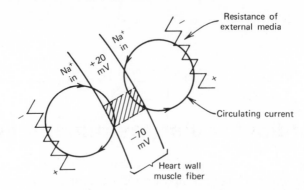

Figure 10.1. **An action potential propagates downward in a heart muscle fiber. Na⁺ ions move into the shaded region and cause the −70 mV resting potential to change to +20 mV. Their movement constitutes a current, which flows in a circulating path through the resistance of the external media, thus causing a voltage drop with the polarity shown. The region ahead of a propagating action potential is positive with respect to the region behind it.**

movements in heart muscle cells constitute a current flow, which results in potential differences in the tissue outside the fibers and on the surface of the body (Fig. 10.1). This current only flows while the action potential is propagating (mainly during the QRS wave of the ECG) or during the recovery period (the T wave). At the peak of the R wave, the potentials on the surface of the body are as shown in Fig. 10.2.

We measure these potential differences on the surface of the body by placing electrodes on the skin, amplifying the potentials, and then displaying the result as an ECG. Moving the electrodes to different positions on the body may result in amplitude changes or even inversion of the signal, as Fig. 10.2 shows. For this reason, ECGs are obtained from well-defined anatomical locations.

10.2. ELECTRODES

A major problem of obtaining an accurate ECG involves the metal electrodes. Electrically, the body can be treated as a bag of salt water, and in salt water current flows in the form of moving ions. In wires and the metal from which electrodes are made, however, current flows in the form of moving electrons. At the interface between the body and a metal electrode, ion flow must be converted to electron flow through a chemical reaction.

If ordinary metals are used for electrodes, polarization results from this

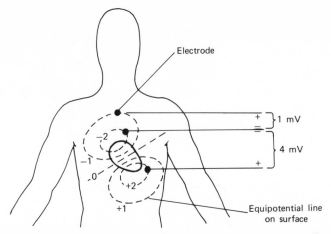

Figure 10.2. At the peak of the R wave of the ECG, most heart wall muscle fiber action potentials are propagating downward. The dashed lines show the resulting surface potentials. Note the difference in the measured potentials caused by using different electrode locations. Both voltage and polarity may change with electrode position.

chemical reaction as shown in Fig. 10.3*a*. At one or both electrodes, gas bubbles form due to electrolysis, and the resulting electrode-to-solution interface is electrically unstable. This instability produces electrical noise and drift which may be much larger than the ECG signal. These problems may be avoided by using silver-silver chloride electrodes, as shown in Fig. 10.3*b*. These electrodes are easily made by electrodepositing a silver chloride coating on pure silver electrodes. Current passes very readily through silver-silver chloride electrodes. The coating merely depletes on one electrode and builds up on the other. There is no formation of gas, and there is no electrical noise from the electrode-to-solution interface.

For these reasons the silver-silver chloride electrode is the natural choice for the typical patient monitoring electrode shown in Fig. 10.4. At the electrode-to-solution interface, complex layers of positive and negative charge form. This electrical double layer should not be disturbed by patient movement, which might cause artifacts (undesirable voltage changes). Hence the metal electrode is recessed from the skin, and the space between the electrode and the skin is filled with a conductive paste. The plastic electrode case is attached to the skin by a pressure-sensitive adhesive, similar to that used on masking tape.

A large motion artifact can come from the skin itself. There is a metabolically generated potential of about 40 mV between the inside and

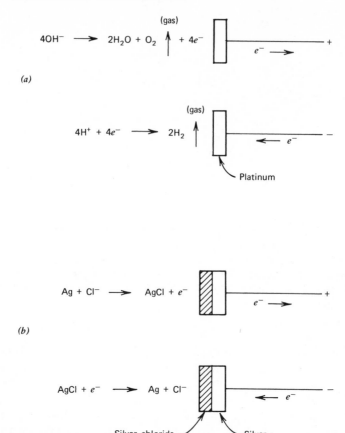

Figure 10.3. The chemical reactions at a skin-electrode interface are determined by the electrode composition. (a) Electrodes of platinum, an inert metal, cause gas bubbles to form (O_2 at the + electrode and H_2 at the − electrode), producing a high resistance and polarization at the interface. (b) Electrodes of silver-silver chloride enter into the chemical reaction. Thus no gas bubbles are formed, the resistance at the interface remains low, and the interface does not become polarized.

outside of the skin. If this potential were stable it would cause no problem. However, it varies with skin movement. This potential causes little problem when a resting ECG is taken; however, it can cause large artifacts in an ECG taken when a patient is exercising or rolls over in bed. These artifacts can be minimized by vigorously abrading the skin under the electrodes or lightly sanding it with very fine sandpaper. To avoid irritation of the sanded skin, a very mild paste must be used.

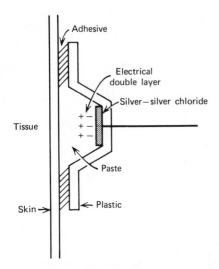

Tissue

Skin

Adhesive

Electrical double layer

Silver—silver chloride

+ −
+ −
+ −

Paste

Plastic

Figure 10.4. Silver-silver chloride electrodes are used for patient monitoring to prevent polarization. The electrode is recessed from the skin to prevent skin motion from disturbing the electrical double layer. A conductive paste fills the space between the electrode and the skin.

10.3. AMPLIFIERS

The design of the amplifier used to record ECG signals is important. The amplitude of the typical ECG signal is only about 1 mV. However, in a typical building the current capacitively coupled into the body from the 120 V power lines can produce a much larger potential. The amplifier used to record the ECG must be able to eliminate interference from voltages induced in the body from such external sources.

Consider the problem caused when the amplifier in an ordinary single-ended laboratory oscilloscope is used to observe the ECG (Fig. 10.5). Since this type of oscilloscope requires power from the wall, one of the input terminals is grounded through the power plug. The capacitively coupled current I from the surrounding power lines, typically about 1 μA, flows to ground through B, because of its lower resistance. From Ohm's law, the voltage drop V across this resistance R is

$$V = IR = (10^{-6} \text{ A}) (10^4 \text{ } \Omega) = 0.010 \text{ V} = 10 \text{ mV}$$

This undesirable 60 Hz interference voltage is ten times larger than the ECG signal and is added to the ECG potential, making the ECG nearly unobservable and useless.

To avoid this interference voltage, we may use the differential amplifier found on some oscilloscopes (Fig. 10.6). This amplifier measures the difference in the two signals appearing at A and B in Fig. 10.6 and is

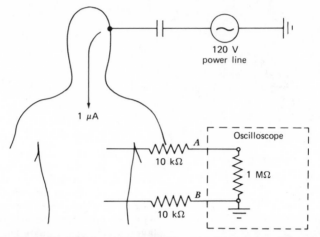

Figure 10.5. When the ECG is measured with a single-ended oscilloscope there is much interference from the 120 V power line that capacitively couples about 1 μA into the body. This current seeks the path of least resistance to ground through *B* to produce an extraneous signal about 10 times larger than the ECG signal.

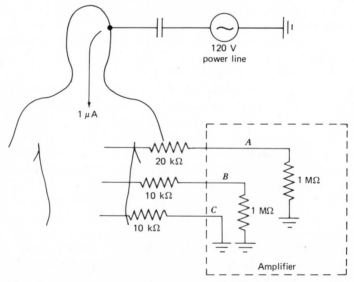

Figure 10.6. Use of a differential amplifier can minimize the interference of capacitively coupled currents from the 120 V power lines. The 1 μA current flows through a separate ground resistance (*C*). However, an unbalance in the skin and electrode resistances to *A* and *B* will cause common mode voltage on the body to be converted to a differential signal in the amplifier.

insensitive to any signal that is common to A and B. The 1 μA current flows to ground through the lowest resistance (C), but the 10 mV is no longer added to the ECG signal. The 10 mV appears everywhere on the body and is known as a common mode voltage (CMV). Since a differential amplifier takes the difference between the voltages at A and B, it would reject the common mode voltage if it were the same at A and B.

However, the resistance of the skin varies in an unpredictable manner and may differ at different electrode locations, as shown in Fig. 10.6. Thus V_A, the interference voltage at point A, is determined by considering the 20 kΩ skin resistance and the 1 MΩ oscilloscope input resistance as a simple voltage divider.

$$V_A = 10 \text{ mV} \left(\frac{1 \text{ M}\Omega}{1 \text{ M}\Omega + 20 \text{ k}\Omega} \right) \simeq 10 \text{ mV}(0.98) = 9.8 \text{ mV}$$

Similarly, the interference voltage at point B is

$$V_B = 10 \text{ mV} \left(\frac{1 \text{ M}\Omega}{1 \text{ M}\Omega + 10 \text{ k}\Omega} \right) \simeq 10 \text{ mV}(0.99) = 9.9 \text{ mV}$$

The differential amplifier measures the difference between these signals

$$V_B - V_A = 9.9 \text{ mV} - 9.8 \text{ mV} = 0.1 \text{ mV}$$

Thus even with a differential amplifier, unbalanced skin resistances can convert common mode voltage into a difference signal at the amplifier input. This 0.1 mV interference is about 10% of the ECG signal and is objectionable. Since it can be reduced by increasing the input resistance, an ECG amplifier is usually designed with an input resistance of 10 MΩ or greater. A 10 MΩ input would reduce the interference caused by common mode voltage to about 1%. An alternate, but less desirable solution is to abrade or lightly sand the skin, which greatly reduces the skin resistance.

Electroencephalogram (EEG) signals are about ten times smaller than ECG signals, or about 0.1 mV. Interference can be such a problem that the scalp is routinely abraded when EEG electrodes are applied and the skin resistance is routinely measured to make sure it is below 5 kΩ.

Another cause of interference in ECGs and in all other biopotential recordings is illustrated in Fig. 10.7. Changing magnetic fields are produced when alternating current flows through wires and are particularly strong near motors and transformers. If the changing magnetic field intersects the loop formed by the ECG wires leading to the amplifier, a voltage will be induced in the loop and it will appear at the amplifier input (Fig. 10.7a). Since the induced voltage is proportional to the area of the loop, this type of interference can be easily minimized by keeping the area of

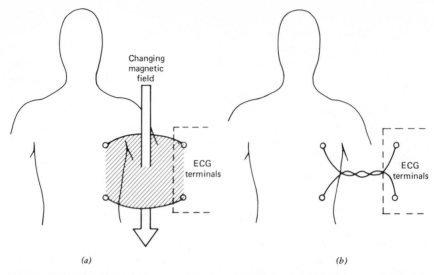

Figure 10.7. An alternating magnetic field that cuts through a loop of wire will induce a voltage in the loop. (*a*) Bad placement of wires for recording the ECG results in a large area in the loop (shown shaded) and a large induced interference voltage. (*b*) Good placement of the wires results in a very small area in the loop and a very small induced interference voltage.

the loop small. The wires can be either twisted together or run parallel and close together in a bundle (Fig. 10.7*b*).

10.4. PATIENT MONITORING

After amplification, the ECG signal must be displayed. When a routine diagnostic ECG is taken, a permanent record is required for analysis and a pen recorder is usually used (Fig. 10.8). In some recorders ink fed through a small capillary tube writes on ordinary paper. Other recorders have a heated stylus that melts off a thin, white wax coating that is on black paper, thus producing a black trace on a white background.

Such a permanent record is impractical for continuous monitoring of a heart attack patient. In a short time the paper would fill up the patient's room, and even if it could be stored no one would have time to examine it. A more convenient approach is to use the ECG oscilloscope shown in Fig. 10.9*a* to continually monitor a patient. It shows one to two cycles of the ECG, which is enough for a doctor or nurse to assess the patient's condition. The viewing screen on the monitoring oscilloscope is coated

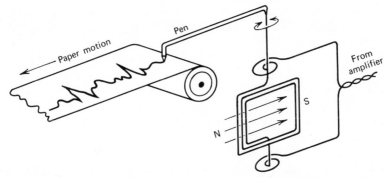

Figure 10.8. In a pen recorder, the amplifier output passes through a coil of wire suspended in a magnetic field. In the same way that a galvanometer twists when current passes through it, the pen twists to write on a moving strip of paper.

with a phosphor that glows for several seconds. Most laboratory oscilloscopes do not use such a long-persistence phosphor and are not useful for ECG monitoring. Modern monitors use a microcomputer to store the ECG information and use it to continually refresh the trace. The trace slowly moves across the screen and never fades. In many intensive care units (ICUs) several patients are monitored by the use of multiple traces on a single large oscilloscope (Fig. 10.9*b*). Several times a day, a short section is taken on a pen recorder and placed in each patient's medical record.

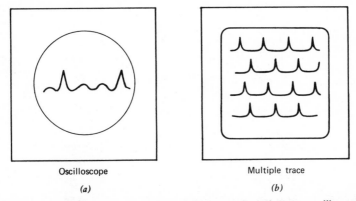

Oscilloscope

(a)

Multiple trace

(b)

Figure 10.9. In an intensive care unit the ECG is monitored on an oscilloscope. The spot on an oscilloscope moves at about 5 cm/sec. A long-persistence phosphor is used. (*a*) A single input oscilloscope is used to monitor one patient. (*b*) Four to eight traces may be displayed on a single large oscilloscope face so that the nurse can monitor a number of patients.

It is very difficult to maintain vigilance while watching ECG tracings on a scope. In order to free the nurse from this monotonous task and allow her or him to perform other tasks, automatic alarm devices have been developed. Electronic devices measure the time between successive R waves. The inverse of this R-R interval is the heart rate, and this rate is indicated on a meter (Fig. 10.10). If the heart stops or if its rate drops below a preset rate, for example, 50 beats/min, the meter needle contacts the low alarm contact, thus sounding an alarm that alerts the nurse to respond immediately.

It is important that the number of false alarms be kept to a minimum, since after a few false responses, the nurse might turn off the alarm and thus not be alerted when a life-threatening condition arises. Therefore, the electronic circuit is designed not to signal an alarm until the preset condition lasts for at least 5 sec. This circuit prevents alarms due to artifacts arising from transient conditions, such as the patient turning over.

Another problem that occurs frequently in monitoring is that after a day or two, the paste may dry out or the electrode may fall off. Because the resulting straight line trace on the oscilloscope and zero reading on the heart rate meter are indistinguishable from the results of the heart stopping, a special circuit is employed to identify this condition. A small, high-frequency current is passed from one wire through one electrode, through the body, and back through the other electrode. If either electrode falls off, this continuous circuit is broken, a special circuit disables the alarm, and an "electrode off" light comes on to alert the nurse.

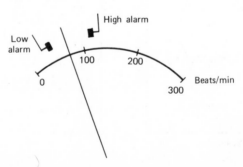

Figure 10.10.   A cardiotachometer indicates the heart rate in beats per minute. If the heart rate exceeds a preset rate, for example, 120 beats/min, the meter needle will contact the high alarm contact, thus sounding an alarm. If it drops below a preset rate, for example, 50 beats/min, the alarm will also be sounded.

10.5. DEFIBRILLATORS

The reason for continuously monitoring the ECG is that should problems arise, prompt therapeutic action can be taken to save the patient's life. Many heart attack patients undergo sudden changes in rhythm. The orderly heart muscle contractions associated with normal heart pumping change to the uncoordinated twitching of ventricular fibrillation, which halts the heart pumping action. Death follows within minutes unless the heart can be defibrillated.

Defibrillation is accomplished as shown in Fig. 10.11. The paddles are metal electrodes 7.5 cm in diameter that are coated with conductive paste and placed above and below the heart. The paddle handles are made of plastic and are electrically insulated to prevent accidental shock to the operator. When the switch is thrown, a current of about 20 A flows through the heart for about 5 msec. This current contracts every muscle fiber in the heart at the same time (Chapter 11). All the muscle fibers then recover at about the same time, and the heart can initiate normal rhythm again.

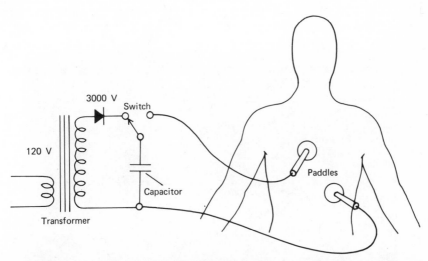

Figure 10.11. A simple defibrillator. The line voltage is stepped up to several thousand volts by a transformer. A diode rectifies the alternating current into direct current to charge up the capacitor. When the switch is thrown, the capacitor discharges through the paddles and the heart.

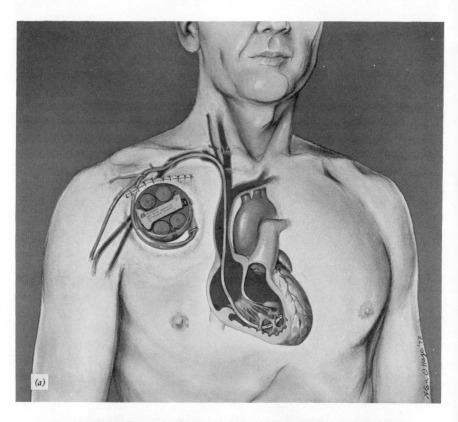

(a)

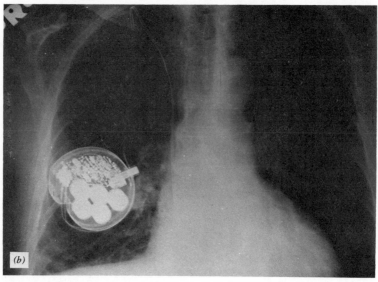

(b)

230

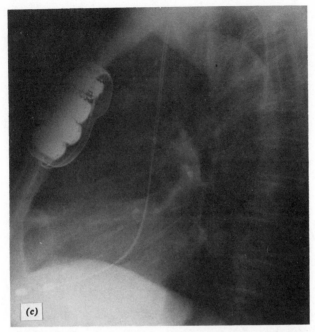

(c)

Figure 10.12. Pacemakers provide electrical pulses to the heart in order to produce a normal heart rate. (a) The pacemaker is usually implanted in a pocket on the right side of the chest. The pacing wire is fed through a vein in the shoulder and advanced through the right atrium and through the valve, until its tip is at the bottom of the right ventricle. (Courtesy of Medtronic, Inc., Minneapolis, Minn.) (b) Chest x-ray of a patient with a pacemaker in place. (c) Lateral chest x-ray showing a pacemaker on the chest wall.

10.6. PACEMAKERS

The atria of the heart are separated from the ventricles by a fatty layer that does not conduct electricity or propagate nerve impulses. At a single location, the atrioventricular node, impulses from the atria are conducted to the ventricles, which perform the heart's pumping action. If this node is damaged, the ventricles receive no signals from the atria. However, the ventricles do not stop pumping; there are natural pacing centers in the ventricles that provide a pulse if none has been recei /ed from the atria for 2 sec. The resulting heart rate, 30 beats/min, will sustain life, but the patient may have to live a life of semi-invalidism.

To improve the quality of life for patients with faulty atrioventricular nodes, artificial pacemakers have been developed. The pacemaker contains a pulse generator that puts out 72 pulses/min. When the pacemaker is put in place, the patient is given local anesthetic and a flap of skin just below the right collarbone is lifted. The pacing wire is fed through a slit in the shoulder vein and advanced under fluoroscopic control until the tip is imbedded in the wall of the right ventricle (Fig. 10.12). Then the pacemaker is placed in the pocket under the skin, and the flap is replaced. The pacemaker runs on batteries that last about 2 years. It is made of materials that are impervious to body fluids and do not cause tissue reaction. In 1976 about 150,000 patients were wearing pacemakers and living near-normal lives.

BIBLIOGRAPHY

Geddes, L. A., *Electrodes and the Measurement of Bioelectric Events,* Wiley-Interscience, New York, 1972.

Huhta, J. C., and J. G. Webster, "Sixty Hertz Interference in Electrocardiography," *IEEE Trans. Biomed. Eng.,* **BME-20,** 91–101 (1973).

Schaldach, M., and S. Furman, *Advances in Pacemaker Technology,* Springer-Verlag, New York, 1975.

Strong, P., *Biophysical Measurements,* Tektronix, Beaverton, Ore., 1970.

Tam, H. W., and J. G. Webster, "Minimizing Electrode Motion Artifact by Skin Abrasion," *IEEE Trans. Biomed. Eng.,* **BME-24,** 134–139 (1977).

Webster, J. G. (Ed.), *Medical Instrumentation: Application and Design,* Houghton Mifflin, Boston, 1978.

REVIEW QUESTIONS

1. Why is the location of the electrodes for obtaining an ECG important?
2. Why are metal electrodes unsatisfactory for obtaining the ECG?
3. What is the advantage of silver-silver chloride electrodes for ECG monitoring?
4. What is the typical potential between the inside and outside of the skin?
5. What is the source of the 60 Hz interference in the ECG signal?
6. How does a differential amplifier reduce the 60 Hz interference of the ECG signal?
7. What is common mode voltage?
8. Why do differential amplifiers used for ECG recording have a large input resistance?
9. How is the interference of alternating magnetic fields minimized through the arrangement of ECG leads?
10. Describe two techniques used to reduce the number of false alarms from automatic alarm devices for monitoring a patient's heart rate.
11. If a defibrillator with a potential of 3000 V produces a current of 20 A, what is the resistance of the body between the electrodes?

CHAPTER 11

Applications Of Electricity And Magnetism In Medicine

This chapter deals with the reaction of the body to externally applied electric and magnetic fields. Fraudulent claims regarding the curative powers of electricity and magnetism have appeared often in the history of medicine. The early medical "specialists" who promoted these cures usually tried to convince others that all illnesses were related to electricity and that they could be cured by the proper application of it. One of the better documented fraudulent cases was the doctrine of "animal magnetism" proposed by Franz Mesmer in the last half of the eighteenth century and called *mesmerism*. Mesmer said he could cure all kinds of sickness through the use of a *baquet* (a tub of soft water with magnetic material in it). His deceptive methods caused him to be run out of both Vienna and Paris. By 1785 this treatment was on the decline; however, in 1842, James Braid of Manchester, England, published a work that verified parts of this doctrine while exposing its fallacies.

A type of electricity many people are familiar with is static electricity. When the atmosphere is dry, it is often possible to generate static sparks many millimeters long corresponding to 15 kV potential. These sparks serve no useful purpose in medicine, but they can cause a serious explosion in an operating room where flammable anesthetics are being used. It is standard policy to guard against such sparks by using conducting floors and conducting footwear in operating rooms.

Recent studies by scientists from many disciplines have helped us to better understand electricity and its use in medicine, and there are now many situations in diagnosis and therapy in which electricity is applied in a controlled fashion. The uncontrolled application of electricity (electrical shock) may be lethal, however, and we begin this chapter by discussing electrical shock and shock hazards that might exist in a clinic or hospital. Some of the many controlled applications of electricity in medicine are then covered. We discuss the application of high-frequency electricity to the body (diathermy and electrosurgery), the measurement of skin resistance, magnetic methods of monitoring blood flow, electroanesthesia, Kirlian photography, and acupuncture.

The physician specialist most closely associated with diathermy is the *physiatrist*—an M.D. who has taken a three-year residency in physical medicine. Electricity is usually applied to the body for therapeutic purposes by a physical therapist (PT). A PT is a paramedical specialist who has earned a bachelor's degree after 4 or 5 years of specialized training followed by an 18-week internship. Physical therapists are state licensed. Their work in rehabilitation involves much more than the application of electricity. Physical therapists can advise on treatment but cannot prescribe treatment.

11.1. ELECTRICAL SHOCK

When an electrode is connected to each hand and 60 Hz currents of different levels are passed through the body, various reactions are produced. As the current is increased from zero, the level at which we can just feel the current—the perception level—is reached. About 50% of adult men feel a 60 Hz current of about 1.0 mA. Women perceive lower levels—about one-third lower than those felt by men. The perception level is frequency dependent; it rises as the frequency increases above 100 Hz.

As a 60 Hz current is increased above the perception level it causes a tingling sensation in the hands or body; at currents of 10 to 20 mA, a sustained muscular contraction takes place in the hands and many subjects cannot let go of the electrodes. The curve in Fig. 11.1 shows the current above which 50% of adult men do not have the ability to control their muscle actions (cannot let go of the electrodes). Note that this current is higher at both low and high frequencies.

As the current is increased still further, pain and in some cases fainting occur; near the 100 mA level the portion of the 60 Hz current passing through the heart is sufficient to cause ventricular fibrillation (rapid, irregular, and ineffectual contraction of the ventricles). We have seen in Chapter 9 that the function of the heart depends upon the proper sequence of electrical activity in the muscle fibers of the heart. Ventricular fibrillation replaces these normal patterns and is fatal if not corrected. The heart is especially vulnerable to fibrillation during one portion of its cycle—the

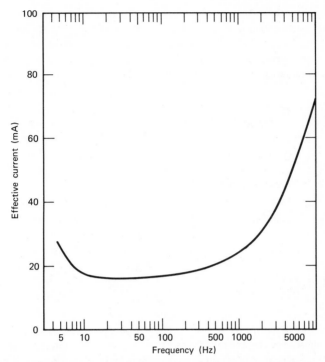

Figure 11.1. **Curve of current versus frequency above which one-half of all men are not able to let go of the electrodes. The values for women are two-thirds of those shown. The 60 Hz current level at which 99.5% of all men can let go of the electrodes is about 9 mA. (Adapted by permission from C.F. Dalziel,** *IEEE Spectrum,* **Vol. 9, No. 2, February 1972, p. 44. Copyright 1972, by the Institute of Electrical and Electronics Engineers, Inc.)**

beginning repolarization of the ventricle (the upswing of the T wave, Fig. 9.21). The current level that will induce fibrillation decreases as the duration of the shock increases. In one set of experiments on sheep, currents of 1000 mA for 0.03 sec and 100 mA for 3 sec both caused fibrillation. Information of this type for several kinds of animals has been used to estimate the fibrillating current levels for humans.

For a 60 Hz shock, the estimated maximum current that will not induce fibrillation in man is given by $(116/t^{1/2})$ mA, where t is the time (in seconds) the shock lasts. For example, if $t = 1$ sec, the safe current is 116 mA; if $t = 4$ sec, the safe current is 58 mA.

Current levels of 6 A and above cause sustained muscular contraction of the heart similar to the "cannot let go" behavior of the hands shown in Fig. 11.1. Defibrillators make use of such current levels; if a patient has ventricular fibrillation, a brief shock from a defibrillator usually restores normal coordinated pumping in the heart (see Chapter 10, p. 229). The defibrillator uses a brief pulse of up to 10 kV. A defibrillator can also be used to synchronize the heart to its normal rhythm when a patient has atrial fibrillation. In this case the electrical pulse is applied after the R wave (depolarization of the ventricles) but before the upswing of the T wave (repolarization of the ventricles).

Continuous currents above 6 A can cause temporary respiratory paralysis and serious burns. The damage depends upon the individual, the dampness of the skin, and the contact of the skin with the conductor.

Our discussion to this point has been concerned with macroshock, which occurs when electrical contact is made on the surface of the body. When the current is applied inside the body, microshock results. In microshock, the current does not have to pass through the high resistance of the skin; it instead often follows the arteries and passes directly through the heart. Obviously, ventricular fibrillation can be induced with microshock current levels that are much smaller than the current levels needed to induce it under macroshock conditions. Experiments on dogs have indicated that 17 μA applied directly to the heart can cause ventricular fibrillation. From these results, it has been estimated that about 30 μA through the human heart would cause ventricular fibrillation.

It is possible for microshock to occur in a medical situation. In the last 10 to 15 years there has been extensive growth in electronic monitoring of patients in intensive care units (ICUs). Many of these patients require catheters for ECG recording, for injection of dyes for radiography, or for internal blood pressure measurements. A patient in an ICU may have a pacemaker catheter running through a major vessel and touching the heart muscle to stimulate the heart if its own timing mechanism should fail. Some catheters contain wires or electrically conducting fluids and, therefore, provide low-resistance electrical paths directly to the heart. These internal electrical paths greatly increase the possibility of microshock.

It has been suggested that microshock via internal electrodes may be a significant factor in fatal accidents in operating rooms and ICUs. Microshocks of this nature involve very small currents that are well below the perception levels of staff members and thus would not be felt by them.

Obvious electrical hazards are faulty lamp sockets, frayed electrical wires, and broken connector plugs. Hazards of this sort are usually easy to find and correct. A more subtle hazard may exist in the power cord of an instrument. The cord is often mistreated; it may be yanked from the wall or the wheel of an instrument or bed may be rolled over it. Eventually some of the wiring within it may fail. Modern power cords have three wires—two that supply the ac power and one that serves as a ground wire.

If either of the power wires breaks the equipment will not operate, and if these wires touch (short) a fuse will blow and the failure will be obvious. However, a break in the ground wire may go undetected and present a serious electrical hazard to patients with internal electrodes.

To understand the hazard of a broken ground wire we need to understand *leakage current*. In all electrical or electronic equipment there is some current flow from the ac power parts to the metal case of the instrument or appliance. This leakage current usually flows to ground through the ground wire in the power cord. The main source of the leakage current is the capacitance between the ac power wires and ground or between the power transformer and its case (Fig. 11.2). The impedance X_c of a capacitance C for an applied voltage of frequency f is

$$X_c = \frac{1}{2\pi f C} \qquad (11.1)$$

A typical leakage capacitance is 2×10^{-2} μF. If the ac potential V is 110 V at a frequency of 60 Hz, then the capacitive reactance is 1.3×10^5 Ω and the leakage current $I = V/X_c = 110/1.3 \times 10^5 = 8.5 \times 10^{-4}$ A (850 μA).

Let us consider what would happen if this leakage current were in an ECG instrument with a broken ground wire and the unit were connected to a patient in an ICU who also had a pacemaker connected as shown in Fig. 11.3. Since the leakage current could not flow to ground through the broken ground wire it would flow through the implanted cardiac pacemaker to ground. The microshock current could result in ventricular fibrillation and death.

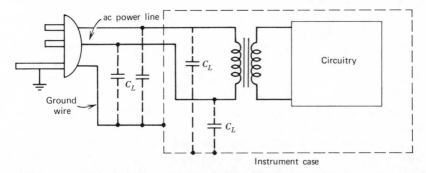

Figure 11.2. Schematic of an electrical instrument showing the leakage capacitances, C_L. The capacitances occur between the ac power wires and ground and between the power transformer and its case, which is usually grounded. The total leakage capacitance C is typically about 2×10^{-2} μF.

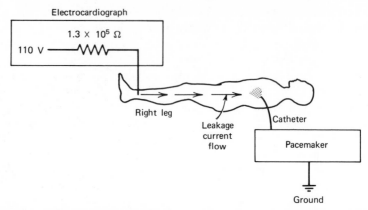

Figure 11.3. An ECG instrument with a broken ground wire on its power cord would place 110 V through a capacitive reactance of 1.3×10^5 Ω and the resistance of the body. An internal pacemaker electrode would provide a relatively low impedance path for the leakage current. At current levels above 30 μA ventricular fibrillation could occur.

A related hazard stems from the fact that many older hospitals were wired without protective grounding systems, and when instruments with three-pin connectors appeared, hospital personnel had to either tear off the third pin of the electrical equipment or modify the sockets. Some of the socket modifications did not provide a good ground. Later more electrical outlets and separate circuits with adequate grounds were added in various areas of many of these hospitals. When separate circuits are used, however, it is possible for a patient to be connected to two grounds at slightly different voltages. If one instrument connects to an internal electrode, hazardous leakage currents can flow through the patient between the two grounds.

Presently electrical safety in hospitals is achieved through periodic inspections (and corrections if necessary) of all electrical equipment. However, there are a number of ways that shock hazards around hospitals could be reduced. The body is less sensitive to direct current than to 60 Hz current (Fig. 11.1). Since $X_c = \infty$ if $f = 0$ (Equation 11.1), there would be no leakage current due to stray capacitance if we operated our electrical equipment with direct current. Hazards could also be reduced by operating electrical equipment at frequencies much higher than 60 Hz where the sensitivity of the heart to ventricular fibrillation is much less.

One proposed means of reducing hazards is to use rechargeable, battery-powered instruments in diagnostic, therapeutic, and monitoring situations. The output would be coupled optically to a conventional display system so that there would be no contact between the patient and the display system. Under these conditions, improper grounds could not exist. While this approach is expensive, it would reduce hazards.

11.2. HIGH-FREQUENCY ELECTRICITY IN MEDICINE

In Chapter 4 we discuss the heating effects produced by electrical diathermy. In this section we discuss electrical diathermy in terms of the physical interaction of the electricity with body tissue. We also discuss electrosurgery, another application of high-frequency electricity.

The heating effects of high-frequency current (10 kHz) were first observed by Jacques A. D'Arsonval in 1890, and by 1929 frequencies near 30 MHz were being used for therapy. The use of frequencies near 30 MHz for heating is called *short-wave diathermy. Long-wave diathermy,* at frequencies near 10 kHz, is no longer used. In 1950 a mode of diathermy that uses microwaves of a frequency of 2450 MHz was introduced. Microwaves are also used in radar and in microwave ovens. Concern about possible hazards to man from these uses has led to federal regulations setting limits of exposure for all forms of nonmedical electromagnetic radiation. We discuss these limits later in this section.

In short-wave diathermy two methods are used to get the electromagnetic energy into the body: the *capacitance method* and the *inductance method.* In both methods the body part to be heated becomes a part of a resonant electrical circuit. A simple resonant circuit consists of a capacitor and an inductor. Electrical energy from a power supply flows back and forth between the capacitor and the inductor, thus providing an alternating electric field (or current).

In the capacitance method of short-wave diathermy, the tissue to be heated is placed between two capacitor plates that have an oscillating electric field across them (Fig. 11.4). The changing electric field forces the ions in the tissue to move back and forth; they thus acquire kinetic energy, part of which is dissipated when the ions collide with molecules in the tissue. The heat produced when the energy is dissipated depends approximately on the square of the current times a constant determined by the tissue properties. This type of energy loss is called *joule* (resistive) *heating.*

In inductance diathermy the portion of the body to be heated is placed within or near the inductor (Fig. 11.5). A 30 MHz current in the coil produces an alternating magnetic field in the tissue that produces eddy currents in it. The energy lost by the eddy currents appears as heat in the tissue.

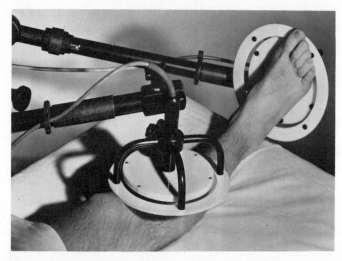

Figure 11.4. In the capacitance method of diathermy, the tissue to be treated is placed between two capacitor plates. (Courtesy of The Burdick Corporation, Milton, Wisc.)

Short-wave diathermy is used in the treatment of bursitis, arthritis, traumatic injuries, strains, and sprains. However, it does have limitations. When short-wave diathermy is used on muscle tissue surrounded by a fatty layer, a disproportionate amount of energy is lost in the fat. While short-wave diathermy is a much better heater of deep tissue than hot packs (see Chapter 4) or infrared light, it is far from ideal because of the large amount of energy deposited in surface fatty layers. For this reason microwave diathermy is frequently used.

Microwave diathermy is fundamentally different from short-wave diathermy. In short-wave diathermy the tissue to be heated is part of a resonant circuit, while in microwave diathermy the tissue absorbs electromagnetic waves that are incident upon it. The radiation is produced in a special high-frequency tube called a *magnetron*. The output of the magnetron is fed to an antenna, and the antenna emits the microwaves. A frequency of 2450 MHz with a wavelength of about 12 cm is usually used. Figure 11.6 shows a patient being treated with microwave diathermy.

Like light waves, microwaves can be transmitted, reflected or refracted at a surface, and absorbed by a medium. Several of the standard antenna arrangements for microwave diathermy make use of the reflection property to direct the radiation to the tissue, where part of it is reflected and part is transmitted. For 2450 MHz radiation the energy

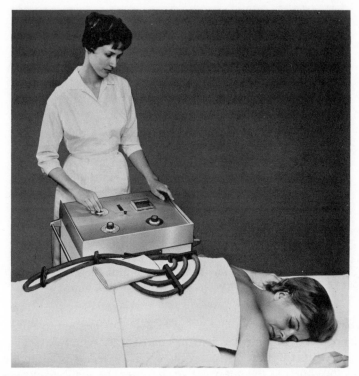

Figure 11.5. The inductance form of diathermy. (Courtesy of The Burdick Corporation, Milton, Wisc.)

reflected from the skin may be over 50%. A good impedance match between the antenna and the tissue increases the amount of radiation that is transmitted. (We discuss impedance matching in Chapter 12.) The transmitted radiation is absorbed by the body and produces heat. For homogeneous tissue, this absorption can be described by an exponential equation. The radiation intensity I at a depth x in the tissue is

$$I = I_0 e^{-x/D} \tag{11.2}$$

where I_0 is the radiation intensity at the surface and D is the tissue thickness that absorbs 63% of the beam, that is, where I is 37% of I_0 (see Appendix A for a general discussion of the exponential equation).

Experiments have shown that the absorption is linked to the amount of water in the tissue and that the heat producing interaction occurs between the electric field in the microwave radiation and the electric dipole mo-

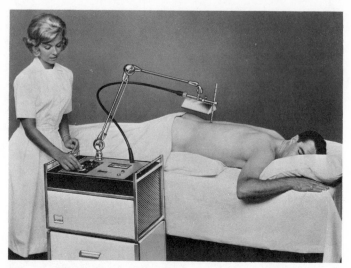

Figure 11.6. Microwave diathermy being applied to the lower back of a patient. (Courtesy of The Burdick Corporation, Milton, Wisc.)

ment of the water molecules in the body. The water molecule has a permanent electric dipole because the center of the net positive charges in the nuclei of the three atoms that make up the molecule is not in the same place as the center of the net negative charges. The slight displacement of the centers of charge in the molecule results in a permanent electric dipole in the water molecule. The electric field from the microwaves tries to align the electric dipole of the water molecule with it. In the alignment process work is done and energy is absorbed by the tissue, thus producing heat. The amount of energy absorbed depends upon the frequency of the microwaves; the energy is absorbed best at frequencies near 20 GHz (1 GHz $= 10^9$ Hz) and poorly at lower frequencies near 100 MHz and at higher frequencies around 1000 GHz.

Because the energy is deposited more effectively in tissue with high water content, microwave energy is absorbed better in muscle tissue than in fatty tissues, which have less water. Figure 11.7 shows the penetration of microwave radiation plotted against the frequency of the radiation for fatty tissue and tissue with a high water content. As the absorption increases D gets smaller (Equation 11.2). Microwave diathermy is used to heat joints, tendon sheaths, and muscles.

Damage can result from overexposure to electromagnetic radiation. The testes and eyes are more sensitive to high temperature than other parts of the body. Overheating of the testes can cause temporary sterility;

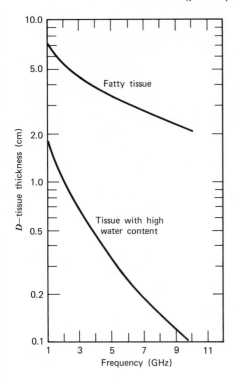

Figure 11.7. Penetration of microwave radiation versus the frequency of the radiation. *D* is the tissue thickness needed to reduce the initial intensity to 37% (1/e). (Adapted from B.D. McLees and E.D. Finch in J.H. Lawrence and J.W. Gofman (Eds.), *Advances in Biological and Medical Physics,* Vol. 14, Academic Press, New York, 1973, p. 169, as calculated from data of Schwan, *J. Non-Ionizing Rad.,* 296, 1973, pp. 485–497.)

it is possible that overheating the testes may someday be used as a means of birth control. Overheating of the eyes can cause irreversible changes in the lens opacity (cataracts).

Because of the hazards of electromagnetic radiation, a maximum long-term microwave exposure level of 10 mW/cm² has been set by the federal government. The Bureau of Radiological Health-FDA enforces the regulation. If one-half of the body's surface (9×10^3 cm²) is exposed to radiation at this maximum level and all of the incident power is absorbed, 90 W of power is produced; this is approximately the rate of energy production of the body when it is inactive, or the basal metabolic rate (see Chapter 5). This long-term radiation exposure level is only one-tenth of the maximum radiation power that could be absorbed from the sun striking the body (100 mW/cm²).

High-frequency electricity is also used to control hemorrhage during surgery. Searing (cauterizing) open wounds has been used to stop bleeding for over 2000 years. In times past hot oil was poured into wounds and hot irons were pressed on wounds without the use of anesthetics!

Currently, searing is done with a high frequency (> 2 MHz), high voltage (≤ 15 kV) arc. The basic device used is shown in Fig. 11.8. The butt plate electrode provides an electrical contact with a large area. The current density at the butt plate is small compared to the current density at the probe electrode, which has a very small tip. By controlling the shape of the probe and the current density, it is possible to deliver different amounts of heat to the tissue by means of the electrical arc. At different instrument settings, the probe can be used to either coagulate small to moderate-size blood vessels that are too small to tie (electrocautery) or cut through tissue (electrosurgery). Care must be taken that the butt plate electrode has adequate contact so that burning does not take place and that the unit does not present a shock hazard. Good electrical contact with the skin is ensured by using an electrically conducting paste on the butt plate.

What makes electrosurgery work? Current flows when a high-frequency probe is immersed in tissue, and under certain conditions a high power density will exist around the probe. For example, if 15 W is dissipated by a 0.25 mm diameter, straight wire probe used at 5 MHz, a direct application of electrical principles would show that the power density at the probe is 3.3×10^3 W/cm^3 and the power density 1.25 cm

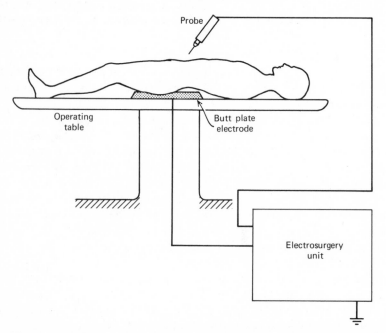

Figure 11.8. Basic arrangement for electrosurgery and electrocautery.

from it is 0.3 W/cm^3. These power densities would cause rapid temperature rises of about 800°C at the probe and about 0.1°C at 1.25 cm from the probe. The "cutting" of electrosurgery is thought to be the physical rupturing of tissue due to rapid boiling of the fluids from the intense local heat. Since the cutting of tissue takes place rapidly, the probe must be moved rapidly (5 to 10 cm/sec) to reduce the destruction of surrounding tissues. With proper control, the destruction can be limited to a depth of about 1 mm from the probe. Electrosurgery is often used in operations on the brain, spleen, bladder, prostate, and cervix.

11.3. LOW-FREQUENCY ELECTRICITY AND MAGNETISM IN MEDICINE

In this section we consider the measurement of blood flow by electromagnetic methods and the measurement of skin resistance to monitor psychological changes.

When an electrical conductor is moved perpendicular to a magnetic field, a voltage is induced in the conductor proportional to the product of the magnetic field and the velocity of the conductor (Faraday's law). This law, which also holds for a conducting fluid moving perpendicular to a magnetic field, is the basis of magnetic blood flow meters.

Blood acts as a conducting fluid. If it passes with mean velocity v through a magnetic field B as shown in Fig. 11.9, a voltage V is induced between the electrodes such that

$$V = Bdv$$

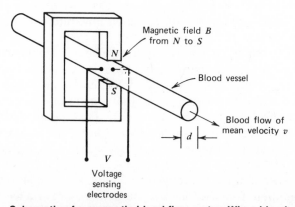

Magnetic field B
from N to S

N

S

Blood vessel

Blood flow of mean velocity v

d

V

Voltage sensing electrodes

Figure 11.9. Schematic of a magnetic blood flow meter. When blood, a conducting fluid, passes through the magnetic field, positive and negative charges are separated, producing the voltage V.

where d is the diameter of the blood vessel. Since V, B, and d can all be measured, the mean velocity can be obtained. The volume flow of blood Q through the vessel can then be calculated, since Q is the product of the mean velocity times the area of the vessel $\pi d^2/4$, or

$$Q = \frac{\pi d^2}{4} \frac{V}{Bd}$$

See Example 11.1.

Example 11.1

A magnetic blood flow meter is positioned across a blood vessel 5×10^{-3} m in diameter. With a magnetic field of 3×10^{-2} T (300 gauss), an induced voltage of 15×10^{-6} V is measured.

a. Find the mean velocity in the vessel.

$$V = Bdv$$

$$v = \frac{V}{Bd} = \frac{1.5 \times 10^{-5}}{(3 \times 10^{-2})(5 \times 10^{-3})} = 0.1 \text{ m/sec}$$

b. Assuming all the blood travels at the mean velocity, what is the volume flow rate?

$$Q = \frac{\pi d^2}{4} \frac{V}{Bd}$$

$$Q = \frac{\pi (5 \times 10^{-3})^2}{4} (0.1) = 1.9 \times 10^{-6} \text{ m}^3/\text{sec} = 1.9 \text{ cm}^3/\text{sec}$$

While this technique is not practical for general clinical use since the blood vessel must be surgically exposed, it is a useful research tool. A noninvasive technique for measuring blood flow rate with ultrasound is described in Section 12.5.

Many body functions such as regulating the heart rate and maintaining a constant temperature and many body processes such as respiration and perspiration (sweating) are controlled by the autonomic nervous system. While this control is involuntary, external stimuli or emotional activity can influence these functions and processes. For example, elevated psychological activity can cause a person to perspire. Perspiration is thus one process that can be monitored to reveal changes in a subject's emotional or psychological state.

Changes in perspiration (sweat gland activity) are related to skin resistance; the variation from the basal, or normal, skin resistance (BSR) due to psychological changes or external stimuli is called the *galvanic skin response* (GSR). A decrease in skin resistance indicates increased sweat gland activity, while an increase in skin resistance indicates reduced sweat gland activity. The GSR can be easily measured where there is a concentration of sweat glands, such as the palm of the hand or sole of the foot. The GSR depends upon the activity of the sweat glands only, and not upon the amount of visible perspiration. The details of this response are still not completely understood.

Several means of measuring the GSR exist. In one, the resistance between two electrodes attached to different fingers is measured. In another, an active electrode is placed on the palm of the hand and a second neutral electrode is placed on the wrist or the back of the hand (Fig. 11.10). Usually a constant tiny direct current ($\sim 10~\mu A/cm^2$) is passed through the electrodes; the resulting voltage indicates the GSR since voltage is proportional to resistance. The problem with using a direct current is that a polarizing voltage is developed between the electrodes and the skin that is indistinguishable from the potential produced by a change in the subject's skin resistance. Using silver-silver chloride surface electrodes with an electrode jelly between the electrodes and the skin reduces this problem (see Chapter 10, p. 221), as does using a pulsed

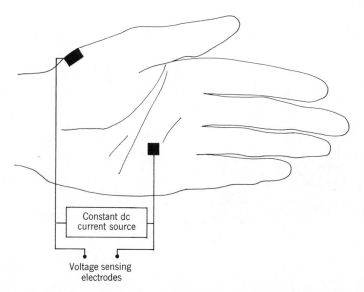

Figure 11.10. Method of measuring the galvanic skin response (GSR). The palm of the hand has a high sweat gland concentration, while the wrist region contains few sweat glands. The constant current source passes a current between the electrodes at these locations. The voltage produced depends upon the resistance (sweat gland activity) between the electrodes.

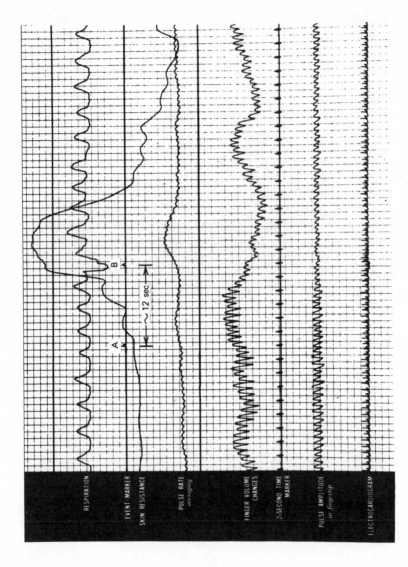

Figure 11.11. Polygraph recording of several physiological phenomena taken during an emotion-charged conversation. The conversation began at A; just before B (the termination of the conversation) the skin resistance, pulse rate, and respiration pattern began to change. In this recording a rise in the GSR curve shows a decrease in the skin resistance. (Courtesy of Gilson Medical Electronics, Inc., Middleton, Wisc.)

current source, which reduces the time the current is on. The range in resistance for the method shown in Fig. 11.10 is from about 20 to 200 kΩ. If for some reason the sweat glands are not functioning, the resistance may be as high as 1 MΩ.

One familiar use of GSR measurements is in lie detection. A polygraph (lie detector) simultaneously measures several body functions including GSR to detect changes in a subject's emotional state resulting from questioning. Figure 11.11 shows the recordings of an individual who had a change in anxiety level. The emotionally disturbing event started at A and terminated at B. Note that it took only a few seconds for the skin resistance to drop sharply.

Some neurological diseases affect the function of sweat glands on one side of the face and not on the other. Measurement of the GSR on the two sides aids in the diagnosis of these diseases.

11.4. CURRENT RESEARCH INVOLVING ELECTRICITY APPLIED TO THE BODY

Various applications of electricity and magnetism to medicine are under investigation, and many more undoubtedly await discovery. In this section we discuss several areas of current research (and controversy) involving the application of electricity to the body.

Considerable interest exists in a type of high-voltage photography called Kirlian photography, named after its primary developer, a Russian named Semyon Kirlian. This photographic technique consists of capturing on film the "aura" or corona from a person subjected to a high frequency ($\sim$ 1 MHz) electromagnetic voltage. The spectacular corona around the fingers appears to change with a person's emotional, mental, and physical state. The corona has caught the attention of a wide range of people including parapsychologists and medical practitioners looking for new medical tools, and there has been a good deal of speculation about what causes it.

Some researchers believe that Kirlian photographs are pictures of the light from the electrical discharge (corona) at the skin. The colors of a corona depend on the composition of the gases present, and supporters of this theory believe that the colors in a Kirlian photograph depend upon the surface chemistry of the skin, the amount of moisture present, and the composition of the surrounding gases. It is clear that more experiments are needed to determine all the factors involved in this phenomenon.

Electroanesthesia can be induced by the application to the skull of direct currents, alternating currents (100 to 1500 Hz), or combinations of the two at levels of a few milliamperes. Electrodes are applied as in Fig. 11.12. While it is obvious that the anesthetic effect is related to a change in the electrical behavior of the brain, sufficient controversy exists on the exact mechanism involved that we will not attempt an explanation (for information on electroanesthesia see the bibliography).

Since limited use on man has indicated that the anesthetic effect produced can be maintained for several hours, electroanesthesia holds promise for use in surgery, particularly in those situations in which chemical anesthetics cannot be used. It also has the decided advantages over chemical anesthetics of a rather fast induction time, a rapid recovery time, and no observable aftereffects. However, some characteristics of electrical shock such as respiratory and cardiac disturbances have been observed early in the application of the current.

At lower current levels than those used for electroanesthesia, electrosleep can be induced. A 100 Hz signal of 1 mA average current used with electrodes placed over each eye and each mastoid (the bony protuberance behind the ear) is effective. Electrosleep has been discussed in the foreign literature for a number of years, but there has been limited interest in the United States. Recent research on this phenomenon in the United States has dealt with its basis and its usefulness in psychiatry.

The Eastern art of acupuncture and its medical applications have aroused considerable interest in the United States as communication channels with China have been renewed. The origin of acupuncture dates back several thousand years in China's history. Acupuncture is used today to reduce or prevent pain associated with surgery and dental work. Stainless steel needles are inserted in one or more different acupuncture

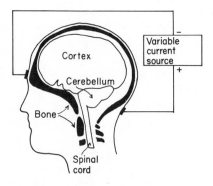

Figure 11.12. Electrode attachment for electroanesthesia.

sites (there are about 1000) depending upon the area to be anesthetized. The needles are then either twisted, moved up and down, or connected to small electrical currents. The effectiveness of acupuncture varies with its intended use, and it is more effective on some subjects than on others.

The reason that acupuncture appears to control pain is not understood. Theories range from hypnosis and autosuggestion to physiological blockage of pain. However, this last theory is in conflict with our present understanding of the nervous system. More studies will have to be done before the mechanisms of acupuncture are understood.

BIBLIOGRAPHY

Becker, R. O., "Electromagnetic Forces and Life Processes," *Technology Review*, 75, 32–38 (1972).

Biologic Effects and Health Hazards of Microwave Radiation, Proceedings of an international symposium in Warsaw, 15–18 October, 1973, Polish Medical, Warsaw, 1974.

Boyers, D. G., and W. A. Tiller, "Corona Discharge Photography," *J. Appl. Phys.*, 44, 3102–3112 (1973).

Dalziel, C. F., "Electric Shock Hazard," *IEEE Spectrum*, 9, 41–50 (1972).

Geddes, L. A., "Electronarcosis," *Med. Biol. Eng.*, 3, 11–26 (1965).

Hopps, J. A., "Electrical Hazards in Hospitals," *Med. Biol. Eng.*, 9, 549–556 (1971).

Lehmann, J. F., "Diathermy," in F. H. Krusen, F. J. Kottke, and P. M. Ellwood, Jr. (Eds.), *Handbook of Physical Medicine and Rehabilitation*, 2nd ed., Saunders, Philadelphia, 1971, pp. 273–345.

McLees, B. D., and E. D. Finch, "Analysis of Reported Physiological Effects of Microwave Radiation," *Advan. Biol. Med. Phys.*, 14, 163–223 (1973).

Sances, A., Jr., and S. L. Larson, "Electroanesthesia Research," in M. Clynes and J. H. Milsum (Eds.), *Biomedical Engineering Systems*, McGraw-Hill, New York, 1970, Chapter 8.

Schwan, H. P., "Microwave Radiation: Biophysical Considerations and Standards Criteria," *IEEE Trans. Biomed. Eng.*, (BME)-19, 304–312 (1972).

Stoyva, J., T. X. Barber, L. V. Dicara, J. Kamiya, N. E. Miller, and D. Shapiro, *Biofeedback and Self-Control–1971*, Aldine Atherton, Chicago, 1972.

Wulfsohn, N. L., and A. Sances, Jr. (Eds.), *The Nervous System and Electrical Currents*, Vol. 1 and 2, Plenum, New York, 1970 and 1971.

REVIEW QUESTIONS

1. How large is the average (50% of a population) "let-go" current for men and women at 60 Hz? How large is it at 10^4 Hz?
2. What is ventricular fibrillation?

3. Which of the electrical sources in the figure is the most hazardous to touch? Assume the resistance of the body to ground is 10^5 Ω.

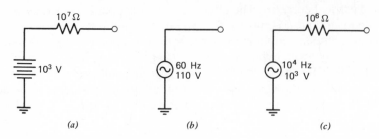

(a)   (b)   (c)

4. Estimate how long the body could safely tolerate a macroscopic shock of 50 mA.
5. What is the physical interaction that leads to tissue heating for each of the following?
   (a) capacitance short-wave diathermy.
   (b) inductance short-wave diathermy.
   (c) microwave diathermy.
6. What organs are the most sensitive to electromagnetic radiation?
7. What is the maximum limit on long-term exposure to microwave radiation? How does this compare to the maximum radiation power that could be absorbed from the sun?
8. What frequency and voltage are used for electrosurgery?
9. What is the GSR?

CHAPTER 12

Sound in Medicine

While we usually consider sound in terms of its physical effect on our ears, this view is much too limited for the purposes of this book. In this chapter we discuss the physical properties of sound and the applications of sound in medicine. These applications range from the use of the stethoscope to the use of modern ultrasonic techniques to study heart valve motion and to "look" at the unborn child (fetus). We discuss the physics of the ear and hearing in Chapter 13.

Sound is a major method of communication and gives enjoyment in the form of music. However, sound pollution, or noise of undesirable levels, is a growing problem in modern society. Federal regulations now limit the permissible noise levels caused by cars and trucks on the highways. Typical limits are 70 decibels (dB) for areas adjacent to a highway and 55 dB for inside work areas. (The decibel, the common unit of sound pressure or intensity, is discussed in Section 12.1 and Appendix A.)

Infrasound refers to sound frequencies below the normal hearing range, or less than 20 Hz. It is produced by natural phenomena like earthquake waves and atmospheric pressure changes; it can also be produced mechanically, such as by a blower in a ventilator system. A typical ventilator system produces frequencies of about 10 Hz. These frequencies cannot be heard, but they can cause headaches and physiological disturbances (see Chapter 2).

The audible sound range is usually defined as 20 Hz to 20,000 Hz (20 kHz). However, relatively few people can hear over this entire range. Older people often lose the ability to hear the frequencies above 10 kHz.

The frequency range above 20 kHz is called *ultrasound*. (Ultrasound should not be confused with supersonic, which refers to velocities faster than the velocity of sound in a medium.) Ultrasound is used clinically in a number of specialties. There is a growing trend to locate ultrasound equipment in the diagnostic radiology area, and some diagnostic radiologists specialize in ultrasonic imaging of the body. Ultrasound is also used by obstetricians to examine the unborn child. It often gives more information than an x-ray, and it is less hazardous for the fetus.

In this chapter we discuss methods of getting information from sounds that are produced within the body and sounds that are made to pass through the body. In order to understand how these sounds can be used medically, we need to discuss and define some of the properties of sound.

12.1. GENERAL PROPERTIES OF SOUND

A sound wave is a mechanical disturbance in a gas, liquid, or solid that travels outward from the source with some definite velocity. We can use a loudspeaker vibrating back and forth in air at a frequency f to demonstrate the behavior of sound. The vibrations cause local increases and decreases in pressure relative to atmospheric pressure (Fig. 12.1). These pressure increases, called *compressions,* and decreases, called *rarefactions,* spread outward as a longitudinal wave, that is, a wave in which the pressure changes occur in the same direction the wave travels. The compressions and rarefactions can also be described by density changes and by displacement of the atoms and molecules from their equilibrium positions.

The relationship between the frequency of vibration f of the sound wave, the wavelength λ, and the velocity v of the sound wave is

$$v = \lambda f$$

For example, for a sound wave with a frequency of 1000 Hz, $v = 344$ m/sec in air at 20°C and $\lambda = 0.344$ m.

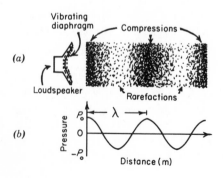

Figure 12.1. Schematic representation of a sound wave from a loudspeaker. (a) A diaphragm vibrates at a frequency f and produces compressions (increased pressure) and rarefactions (decreased pressure) in air. (b) The pressure relative to atmospheric pressure versus distance. P_0 is the maximum pressure variation from atmospheric, and λ is the wavelength.

Table 12.1. Values of ρ, ν, and Z for Various Substances at Clinical Ultrasound Frequencies

	ρ (kg/m³)	ν (m/sec)	Z (kg/m² · sec)
Air	1.29	3.31×10^2	430
Water	1.00×10^3	14.8×10^2	1.48×10^6
Brain	1.02×10^3	15.3×10^2	1.56×10^6
Muscle	1.04×10^3	15.8×10^2	1.64×10^6
Fat	0.92×10^3	14.5×10^2	1.33×10^6
Bone	1.9×10^3	40.4×10^2	7.68×10^6

Energy is carried by the wave as potential and kinetic energy. The intensity I of a sound wave is the energy passing through 1 m²/sec, or watts per square meter. For a plane wave I is given by

$$I = \frac{1}{2} \rho v A^2 (2\pi f)^2 = \frac{1}{2} Z(A\omega)^2 \qquad (12.1)$$

where ρ is the density of the medium; v is the velocity of sound; f is the frequency; ω is the angular frequency, which equals $2\pi f$; A is the maximum displacement amplitude of the atoms or molecules from the equilibrium position; and Z, which equals ρv, is the acoustic impedance. Some typical values of ρ, v, and Z are given in Table 12.1. The intensity can also be expressed as

$$I = \frac{P_0^2}{2Z} \qquad (12.2)$$

where P_0 is the maximum change in pressure (see Example 12.1).

Example 12.1

a. The maximum sound intensity that the ear can tolerate at 1000 Hz is approximately 1 W/m². What is the maximum displacement in air corresponding to this intensity?

From Equation 12.1 and Table 12.1,

$$A = \frac{1}{2\pi f} \left(\frac{2I}{Z} \right)^{1/2} = \frac{1}{6.28 \times 10^3} \left(\frac{2 \times 1}{4.3 \times 10^2} \right)^{1/2} = 1.1 \times 10^{-5} \text{ m}$$

b. The faintest sound intensity the ear can hear at 1000 Hz is approximately 10^{-12} W/m². What is A under these conditions?

We can use ratios between case a and this case

$$\frac{A_b}{A_a} = \left(\frac{I_b}{I_a}\right)^{1/2}$$

$$A_b = A_a \left(\frac{I_b}{I_a}\right)^{1/2} = 1.1 \times 10^{-5} \left(\frac{10^{-12}}{10^0}\right)^{1/2} = 1.1 \times 10^{-11} \text{ m}$$

This displacement is smaller than the diameter of the hydrogen atom!

c. Calculate the sound pressures for cases *a* and *b* using Equation 12.2.

$$P_0 = \sqrt{2ZI}$$

$$P_{0a} = [(2)\, 4.3 \times 10^2 (1)]^{1/2} \left|\frac{\text{kg}}{\text{m}^2 \text{ sec}} \quad \frac{\text{kg}}{\text{sec}^3}\right|^{1/2} = 29 \left|\frac{\text{kg m}}{\text{m}^2 \text{ sec}^2}\right|$$

$$= 29 \text{ N/m}^2 \simeq 0.0003 \text{ atmospheres} \qquad \text{(case } a\text{)}$$

$$P_{0b} = 2.9 \times 10^{-5} \text{ N/m}^2 \qquad \text{(case } b\text{)}$$

For comparison, atmospheric pressure is about 10^5 N/m².

For many purposes it is not necessary to know the absolute pressure or the absolute intensity of a sound wave. A special unit, the *bel,* has been developed for comparing the intensities of two sound waves (I_2/I_1). This unit was named after Alexander Graham Bell, who invented the telephone and did research in sound and hearing. The intensity ratio in bels is equal to $\log_{10} (I_2/I_1)$. Thus if one sound is ten times more intense than another, $I_2/I_1 = 10$; since $\log_{10} 10 = 1.0$, the two sound intensities differ by 1 bel. Because the bel is a rather large unit, it is common to use the decibel (dB) in comparing two sound intensities (1 bel = 10 dB). This unit is often used to describe the performance of hi-fi systems.

Since *I* is proportional to P^2, the pressure ratio between two sound levels can be expressed as $10 \log_{10} (P_2^2/P_1^2)$, or $20 \log_{10} (P_2/P_1)$. This expression for the number of decibels can be used to compare any two sound pressures in the same medium. For two sounds with pressures that differ by a factor of 2 we get $20 \log_{10} (P_2/P_1) = 20 \log_{10} 2 = 20(0.301) \simeq 6$ dB. Thus, a hi-fi set that gives a sound output uniform to ± 3 dB (a total variation of 6 dB) from 30 to 15,000 Hz has a sound pressure variation over its frequency range of 2. This variation would not be noticed by the average ear except under controlled laboratory conditions.

For hearing tests, it is convenient to use a reference sound intensity (or sound pressure) to which other sound intensities can be compared. The reference sound intensity I_0 is 10^{-16} W/cm² (10^{-12} W/m²); $P_0 \simeq 2 \times 10^{-4}$ dyne/cm². A 1000 Hz note of this intensity is barely audible to a person with good hearing.

Table 12.2. Approximate Intensities of Various Sounds

	Intensity (W/m²)	Level (dB)
Sound that is barely perceptible	10^{-12}	0
Whisper	10^{-10}	20
Average dwelling	10^{-9}	30
Business office	10^{-7}	50
Speech at 1 m	10^{-6}	60
Busy street	10^{-5}	70
Subway or automobile	10^{-3}	90
Sound that produces pain	10^{0}	120
Jet aircraft	10^{1}	130
On rocket launch pad	10^{5}	170

If a sound intensity is given in decibels with no reference to any other sound intensity, you can assume that I_0 is the reference intensity. Table 12.2 gives the intensities of some typical sounds in terms of this reference value. The most intense sound that the ear can tolerate without pain is about 120 dB. For 120 dB, $P/P_0 = 10^6$ and $I/I_0 = 10^{12}$.

When a sound wave hits the body, part of the wave is reflected and part is transmitted into the body (Fig. 12.2). The ratio of the reflected amplitude R to the incident amplitude A_0 depends on the acoustic impedances of the two media, Z_1 and Z_2. The relationship is

$$\frac{R}{A_0} = \frac{Z_1 - Z_2}{Z_1 + Z_2} \qquad (12.3)$$

For a sound wave in air hitting the body, Z_1 is the acoustic impedance of air and Z_2 is the acoustic impedance of tissue. Note that if $Z_1 = Z_2$, there is

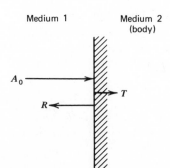

Medium 1   Medium 2 (body)

A_0

R

T

Figure 12.2. A sound wave of amplitude A_0 incident upon the body. Part of the wave, of amplitude R, is reflected and part, of amplitude T, is transmitted.

no reflected wave and transmission to the second medium is complete. Also, if $Z_2 > Z_1$, the sign change indicates a phase change of the reflected wave.

The ratio of the transmitted amplitude T to the incident wave amplitude A_0 is

$$\frac{T}{A_0} = \frac{2 Z_1}{Z_1 + Z_2} \qquad (12.4)$$

Equations 12.3 and 12.4 are for sound waves striking perpendicular to the surface. The equations for sound waves hitting from various angles are more complicated and are not needed in this book.

It is obvious from Example 12.2 that whenever acoustic impedances differ greatly there is almost complete reflection. This is true regardless of which medium the sound originates in and is the reason heart sounds are poorly transmitted into the air adjacent to the chest. The large impedance difference between air and tissue accounts for the large amount of reflection.

Example 12.2
Calculate the amplitudes of the reflected and transmitted sound waves from air to muscle.

Using the values of Z from Table 12.1 in Equations 12.3 and 12.4 and ignoring the negative sign due to a phase change on reflection, we get for the fraction reflected

$$\frac{R}{A_0} = \frac{1.64 \times 10^6 - 430}{1.64 \times 10^6 + 430} = 0.9995 \text{ (or } 99.95\%)$$

and for the fraction transmitted

$$\frac{T}{A_0} = \frac{2(430)}{1.64 \times 10^6 + 430} \simeq 5 \times 10^{-4} \text{ (or } \sim 0.05\%)$$

Example 12.3 shows that when the acoustic impedances of the two media are similar almost all of the sound is transmitted into the second medium. Choosing materials with similar acoustic impedances is called *impedance matching*. Getting sound energy into the body requires impedance matching.

Example 12.3
Calculate the amplitudes of the reflected and transmitted sound waves from water to muscle.

Using the values from Table 12.1 in Equations 12.3 and 12.4 we obtain

$$\frac{R}{A_0} = \frac{(1.64 - 1.43) \times 10^6}{(1.64 + 1.43) \times 10^6} = 0.0684 \text{ (or } 7\%)$$

$$\frac{T}{A_0} = \frac{2(1.43) \times 10^6}{(1.64 + 1.43) \times 10^6} = 0.9316 \text{ (or } 93\%)$$

Since the intensity of a wave in a medium is proportional to ZA^2 (Equation 12.1), the ratio of the reflected intensity to the incident intensity (both in medium 1) is

$$\frac{Z_1(R)^2}{Z_1(A_0)^2} = \left(\frac{R}{A_0}\right)^2 = (0.0684)^2 = 0.0047 \text{ (or } 0.47\%)$$

and the ratio of the transmitted intensity to the incident intensity is

$$\frac{Z_2(T)^2}{Z_1(A_0)^2} = 0.9953 \text{ (or } 99.5\%)$$

In our discussion of the reflection of a sound wave we assumed that the wave was perpendicular to the surface. Thus the transmitted wave went straight in and the reflected wave went straight back. How do the reflected and transmitted sound waves behave when a wave hits at an angle θ_i to a boundary between two media (Fig. 12.3)?

The geometric laws involving the reflection and refraction (bending) are the same as for light. This means that θ incident = θ reflected, or $\theta_i = \theta_r$. The angle of the refracted sound wave θ_2 is determined by the velocities of sound in the two media v_1 and v_2 from the equation

$$\frac{\sin \theta_i}{v_1} = \frac{\sin \theta_2}{v_2}$$

Because sound can be refracted, acoustic lenses can be constructed to focus sound waves.

When a sound wave passes through tissue, there is some loss of energy due to frictional effects. The absorption of energy in the tissue causes a reduction in the amplitude of the sound wave. The amplitude A at a depth x cm in a medium is related to the initial amplitude A_0 ($x = 0$) by the exponential equation*

$$A = A_0 e^{-\alpha x}$$

where α, in cm^{-1}, is the absorption coefficient for the medium at a particular frequency. Table 12.3 gives some typical absorption coefficients. Figure 12.4 illustrates the exponential absorption of ultrasound in oil.

* Appendix A discusses exponential behavior. Review it if you need help on exponentials.

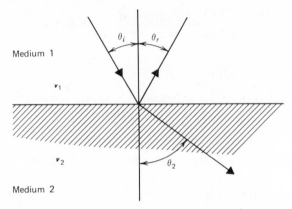

Figure 12.3. **Behavior of a sound wave at an interface between two substances where reflection and refraction (bending) take place. The velocity of sound in medium 2 (v_2) is greater than that in medium 1 (v_1).**

Since the intensity is proportional to the square of the amplitude (Equation 12.1), its dependence with depth is

$$I = I_0 e^{-2\alpha x} \qquad (12.5)$$

where I_0 is the incident intensity at $x = 0$ and I is the intensity at a depth x in the absorber. Since the absorption coefficient in Equation 12.5 is 2α, the intensity decreases more rapidly than the amplitude with depth.

Table 12.3. **Absorption Coefficients and Half-Value Thicknesses for Various Substances**

Material	Frequency (MHz)	α (cm^{-1})	Half-Value Thickness (cm)[a]
Muscle	1	0.13	2.7
Fat	0.8	0.05	6.9
Brain (ave)	1	0.11	3.2
Bone (human skull)	0.6	0.4	0.95
	0.8	0.9	0.34
	1.2	1.7	0.21
	1.6	3.2	0.11
	1.8	4.2	0.08
	2.25	5.3	0.06
	3.5	7.8	0.045
Water	1	2.5×10^{-4}	1.4×10^3

[a]The intensity half-value thickness layer $= \ln 2/2\alpha$.

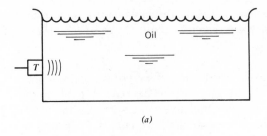

(a)

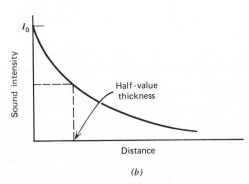

(b)

Figure 12.4. Absorption of ultrasound. (a) An ultrasonic beam is sent from the transducer T into a long oil bath. (b) It is absorbed exponentially. The half-value thickness is the path length through oil that reduces the beam to 50% of its original intensity I_0.

The half-value thickness (HVT) is the tissue thickness needed to decrease I_0 to $I_0/2$. Table 12.3 gives typical HVTs for different tissues. Note the high absorption in the human skull and that the absorption increases as the frequency of the sound increases. This increasing absorption with frequency also occurs for other body tissues and limits the maximum frequencies that can be used clinically (see Example 12.4).

Example 12.4
What is the attenuation of sound intensity by 1 cm of bone at 0.8, 1.2, and 1.6 MHz?
    While Equation 12.5 could be used, a quick estimate can be made using the HVTs given in Table 12.3. At 0.8 MHz, the HVT is 0.34 cm; therefore, 1 cm is about 3 HVT, and the intensity is reduced by 2^3, or by a factor of 8 (i.e., $\sim$ 12% remains). At 1.2 MHz, the HVT is 0.21; 1 cm is nearly 5 HVT, and the intensity is reduced by almost 2^5, or by a factor of 32 (i.e.,

~ 3% remains). At 1.6 MHz the HVT is 0.11 cm; 1 cm is about 9 HVT, and the intensity is reduced by about 2^9, or by a factor of 512 (i.e., ~ 0.5% remains).

In addition to the absorption of sound, the spreading out, or divergence, of sound causes the intensity to decrease. If the sound is from a small source (point source) the divergence causes the intensity to decrease according to the inverse square law. That is, I is proportional to $1/r^2$ where r is the distance from the source to the measuring point. The typical clinical ultrasound source is small (~ 1 cm), but it is not an ideal point source.

12.2. THE BODY AS A DRUM (PERCUSSION IN MEDICINE)

Percussion (tapping) has been used since the beginnings of civilization for various purposes such as testing whether walls are solid or coverings for hiding places and whether wine barrels are empty or full. The first recorded use of percussion on the human body as a means of diagnosis occurred in the eighteenth century. In 1761, L. Auenbrugger published a short book, *On Percussion of the Chest,* which was based on his clinical observations over seven years of the different sounds he produced by striking the chests of patients in various places. It should be mentioned

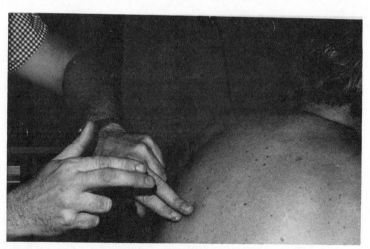

Figure 12.5. A method of inducing percussion of the chest from the back. The fingers of one hand are held against the skin and tapped with the fingers of the other hand.

that Auenbrugger was an accomplished musician and that his father had been an innkeeper. He probably learned the technique of percussion by tapping wine barrels at his father's inn, and his musical ear probably helped him in interpreting the sounds.

In his book, Auenbrugger described how to strike the chest with the fingers (Fig. 12.5), and he stated, "the sound thus elicited from the healthy chest resembles the stifled sound of a drum covered with a thick woolen cloth or other envelope." He discussed both the sounds heard from healthy subjects and the sounds heard from patients with various pathological conditions. Auenbrugger stated that with percussion he could diagnose cancer, the presence of abnormal cavities in an organ, and those diseases involving fluids in the chest region. He confirmed many of these diagnoses by examining bodies after death.

Auenbrugger's discovery was largely ignored until 1808 when his work, originally published in Latin, was translated to French. Percussion has since become an important technique in the detection of disease.

12.3. THE STETHOSCOPE

Perhaps no symbol is more associated with the physician than the stethoscope hanging around his neck or protruding from his pocket. This simple "hearing aid" permits a physician or nurse to listen to sounds made inside the body, primarily in the heart and lungs. The act of listening to these sounds with a stethoscope is called *mediate auscultation,* or usually just *auscultation.* (We should be careful to distinguish auscultation from osculation, which is an entirely different activity.)

Many sounds from the chest region can be useful in the diagnosis of disease. Prior to 1818, the only methods available for examination of the chest were feeling with the hand, percussion, and immediate auscultation with the ear directly on the chest. In *A Treatise on the Diseases of the Chest and on Mediate Auscultation* (1818), R. T. H. Laennec described the objections to putting the ear directly on the chest: "it is always inconvenient, both to the physician and the patient; in the case of females it is not only indelicate, but often impracticable; and for that class of persons found in the hospitals it is disgusting." It should be mentioned that at that time, physicians routinely made house calls and examined and treated almost everyone of any means in his home. Only charity patients went to the hospital.

Laennec used immediate auscultation until 1816 when he was examining a girl with general symptoms of a diseased heart. Because she was fat, young, and female, he felt that the usual examination methods were inappropriate. However, he recalled that if one end of a piece of wood is scratched with a pin, the sound can be heard well when the other end is held to the ear. He immediately rolled several pieces of paper into a cylinder and held one end to his ear and the other to the girl's chest above her heart. The results were dramatic and encouraged Laennec to improve this instrument. Eventually he developed a hollow wood cylinder 30 cm long with an inner diameter of approximately 1 cm and an outer diameter of about 7.5 cm. He named it the *stethoscope.* In his book he described his research on the stethoscope and his interpretation of the natural and pathological sounds of the lungs, heart, and voice.

The stethoscope currently in use is based on Laennec's original work. The main parts of a modern stethoscope are the bell, which is either open or closed by a thin diaphragm, the tubing, and the earpieces (Fig. 12.6).

The open bell is an impedance matcher between the skin and the air and accumulates sounds from the contacted area. The skin under the open bell behaves like a diaphragm. The skin diaphragm has a natural resonant frequency at which it most effectively transmits sounds; the factors controlling the resonant frequency are similar to those controlling the frequency of a stretched vibrating wire. The tighter the skin is pulled, the higher its resonant frequency. The larger the bell diameter, the lower the skin's resonant frequency. Thus it is possible to enhance the sound range of interest by changing the bell size and varying the pressure of the bell against the skin and thus the skin tension. A low-frequency heart murmur will appear to go away if the stethoscope is pressed hard against the skin!

A closed bell is merely a bell with a diaphragm of known resonant frequency, usually high, that tunes out low-frequency sounds. Its resonant frequency is controlled by the same factors that control the frequency of the open bell pressed against the skin. The closed-bell stethoscope is primarily used for listening to lung sounds, which are of higher frequency than heart sounds. Figure 12.7 shows the typical frequency ranges of heart and lung sounds.

What is the best shape for the bell? Since we are dealing with a system

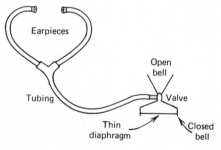

Figure 12.6. A schematic of a modern stethoscope.

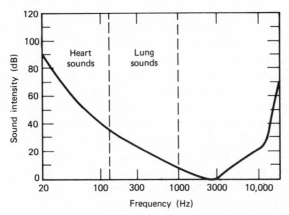

Figure 12.7. Most of the heart sounds are of low frequency in the region where the sensitivity of the ear is poor. Lung sounds generally have higher frequencies. The curve represents the threshold of hearing for a good ear. Some of the heart and lung sounds are below this threshold.

closed at the far end by a pressure sensitive diaphragm, the eardrum, it is desirable to have a bell with as small a volume as possible. The smaller the volume of gas, the greater the pressure change for a given movement of the diaphragm at the end of the bell.

The volume of the tubes should also be small, and there should be little frictional loss of sound to the walls of the tubes. The small volume restriction suggests short, small diameter tubes, while the low-friction restriction suggests large diameter tubes. If the diameter of the tube is too small, frictional losses occur, and if it is too large, the moving air volume is too great; in both cases the efficiency is reduced. Below about 100 Hz tube length does not greatly affect the efficiency, but above this frequency the efficiency decreases as the tube is lengthened. At 200 Hz 15 dB is lost in changing from a tube 7.5 cm long to a tube 66 cm long. A compromise is a tube with a length of about 25 cm and a diameter of 0.3 cm.

The earpieces should fit snugly in the ear because air leaks reduce the sounds heard. The lower the frequency, the more significant the leak. Leaks also allow background noise to enter the ear. The earpieces are usually designed to follow the slightly forward slant of the ear canals.

12.4. ULTRASOUND PICTURES OF THE BODY

Human ears respond to sound in the frequency range of about 20 to 20,000 Hz, although many animals can produce and hear sounds of considerably higher frequencies. For example, bats emit blips of ultrasonic frequencies—30 to 100 kHz—and navigate by listening to the echoes. It was discovered during World War II that man can use ultrasound in much the same way bats can. The navy developed sonar (*SO*und *NA*vigation and *R*anging), a method of locating underwater objects, such as submarines, with ultrasound echoes. After World War II medical engineers developed techniques for using ultrasound in diagnosis.

In this section we discuss the use of ultrasound to produce pictures for medical diagnosis. Basically, an ultrasound source sends a beam of pulses of 1 to 5 MHz sound into the body. The time required for the sound pulses to be reflected gives information on the distances to the various structures or organs in the path of the ultrasound beam.

While there are several methods of generating ultrasound, the most important for medical applications involves the piezoelectric effect. This effect was discovered by Jacques and Pierre Curie in about 1880. Many crystals can be cut so that an oscillating voltage across the crystal will produce a similar vibration of the crystal, thus generating a sound wave (Fig. 12.8).

A device that converts electrical energy to mechanical energy or vice versa is called a *transducer*. Ultrasound generators are often simply referred to as transducers.

Each transducer has a natural resonant frequency of vibration. The thinner the crystal, the higher the frequency at which it will oscillate. For a quartz crystal cut along a certain axis (*x*-cut), a thickness of 2.85 mm gives a resonant frequency of about 1 MHz. Typical frequencies for medical work are in the 1 to 5 MHz range. An average power level for diagnostic applications is a few milliwatts per square centimeter.

Pulses of ultrasound are transmitted into the body by placing the vibrating crystal in close contact with the skin, using water or a jelly paste to eliminate the air. This gives a good impedance match at the skin and greatly increases the transmission of the ultrasound into the body and of the echoes back to the detector.

The same transducer that produced the pulse serves as the detector. The vibrations of the crystal produced by the echoes generate a voltage across it—just the reverse of what happens in the production of ultrasound. The weak signals are then amplified and displayed on an oscilloscope.

Many of the applications of ultrasound in medicine are based on the principles of sonar. In sonar a sound wave pulse is sent out and is reflected from an object; from the time required to receive the echo and the known velocity of sound in water, the distance to the object can be determined. Bats and porpoises use the sonar principle for navigating and

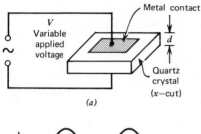

(a)

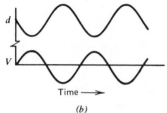

Time ———➤

(b)

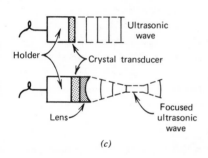

(c)

Figure 12.8. Behavior of a quartz crystal used in the production of ultrasound. (a) Attachment of electrodes; (b) change in crystal thickness d (greatly exaggerated) due to an applied alternating voltage V; (c) the crystal mounted in a holder to produce a beam of ultrasound. A focused beam is produced when an acoustic lens is attached to the crystal.

finding food (Fig. 12.9). Blind humans use the same principle when they listen to echoes from the tap of a cane to help avoid large objects.

To obtain diagnostic information about the depth of structures in the body, we send pulses of ultrasound into the body and measure the time required to receive the reflected sound (echoes) from the various surfaces in it. This procedure is called the *A scan* method of ultrasound diagnosis. Pulses for A scan work are typically a few microseconds long. They are usually emitted at 400 to 1000 pulses/sec.

The A scan method is illustrated schematically in Fig. 12.10. In Fig. 12.10*a*, a transducer *T* sends a pulse of ultrasound through a beaker of water of diameter *d*. The sound is reflected from the other side of the beaker and returns to the transducer, which also acts as a receiver. The detected echo is converted to an electrical signal and is displayed as the vertical deflection *R* on the cathode ray tube (CRT) of an oscilloscope (Fig. 12.10*a'*). Since the echo has been attenuated by the water, *R* is

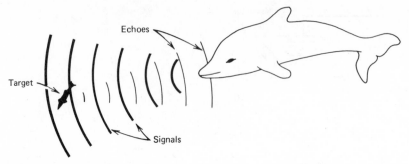

Figure 12.9. The sonar principle. The porpoise sends out ultrasonic waves and uses the echoes to locate food.

smaller in amplitude than the initial pulse shown on the oscilloscope at 0. The time required for the pulse to travel from the transducer to the far side and return to the transducer is indicated on the horizontal scale of the oscilloscope. This time can easily be converted to distance by using the known velocity of sound in water (Table 12.1) to calibrate the scale.

An object in the beaker can be located with ultrasound. In Fig. 12.10b a surface S at a distance d_1 produces an additional echo, which is displayed on the oscilloscope as S at the position d_1 (Fig. 12.10b'). Note that the echo R is now smaller. When the surface vibrates (Fig. 12.10c), the position of the echo on the oscilloscope also moves (Fig. 12.10c').

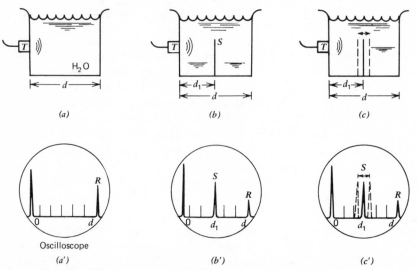

Figure 12.10. Schematic of an A scan. See text for explanation.

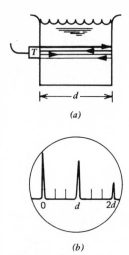

(a)

(b)

Figure 12.11. Schematic of an A scan showing multiple reflections taking place. See text for explanation.

It is also possible to have multiple reflections between surfaces. Figure 12.11*a* illustrates this effect. A pulse emitted at the transducer T is reflected from the far side and returns to the transducer, where part is converted to a signal and part is again reflected to the far side; this part returns to the transducer again and appears as a signal. Such a multiple echo appears in Fig. 12.11*b* as an object at a distance d and a second object at $2d$.

It must be remembered that the basis for the use of ultrasound in medicine is the partial reflection of sound at the surface between two media that have different acoustical properties. The amount of the reflection depends primarily upon the difference in the acoustical impedances of the two materials and the orientation of the surface with respect to the beam. Since the transmitter and detector are the same unit, the most intense detected signals are due to reflections from surfaces perpendicular to the beam. In many diagnostic uses of ultrasound the echoes are very small signals due to weak reflection and the absorption of the sound by tissue. The great loss in the signal due to exponential absorption of sound in the tissue can be compensated for by electronically amplifying the echo by an amount proportional to the body depth from which the sound is reflected (Fig. 12.12).

Another problem is the lack of resolution, or the ability of the equipment to detect separate echoes from two objects close together. In general, structures smaller than the wavelength λ cannot be resolved. Since $\lambda = v/f$, where v is the velocity of sound and f is the frequency, high-frequency sound has shorter wavelengths and allows better resolution

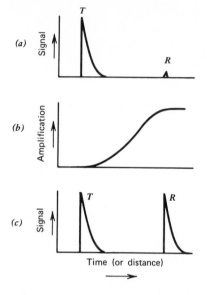

Figure 12.12. Weak echoes from deep structures can be electronically amplified to make them more visible. (a) The transmitted pulse *T* produces a weak echo *R* on the CRT. (b) The amount of amplification increases with time (or distance). (c) The amplification increases the size of the echo *R*.

than low-frequency sound. For example, 1.2 MHz sound has a wavelength in water of $1.5 \times 10^3/1.2 \times 10^6 = 1.2 \times 10^{-3}$ m or 1.2 mm, while 3.5 MHz sound has a wavelength of 0.43 mm. However, a compromise must be made in the choice of the frequency since the absorption increases as the frequency increases. For example, a 3.5 MHz signal does not penetrate through the skull nearly as well as 1.2 MHz sound (Table 12.3), and it has to go through the skull twice for the echo to reach the detector.

One A scan procedure, *echoencephalography,* has been used in the detection of brain tumors. Pulses of ultrasound are sent into a thin region of the skull slightly above the ear and echoes from the different structures within the head are displayed on an oscilloscope (Fig. 12.13). The usual procedure is to compare the echoes from the left side of the head to those from the right side and to look for a shift in the midline structure. A tumor on one side of the brain tends to shift the midline toward the other side (Fig. 12.14). Generally a shift of more than 3 mm for an adult or 2 mm for a child is considered abnormal.

During echoencephalography care must be taken that the instrument can detect the echo immediately after the initial pulse from the near side of the skull. This information is necessary to compare scans from right to left and left to right. It is also necessary to exercise caution in interpreting the echo patterns when the skull is asymmetric to avoid making a false diagnosis.

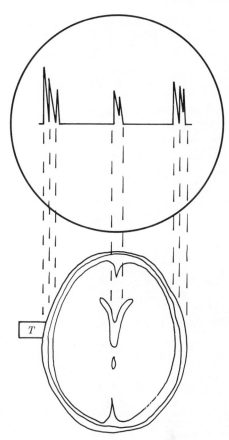

Figure 12.13. The A scan method to locate the midline of the brain. (echoencephalography). Pulses of ultrasound are sent into the brain by the transducer *T*, and the echoes are displayed on the oscilloscope.

Applications of A scans in ophthalmology can be divided into two areas: one is concerned with obtaining information for use in the diagnosis of eye diseases; the second involves biometry, or measurements of distances in the eye. At the low power levels used, there is no danger to the patient's eye. Ultrasound frequencies of up to 20 MHz are used. These high frequencies can be used in the eye to produce better resolution since there is no bone to absorb most of the energy and absorption is not significant because the eye is small.

Ultrasound diagnostic techniques are supplementary to the generally practiced ophthalmological examinations; they can provide information about the deeper regions of the eye and are especially useful when the cornea or lens is opaque. Use of the A scan plus optical and even x-ray information may be necessary for a complete diagnosis. Tumors, foreign bodies, and detachment of the retina (the light-sensitive part of the eye)

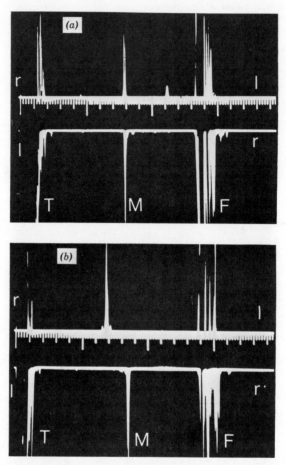

Figure 12.14.   Echoencephalograms of a normal brain and an abnormal brain. (*a*) A pair of scans of a normal brain. The transducer T is on the right side of the head in the top scan and on the left side in the bottom scan. F indicates echoes from the far side of the skull. There is no shift of the midline echo M. (*b*) A pair of scans of an abnormal brain showing a shift of 7 mm toward the right side that could be caused by a tumor on the left side of the brain. (From M.M. Lapayowker in S. Gottlieb and M. Viamonte (Eds.), *Diagnostic Ultrasound,* Committee on New Technology, American College of Radiology, no date, p. 16.)

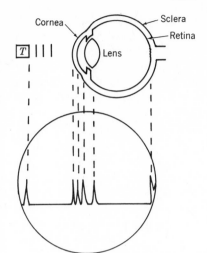

Figure 12.15. An ultrasound transducer *T* transmits sound through water into the eye, and the reflected sound is displayed on an oscilloscope.

are some of the problems that can be diagnosed with ultrasound. Figure 12.15 is a schematic view of a normal A scan of the eye. Figure 12.16 shows an A scan of a severe retinal detachment.

Without ultrasound ophthalmologists can look into the living eye up to the optic nerve (see Chapter 15), but measurements of the eye have been largely confined to the exterior segment. With ultrasound it is possible to measure distances in the eye such as lens thickness, depth from cornea to lens, the distance to the retina, and the thickness of the vitreous humor. This information can be combined with other quantities such as the curvature of the cornea and the prescription for corrective glasses to determine the indexes of refraction of components of the eye.

Figure 12.16. Ultrasound studies of a detached retina. The CRT shows an echo s from the anterior sclera, an echo r from the retina, and an echo s from the sclera at the back of the eye. In a normal eye the echo from the retina would blend with the echo from the posterior sclera. (From K. Ossoinig, in A. Oksala and H. Gernet (Eds.), *Ultrasonics in Ophthalmology*, Karger, Basel/New York, 1967, pp. 116–133.)

For many clinical purposes A scans have been largely replaced by B scans. The B scan method is used to obtain two-dimensional views of parts of the body. The principles are the same as for the A scan except that the transducer is moved. As a result each echo produces a dot on the oscilloscope at a position corresponding to the location of the reflecting surface (Fig. 12.17). A storage oscilloscope is usually used so that a lasting image can be formed and a photograph can be made (Fig. 12.18).

B scans provide information about the internal structure of the body. They have been used in diagnostic studies of the eye, liver, breast, heart, and fetus. They can detect pregnancy as early as the fifth week and can provide information about uterine anomalies (Fig. 12.19). Information on the size, location, and change with time of a fetus is extremely useful in both normal deliveries (Fig. 12.20) and cases such as abnormal bleeding and threatened abortion. In many cases B scans can provide more information than x-rays, and they present less risk (see Chapter 19). For example, x-ray studies can only detect cysts that take up radiopaque solutions, while ultrasound can be used to quantitatively image many types of cysts.

In the early B scan work all the echoes displayed on the CRT were the

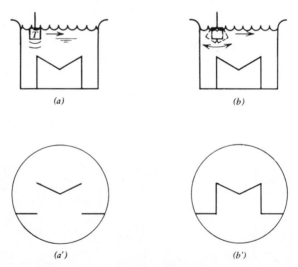

(a)   (b)

(a')   (b')

Figure 12.17.   Schematic of B scan method. (a) As the transducer *T* moves to the right it produces echoes from the submerged object. (a') The storage oscilloscope shows a dot corresponding to the location of each echo received. The dots outline the top surface of the object. (b) When the transducer is rocked as it is moved to the right it produces echoes from other surfaces. (b') The resulting scan shows the sides of the object.

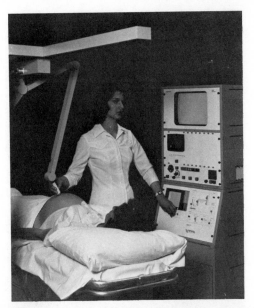

Figure 12.18. An ultrasound technician moves the transducer of a B scan unit over the bare abdomen of a pregnant patient while adjusting the brightness of the image on the monitor. When a satisfactory image is obtained, a copy is made with a camera. (Courtesy of Unirad Corporation, Denver, Colo.)

same brightness. The operator could exclude echoes of low magnitude by setting an electronic control at a chosen threshold value. While this mode, called the *leading-edge display,* is very useful for many purposes, it gives no information on the sizes of the echoes. The improved *gray-scale display* electronically changes the brightness on the CRT so that large echoes appear brighter than weak echoes. Figure 12.21 shows a scan displayed in the two modes. With the gray-scale display, tumors in the liver that might have been missed with the leading-edge display have been easily detected.

The success of the gray-scale display has led to the development of color displays that show a greater range of echoes. A still greater range of echoes can be shown by using a digital display. In this case the ultrasound echoes must be electronically processed and fed to a computer.

12.5. ULTRASOUND TO MEASURE MOTION

Two methods are used to obtain information about motion in the body with ultrasound; the M (motion) scan, which is used to study motion such

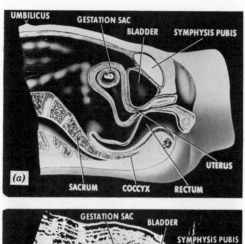

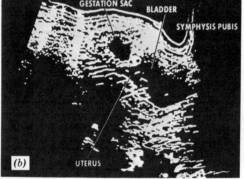

Figure 12.19. B scans have an important use in obstetrics. (a) A schematic cross-section of a woman showing a small fetus in the gestation sac. (b) A B scan of a patient showing the gestation sac, uterus, and bladder. (Courtesy of Unirad Corporation, Denver, Colo.)

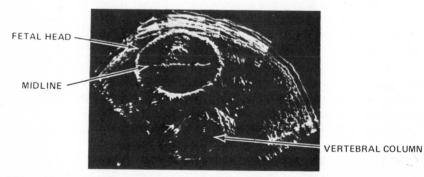

Figure 12.20. A side-to-side B scan of a more developed fetus than that shown in Fig. 12.19. The fetal head and the midline of the brain are clearly seen. (Courtesy of Kenneth Gottesfeld, M.D., University of Colorado Medical Center, Denver, Colo.)

as that of the heart and the heart valves, and the Doppler technique, which is used to measure blood flow.

The M scan combines certain features of the A scan and the B scan. The transducer is held stationary as in the A scan and the echoes appear as dots as in the B scan.

Figure 12.22*a* shows a transducer fixed at one position emitting a pulse of ultrasound into a beaker of water that has a vibrating interface in it. Figure 12.22*b* is a standard B scan showing the motion of the interface on the oscilloscope screen. When the oscilloscope trace is made to move vertically as a function of time, the motion of the interface is displayed as an M scan as seen in Fig. 12.22*c*.

M scans are used to obtain diagnostic information about the heart. The places where the heart can be probed are quite limited because of poor ultrasound transmission through lung tissue and bone. The usual method is to put the transducer on the patient's left side, aim it between the ribs over the heart, and tip it at different angles to explore various regions of the heart (Fig. 12.23). By moving the probe it is possible to obtain information about the behavior of a particular valve or section of the heart. The examiner must be familiar with the patterns of specific cardiac echoes to interpret the information. Several heart conditions can be diagnosed with M scans; we consider here M scans of mitral valves and M scans showing accumulation of fluid in the heart sac (pericardial effusion).

A scan taken from position 1 of Fig. 12.23 shows reflections from the chest wall, the right ventricular cavity, the interventricular septum (IVS), the anterior mitral valve leaflet (AMVL), and the posterior wall of the heart. The part of the scan consisting of the AMVL signal (Fig. 12.24*a*) shows the motion—the opening and closing—of the anterior leaflet of the mitral valve. The motion is correlated to the electrical activity of the heart (ECG), which is recorded simultaneously. The information of interest is the rate of closing of the mitral valve. The rate of closing for a normal valve is indicated by the slope in Fig. 12.24*a;* in this case the rate of closing is 72 mm/sec. Figure 12.24*b* is an M scan showing an abnormality called mitral stenosis (a narrowing of the valve opening). The reduced slope for mitral stenosis is quite different from the normal slope—the slower the rate of closure, the larger the amount of stenosis. Other heart valves can be examined in a similar fashion.

Pericardial effusion can be easily detected with an M scan. Normally the pericardial sac surrounding the heart is in direct contact with it. Thus an M scan of a normal heart taken from position 1 in Fig. 12.23 would show the anterior right ventricular wall in direct contact with the anterior pericardium and the stationary chest wall and the posterior left ventricular wall in contact with the posterior pericardium. When pericardial effusion occurs, the space between the heart and pericardium fills with a fluid that is relatively echo free, and an M scan will show a separation between the sac and the heart both anteriorly and posteriorly (Fig. 12.25). This scan can be repeated during treatment to determine whether progress is being made.

Since the early studies on sound in the 1800s, physicists have realized that a source of sound of frequency f_0 has a higher pitch when it is moving toward a listener and a lower pitch when it is moving away from him (Fig.

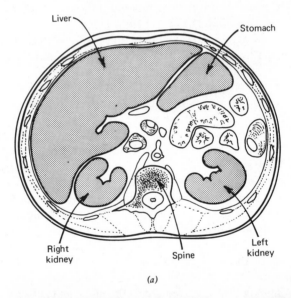

Liver

Stomach

Right kidney

Spine

Left kidney

(a)

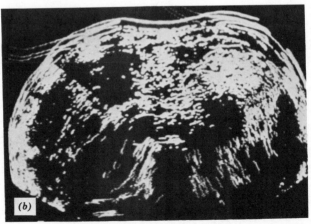

(b)

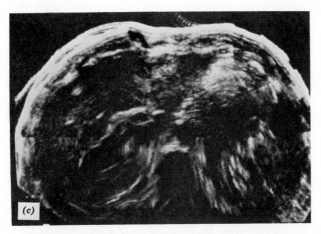

Figure 12.21.   Transverse ultrasound B scans of the upper abdomen. (a) A sketch identifying the various structures in the scans. (b) A leading-edge B scan in which all echoes are equally bright. (c) A gray-scale B scan of the same patient in which large echoes are brighter than small echoes. (Courtesy of James Zagzebski, University of Wisconsin, Madison, Wisc.)

(a)

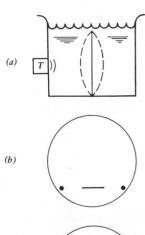

(b)

(c)

Figure 12.22.   A schematic of the M scan method. (a) A vibrating interface in a beaker of water reflects sound pulses from the transducer T. (b) On a stationary B scan the vibrating membrane appears as a line in the middle of the scan. (c) When the electron beam of the CRT is moved vertically, the motion of the vibrating interface is displayed as an M scan.

279

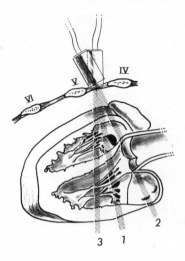

Figure 12.23. Schematic view of the heart being scanned with ultrasound. The ribs are identified with Roman numerals. (Courtesy of Richard D. Spangler, M.D., and Michael Johnson, M.D., University of Colorado Medical Center, Denver, Colo.)

12.26*a*). It also has a higher pitch when the listener is moving toward the source than when he is moving away from it (Fig. 12.26*b*). The frequency change is called the *Doppler shift*. When the sound source is moving toward the listener or when he is moving toward the source, the sound waves are pushed together and he hears a frequency higher than f_0. When the source is moving away from the listener or when he is moving away from the source, he hears a frequency lower than f_0. If we know the frequency of the source f_0 and can measure the frequency that is received by the listener, we can determine how fast the sound source or listener is moving. This technique has been used to measure the velocity of rockets; a rocket receives a radio-frequency signal and then rebroadcasts it to the sender, who compares the received signal to the original to determine the relative velocity of the rocket.

The Doppler effect can be used to measure the speed of moving objects or fluids within the body, such as the blood. When a continuous ultrasound beam is "received" by some red blood cells in an artery moving away from the source, the blood "hears" a slightly lower frequency than the original frequency f_0. The blood sends back scattered echoes of the sound it "hears," but since it is now a source of sound moving away from the detector, there is another shift to a still lower frequency. The detector receives a back-scattered signal that has undergone a double Doppler shift. When the blood is moving at an angle θ from the direction of the sound waves, the frequency change f_d is

$$f_d = \frac{2f_0 V}{v} \cos \theta$$

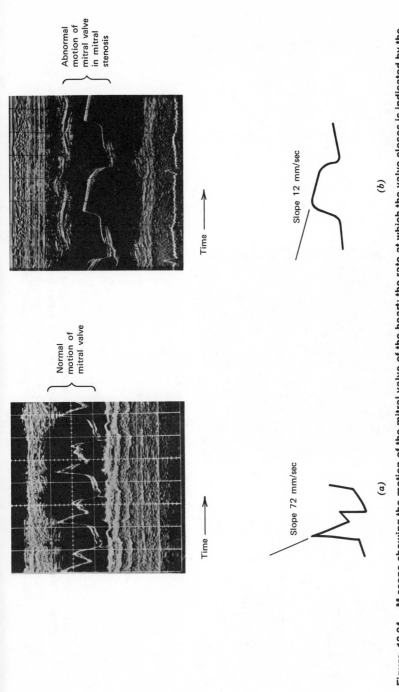

Figure 12.24. M scans showing the motion of the mitral valve of the heart; the rate at which the valve closes is indicated by the slope, which is sketched below each scan. (a) A slope of 72 mm/sec is normal. (b) A slope below 35 mm/sec indicates an abnormality called mitral stenosis (narrowing of the opening). (Scans courtesy of Richard D. Spangler, M.D., and Michael Johnson, M.D., University of Colorado Medical Center, Denver, Colo.)

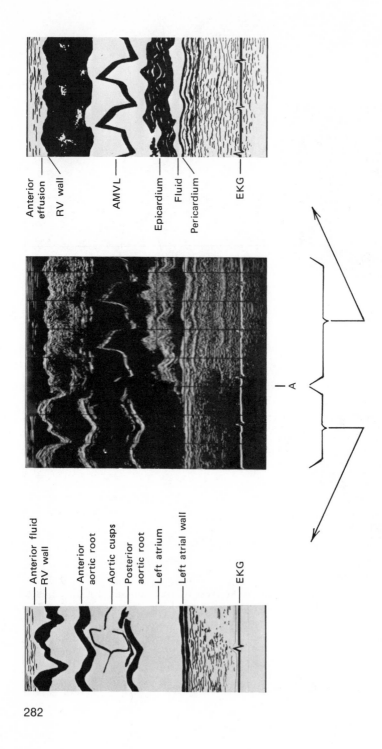

Anterior fluid
RV wall

Anterior aortic root
Aortic cusps
Posterior aortic root
Left atrium
Left atrial wall

EKG

Anterior effusion
RV wall

AMVL

Epicardium
Fluid
Pericardium

EKG

A

Figure 12.25. An M scan showing fluid accumulation in the sac surrounding the heart (pericardial effusion). The probe was moved at point A. The sketch on the left identifies the structures in the left portion of the M scan, up to point A. The sketch on the right, corresponding to the right portion of the M scan, shows the fluid between the heart and pericardium. (Courtesy of Richard D. Spangler, M.D., and Michael Johnson, M.D., University of Colorado Medical Center, Denver, Colo.)

Moving sound source

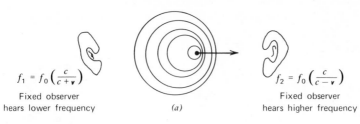

$$f_1 = f_0 \left(\frac{c}{c+v} \right)$$

Fixed observer
hears lower frequency

(a)

$$f_2 = f_0 \left(\frac{c}{c-v} \right)$$

Fixed observer
hears higher frequency

Fixed sound source

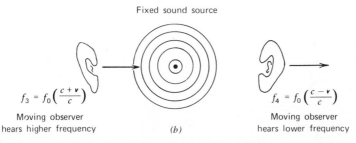

$$f_3 = f_0 \left(\frac{c+v}{c} \right)$$

Moving observer
hears higher frequency

(b)

$$f_4 = f_0 \left(\frac{c-v}{c} \right)$$

Moving observer
hears lower frequency

Figure 12.26. The Doppler effect. (a) The listener hears a higher frequency from a sound source moving toward him and a lower frequency when it is moving away from him. (b) A listener hears a higher frequency when he is moving toward a sound source than when he is moving away from it. Here c is the velocity of sound in air, v is the velocity of the source in a and the listener in b, and f_0 is the frequency in the absence of motion.

where f_0 is the frequency of the initial ultrasonic wave, V is the velocity of the blood, v is the velocity of sound, and θ is the angle between V and v (Fig. 12.27). While there are other ways of measuring blood flow (see Chapter 11, p. 245), this method has the decided advantage of not requiring catheters in the artery or surgery to implant measuring devices.

The Doppler effect is also used to detect motion of the fetal heart, umbilical cord, and placenta in order to establish fetal life during the 12- to 20-week period of gestation when radiological and clinical signs are unreliable. When a continuous sound wave of frequency f_0 is incident upon the fetal heart, the reflected sound is shifted to frequencies slightly higher than f_0 when the fetal heart is moving toward the source of sound and slightly lower than f_0 when the fetal heart is moving away from it. Variations in the frequency give the fetal heart rate. Figure 12.28 shows the instrument arrangement for monitoring the fetal heart. The output can be audible or displayed on an oscilloscope.

Perhaps the most common use of the Doppler effect in obstetrics is in

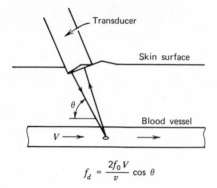

Transducer

Skin surface

θ

Blood vessel

$V \longrightarrow$ $\longrightarrow$

$$f_d = \frac{2f_0 V}{v} \cos \theta$$

Figure 12.27. Schematic arrangement for using the Doppler effect to measure the velocity of blood in a blood vessel. The transducer contains two crystals— one for transmitting the sound wave and one for receiving the echo. A continuous rather than a pulsed sound wave is used.

locating the point of entry of the umbilical cord (artery) into the placenta. This information is very useful if there is bleeding due to a misplaced placenta (*placenta praevia*) or if there is to be an intrauterine transfusion for Rh incompatibility. In one study, location prediction by Doppler ultrasound was checked by other methods and found to be more than 90% accurate. Figure 12.29 shows the frequency shifts from the Doppler effect for the placenta and other regions in the pregnant uterus. Care must be taken to avoid monitoring maternal arteries and veins, which have a much lower rate.

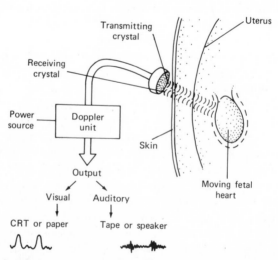

Transmitting crystal

Uterus

Receiving crystal

Power source

Doppler unit

Skin

Output

Visual

Auditory

Moving fetal heart

CRT or paper

Tape or speaker

Figure 12.28. Schematic diagram of the ultrasonic motion sensor for monitoring the fetal heart. (Adapted from Bishop, E.H.: "Obstetric Uses of the Ultrasonic Motion Sensor," *Amer. J. Obstet. Gynecol.*, 96, 1966, pp. 864–867.)

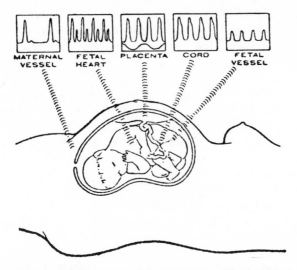

Figure 12.29. Various types of "sounds" obtained from the pregnant uterus. (From Bishop, E.H., "Obstetric Uses of the Ultrasonic Motion Sensor," *Amer. J. Obstet. Gynecol.,* 96, 1966, pp. 864–867.)

12.6. PHYSIOLOGICAL EFFECTS OF ULTRASOUND IN THERAPY

Various physical and chemical effects occur when ultrasonic waves pass through the body, and they can cause physiological effects. The magnitude of the physiological effects depends on the frequency and amplitude of the sound. At the very low intensity levels used for diagnostic work (0.01 W/cm² average power and 20 W/cm² peak power) no harmful effects have been observed. As the power is increased, ultrasound becomes useful in therapy. Ultrasound is used as a deep heating agent at continuous power levels of about 1 W/cm² and as a tissue-destroying agent at power levels of 10^3 W/cm².

The primary physical effects produced by ultrasound are temperature increase and pressure variations. The primary effect used for therapy is the temperature rise due to the absorption of acoustic energy in the tissue.

Ultrasound diathermy complements deep heating electromagnetic diathermy (see Chapter 11). The ultrasound is applied with a piezoelectric crystal transducer with a radiating surface of approximately 10 cm². A gel or mineral oil is used between the transducer and the skin for impedance matching. The probe should be calibrated and tuned in water before the treatment to determine the average intensity and total power output. A treatment plan may use intensities of several watts per square centimeter for periods of 3 to 10 min from once or twice a day to three times a week. Many times the applicator is moved slowly in a back-and-forth stroking motion to avoid forming "hot spots" in the tissue. When a joint is being treated the applicator may be moved over the entire external surface of the joint.

Ultrasound deposits its energy in the deeper muscles and tissues of the body while causing little temperature rise in the soft surface tissue layers. Research suggests that ultrasound is the most effective deep heater of bones and joints. Figure 12.30 shows the change in temperature inside a hip joint as a function of time for ultrasound and microwave diathermy.

Ultrasound diathermy is helpful in the treatment of joint disease and joint stiffness. It has also been used on joints that have calcium deposits; there are some indications that it aids in the removal of the deposits. It is not used on regions of the body such as the eyes and gonads where increased temperatures can cause damage.

Ultrasound waves differ completely from electromagnetic waves; they interact with tissue primarily by microscopic motion of the tissue particles. As a sound wave moves through tissue, the regions of compression and rarefaction cause pressure differences in adjacent regions of tissue. Stretching occurs in these regions; if the stretching exceeds the elastic limit of the tissue, tearing results. This is why an eardrum can be ruptured by a very intense sound source. In physical therapy the typical intensity is

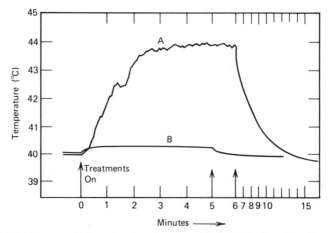

Figure 12.30.   Change in temperature inside a hip joint during (A) ultrasound diathermy and (B) microwave diathermy. The microwaves were turned off at 5 min and the ultrasound at 6 min as indicated by the arrows. (Adapted from J.F. Lehmann, J.A. McMillan, G.D. Brunner, and J.B. Blumberg, *Arch. Phys. Med. Rehabil.,* 40, 1959, p. 511. © 1959 *Arch. Phys. Med. Rehabil.*)

about 1 to 10 W/cm^2 and the frequency is about 1 MHz. Using Equation 12.1, we find that the amplitude of displacement A at 10 W/cm^2 in tissue is about 10^{-6} cm; the maximum pressure amplitude P_0 (Equation 12.2) is approximately 5 atm. Recall that the change from maximum to minimum pressure occurs in a distance of one-half the wavelength; for a 1 MHz wave in tissue, $\lambda/2 = 0.7$ mm. Thus there is a substantial pressure change over a short distance. A beam of ultrasound with an intensity of 35 W/cm^2 can produce pressure changes of approximately 10 atmospheres! At very high frequencies, the energy can be passed to the molecules so quickly that it is impossible for the molecules to disperse the energy to the surrounding tissue through vibrations. The molecules can gain sufficient energy to break their chemical bonds. Intense ultrasound waves can change water into H_2 and H_2O_2 and rupture DNA molecules.

Negative pressure in the tissue during rarefaction can cause dissolved gas to come out of solution and form bubbles. This forming of bubbles, called *cavitation*, can break molecular bonds between the gas and tissue. The collapse of the bubbles releases energy that can also break bonds. Free radicals produced during the breaking of bonds can lead to oxidation reactions.

At power levels of 10^3 W/cm^2 it is possible to selectively destroy tissue at a desired depth by using a focused ultrasound beam. Work on the brains of cats indicates that the mechanism for the destruction of tissue appears to be biochemical and not merely due to local heating.

One might conclude that an intense ultrasound wave would be an ideal agent for destroying cancer tissue. Some studies with ultrasound have shown that cancer cell destruction does occur in some regions of treated tumors; however, other cancer cells in these tumors sometimes show stimulated growth. This method obviously needs further study.

Ultrasound was at one time used successfully on patients suffering from Parkinson's disease. However, it was found that directing the focused sound to the correct region of the brain was difficult. Because of the possibility of complications due to improper aim, the ultrasound treatment is currently not being used.

Meniere's disease, a condition involving dizziness and hearing loss, has been treated with intense ultrasound with nearly 95% success. The ultrasound destroys tissues near the middle ear.

12.7. THE PRODUCTION OF SPEECH (PHONATION)

Normal speech sounds are produced by modulating an outward flow of air. For most sounds the lungs furnish the stream of air, which flows through the vocal folds (cords), sometimes called the glottis, and several vocal cavities and exits from the body through the mouth and to a slight degree through the nostrils (Fig. 12.31). The speech sounds produced in this way are called *voiced sounds*. Some sounds are produced in the oral portion of the vocal tract without the use of the vocal folds—these are called *unvoiced sounds*. Examples are *p, t, k, s, f, th,* and *ch.* The *p, t,* and *k* are plosive sounds; the *s, f,* and *th* are fricative sounds; and the *ch* is a combination of the two types. The unvoiced sounds involve air flow through constrictions or past edges formed by the tongue, teeth, lips, and

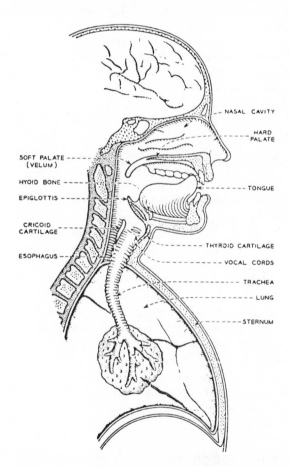

Figure 12.31.   Schematic of human vocal mechanism. (From J.L. Flanagan, *Speech Analysis, Synthesis and Perception,* 2nd ed., Springer-Verlag, Heidelberg, 1972, p. 10.)

palate. Try making some of these sounds and notice how you use your tongue, teeth, and lips in the process.

In this section we consider only the production of voiced sounds. Since the vocal mechanism is far too complex to examine in detail from the acoustic point of view, we use a model of the vocal tract (Fig. 12.32). In this model, the sound is produced at the vocal folds and is selectively modified or filtered by three cavities. (This type of model is sometimes called a source-filter model.)

The vocal folds are located within the larynx, or adam's apple, inside the trachea, or windpipe. Figure 12.33 shows the vocal folds as viewed from above, and Fig. 12.34 shows a vertical cross-section of the larynx as seen from the front. During normal respiration the folds are widely separated, forming a large triangular opening (Fig. 12.33*a*). In the production of the vocal sounds the vocal folds are drawn close together by muscles (Fig. 12.33*b*), the air in the lungs is exhaled, the pressure below the vocal folds rises, and the closed folds are forced apart. The resulting rapid upward flow of air causes a decrease in pressure between the folds due to the Bernoulli effect (see Section 8.6). The decrease in pressure, along with the elastic forces in the tissues, causes the folds to move together, par-

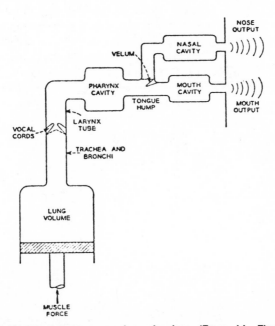

Figure 12.32. Model of human vocal mechanism. (From J.L. Flanagan, *Speech Analysis, Synthesis and Perception,* 2nd ed., Springer-Verlag, Heidelberg, 1972, p. 24.)

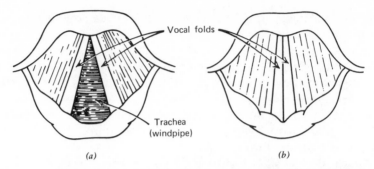

(a)   (b)

Figure 12.33. Sketch of the vocal folds as seen with a mirror held in the back of the throat. (a) The normal opening during inspiration. The dark area is the windpipe below the folds. (b) During phonation (speech). The vocal folds are drawn close together and vibrate as air is forced between them.

tially blocking the passage and thus reducing the air velocity. This reduced air velocity increases the pressure below the folds and causes the process to begin again.

The fundamental frequency of the resulting complex vibration depends on the mass and tension of the vocal folds. Men, who have longer and heavier vocal folds than women, have a typical fundamental frequency of about 125 Hz; the typical fundamental frequency for women is about one octave higher, or 250 Hz. The lowest frequency that can be produced by a bass singer is about 64 Hz (low C), and the highest frequency that a soprano can produce is about 2048 Hz (five octaves above low C).

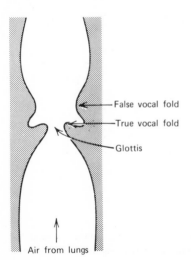

Figure 12.34. A cross-section view of the larynx as seen from the front.

The glottal sound wave passes through several vocal cavities—the pharyngeal (throat), oral, and nasal cavities—that further change the sound of the wave that is emitted. The throat and nasal cavities are pretty well fixed for each individual and to a large extent determine the sound of the voice. They cannot be changed much voluntarily, but the swelling of tissues due to a head cold will alter them and cause a change in the voice. The oral cavity changes shape through the movement of the tongue, lower jaw, soft palate, and cheeks to determine the specific sounds that are emitted. The tongue, palate, and cheeks in particular select the desired sounds out of the complicated periodic wave. You can feel this selection when speaking the vowel sounds and some of the consonant sounds.

Figure 12.35a shows schematically the air velocity in the glottal region. Figure 12.35b shows the vocal tract modification, and Fig. 12.35c shows the resultant radiative sound wave. It is possible to decompose the complex glottal wave into frequency components and to determine the amplitudes of these components by a method called Fourier analysis (Fig. 12.35a'). Figure 12.35b' shows the sound transmission characteristics of

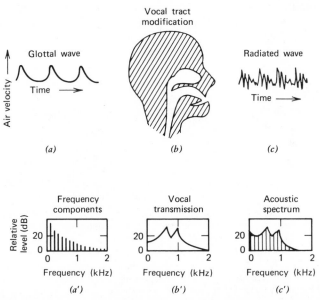

Figure 12.35. Two ways of viewing production of one speech sound. The glottal sound wave (a) is modified by the vocal tract (b) to produce a radiative wave (c). The amplitude of the frequency components of the glottal wave can be obtained (a'). They are modified by a function that represents the characteristics of the vocal tract (b') to produce the acoustic spectrum of the radiated wave (c'). (Adapted from Gunnar Fant, *Acoustic Theory of Speech Production*, © 1970 Mouton, The Hague/Paris, p. 19.)

the vocal tract. The action of the transmission characteristics of the vocal tract on the frequency components of the glottal sound produces the sound spectrum shown in Fig. 12.35c'. Human speech is composed from a rich variety of glottal sounds and vocal tract shapes that are timed by the central nervous system.

When the sentence "Joe took Father's workbench out" is spoken in a normal voice, the kinetic and potential energy in the resultant sound is 3×10^{-5} to 4×10^{-5} J. This is a very tiny amount of energy. The time needed to say the sentence is about 2 sec, and the average power is about 10 to 20 μW. A person could talk continuously for a year and not produce the sound energy equivalent to the heat energy needed to boil a cup of water. We can hear the spoken word even though the energy is small because of the great sensitivity of the ear (see Chapter 13, p. 297). The vowel sounds contain much more power than the consonant sounds. Thus vowel sounds are easier to hear and understand than consonant sounds. In one study the relative power between the vowel sound in *aw*l and the consonant sound *th* in *th*in was found to be 680:1. This corresponds to 29 dB!

We normally think of the sound produced by belching as having no practical value. However, patients who have had their larynx removed can be taught to swallow air and to use controlled belching as an artificial larynx to produce voice sounds.

BIBLIOGRAPHY

Denes, P. B., and E. N. Pinson, *The Speech Chain: The Physics and Biology of Spoken Language,* Doubleday Anchor, Garden City, N.Y., 1963.

Dunn, F., P. D. Edmonds, and W. F. Fry, "Absorption and Dispersion of Ultrasound in Biological Media," in H. P. Schwan (Ed.), *Biological Engineering,* McGraw-Hill, New York, 1969, pp. 205-332.

Feigenbaum, H., *Echocardiography,* Lea and Febiger, Philadelphia, 1972.

Flanagan, J. L., *Speech Analysis, Synthesis and Perception,* Academic, New York, 1965.

Fry, E. (Ed.), *Ultrasound in Biology and Medicine,* A symposium sponsored by the Bioacoustics Laboratory of the University of Illinois and the Physiology Branch of the Office of Naval Research, held at Robert Allerton Park, Monticello, Ill., June 20-22, 1955, American Institute of Biological Science, Washington, D.C., 1957.

Goldberg, R. E., and L. K. Sarin (Eds.), *Ultrasonics in Ophthalmology: Diagnostic and Therapeutic Applications,* Saunders, Philadelphia, 1967.

McDicken, W. N., *Diagnostic Ultrasonics: Principles and Use of Instruments,* Wiley, New York, 1976.

Minifie, F. D., T. J. Hixon, and F. Williams, *Normal Aspects of Speech, Hearing, and Language,* Prentice-Hall, Englewood Cliffs, N.J., 1973.

Weaver, G., and M. Lawrence, *Physiological Acoustics,* Princeton U. P., Princeton, N.J., 1954.

Went, J. M. van, *Ultrasonic and Ultrashort Waves in Medicine,* Elsevier, Amsterdam, 1954.

Wood, A. B., *A Textbook of Sound,* Bell, London, 1955.
Zemlin, W. R., *Speech and Hearing Science: Anatomy and Physiology,* Prentice-Hall, Englewood Cliffs, N.J., 1968.

REVIEW QUESTIONS

1. What is infrasound? Ultrasound?
2. What is the wavelength of a 1000 Hz sound wave in water if its velocity in water is 1480 m/sec?
3. What is the acoustic impedance Z?
4. What is the maximum displacement in air for a 1000 Hz sound wave with an intensity of 50 dB or 10^{-7} W/m²?
5. If your loudest shout is 1000 times more intense than your normal speaking voice, what is the dB difference between them?
6. For hearing tests, what is the reference sound intensity? The reference sound pressure?
7. Calculate the relative amplitudes of the reflected and transmitted sound waves from fat to muscle using the acoustic impedances given in Table 12.1.
8. What is the attenuation of sound intensity in 15 cm of brain tissue?
9. What is the difference between percussion and auscultation?
10. What factors affect the selection of the diameter and the length of the tube for a stethoscope?
11. What is a transducer?
12. What is the difference between an ultrasonic A scan and an ultrasonic B scan?
13. What is the advantage of an ultrasonic gray-scale display over a leading-edge display?
14. How is the ultrasound Doppler effect used to monitor fetal heart rate?
15. What power level of ultrasound is used for deep heating therapy?
16. List five unvoiced sounds.
17. What is the typical fundamental frequency of the vocal folds of men? Of women?
18. What is the typical average power of the human voice?
19. Why are vowel sounds easier to hear than consonant sounds?

CHAPTER 13

Physics of the Ear and Hearing

Speech and hearing are the most important means by which we communicate with our fellow man. Through hearing we receive speech sounds from others and also listen to ourselves! In some ways it is more of a handicap to be born "stone deaf" than to be born blind. Any child who cannot hear the sounds from his own vocal cords cannot learn to talk without special training. In earlier times a child deaf from birth was also mute, or *dumb,* and since so much of our learning takes place through hearing, he often was not educated. (This might be the origin of the use of *dumb* to indicate ignorant or stupid.) It was not until the sixteenth century that people first realized that the inability of a deaf child to talk was fundamentally related to his deafness. In the early nineteenth century special schools were established for deaf-mutes. While deaf people can now be taught to speak, their voices usually sound abnormal since they have no easy way to compare them to the voice sounds produced by other people.

If a sound is loud enough, it can be "heard" by a deaf person through the sense of touch; for example, he may feel vibrations of the exposed hairs on his body and thus "hear" the loud sound through the nerve sensors at the roots of the hairs. We discuss shortly how the ear uses a sophisticated refinement of this technique to hear sounds that are billions of times weaker.

The sense of hearing is in some ways more remarkable than the sense of vision. We can hear a range of sound intensities of over a million million (10^{12}), or 100 times greater than the range of light intensities the eye can handle. The ear can hear frequencies that vary by a factor of 1000, while the frequencies of light that the eye can detect vary by only a factor of 2.

The sense of hearing involves (1) the mechanical system that stimulates the hair cells in the cochlea; (2) the sensors that produce the action potentials in the auditory nerves; and (3) the auditory cortex, the part of the brain that decodes and interprets the signals from the auditory nerves. Deafness or hearing loss results if any of these parts malfunctions. While they all involve physics, we know much more about the physics of the first part than about the physics of the other parts. In this chapter we deal with the sense of hearing only up to the auditory nerve.

The ear is a cleverly designed converter of very weak mechanical waves in air into electrical pulses in the auditory nerve. Figure 13.1 shows most of the structures of the ear that are involved with hearing. The ear is usually thought of as divided into three areas: the outer ear, the middle ear, and the inner ear. What we commonly call the ear (the appendage we use to help hold up our glasses) has no essential role in hearing. The *outer ear* consists of the ear canal, which terminates at the eardrum (tympanic membrane). The *middle ear* includes the three small bones (ossicles) and an opening to the mouth (Eustachian tube). The *inner ear* consists of the fluid-filled, spiral-shaped cochlea containing the organ of Corti. Hair cells in the organ of Corti convert vibrations of sound waves hitting the eardrum into coded nerve pulses that inform the brain of these sound waves.

One of the first medical physicists to study the physics of the ear and hearing was Hermann von Helmholtz (1821–1894). He developed the first

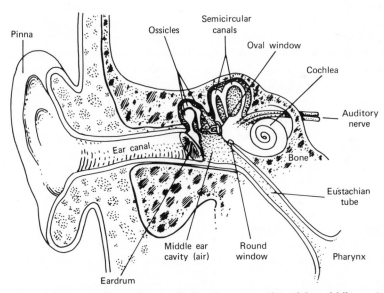

Figure 13.1. Cross-section of the ear. Notice the connection of the middle ear to the pharynx via the Eustachian tube.

modern theory of how the ear works. His work was expanded and extended by Georg Von Bekesey (1900–1970), a communications engineer who became interested in the function of the ear as part of the communications system. Von Bekesey received a Nobel prize in 1961 for his contributions to the understanding of the ear.

The medical specialists concerned with the ear and hearing are the *otologist,* an M.D. who specializes in diseases of the ear and ear surgery; the *otorhinolaryngologist,* an M.D. who specializes in diseases of the ear, nose, and throat (also called an ENT specialist); and the *audiologist,* a non-M.D. who specializes in measuring hearing response, diagnosing hearing disorders through hearing tests, and rehabilitating those with varying degrees of hearing loss.

13.1. THE OUTER EAR

The outer ear does not refer, as you might think, to the visible part of the ear, which in medical jargon is called the external auricle or pinna. The outer ear is the external auditory canal, which terminates at the eardrum. The outer structure, or auricle, is the least important part of the hearing system; it aids only slightly in funneling sound waves into the canal and can be completely removed with no noticeable loss in hearing, although its removal will not help anyone's appearance. Those of us who have control over the movement of our auricles (who can wiggle our ears) find that its primary use is to entertain children.* In some animals the auricle does play a role in collecting sound energy and concentrating it on the eardrum. Man can often get a 6 to 8 dB gain by cupping his hand behind his ear to make up for nature's deficiencies in this regard. The large ears of the elephant and many desert animals also serve an important function in the loss of body heat.

The external auditory canal, besides being a storage place for ear wax, serves to increase the ear's sensitivity in the region of 3000 to 4000 Hz. The canal is about 2.5 cm long and the diameter of a pencil. You can think of the canal as an organ pipe closed at one end (length = $\lambda/4$) with a resonant frequency of about 3300 Hz (λ = 10 cm). You will notice that the sensitivity of the ear is best in this region (Fig. 13.2).

The eardrum, or tympanic membrane, is about 0.1 mm thick (paper thin) and has an area of about 65 mm^2. It couples the vibrations in the air to the small bones in the middle ear. Because of the off-center attachment

---

*One of the authors (JRC) can wiggle each ear independently. Some people think it is his most outstanding accomplishment!

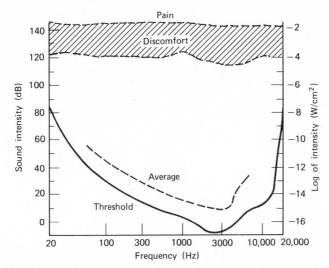

Figure 13.2. The sensitivity of the ear. The solid curve is the threshold of hearing for a young person with good hearing. Zero decibels occurs at 1000 Hz. The "average" curve is the average threshold for all people, young and old. Both axes—horizontal and vertical—are logarithmic scales.

of the malleus (Fig. 13.3), the eardrum does not vibrate symmetrically like a drumhead. The motion of the eardrum was studied extensively by Von Bekesey, who used cadaver ears. Recent studies using a sophisticated nuclear physics technique (the Mössbauer effect) to study the motion of the eardrum in the living ear indicate that many of the values Von Bekesey obtained were in error because of changes in tissue elasticity after death. However, it is clear that the actual movement of the eardrum is exceedingly small since it must be less than the movement of the air molecules in the sound wave. This movement at the threshold of hearing at 3000 Hz is about 10^{-9} cm—less than the diameter of a hydrogen atom! At the threshold of hearing at the lowest frequencies that we can hear (~20 Hz), the motion of the eardrum may be as large as 10^{-5} cm. This is still less than the wavelength of visible light.

It is possible for sound pressures above 160 dB to rupture the eardrum. A ruptured eardrum normally heals just as other living tissue does.

13.2. THE MIDDLE EAR

The dominant features of the middle ear are the three small bones (ossicles), which are shown in Fig. 13.1 and in more detail in Fig. 13.3. These

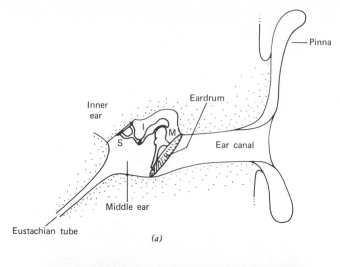

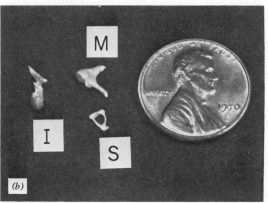

Figure 13.3.   The ossicles of the middle ear. (*a*) A schematic cross-section showing the stapes S, the malleus M, and the incus I. (*b*) Photographed next to a penny. Note the small size of the ossicles.

bones are full adult size before birth. (The fetus can hear while it is still in the womb.) The ossicles play an important role in matching the impedance of the sound waves at the eardrum to the liquid-filled chambers of the inner ear. The ossicles are named after the objects they resemble: the malleus (hammer), the incus (anvil), and the stapes (stirrup). They are arranged so that they efficiently transmit vibrations from the eardrum to the inner ear. They transmit poorly vibrations in the skull—even the large vibrations from the vocal cords.* You hear your own voice primarily by transmission of sound through the air. Try plugging both your ears and listen to the reduction in your sound volume. (It is best to do this while you are alone and not in the library.)

The ossicles amplify the pressure of the sound waves at the entrance to the inner ear. The lever action of the ossicles is such that the motion of the plate of the stapes at the oval window of the inner ear is about 0.7 that of the malleus at the eardrum. Thus the lever action amplifies the force by a factor of about 1.3. A much larger gain in pressure is obtained by the piston action shown schematically in Fig. 13.4. The eardrum, which acts like a large piston, is mechanically coupled to the stapes, which acts like a small piston at the entrance to the inner ear. The ratio of the effective area of the eardrum to that of the base of the stapes is about 15 to 1. This gain combined with the lever gain of 1.3 results in a total gain of about 20!

You have learned in Chapter 12 (we hope) that when a sound wave encounters a very different medium most of the sound energy is reflected. A sound wave in air striking a wall is about 99.9% reflected; that is, only 0.1% or 1 part in 1000 is transmitted! An attenuation of 1000 amounts to a sound loss of (10 log 1000) or 30 dB! The ear is designed to reduce this loss by impedance matching. In the ear, the factors that affect the impedance are primarily the springiness of the eardrum and its mass. The impedance in the ear is fairly well matched from about 400 to 4000 Hz; below 400 Hz the "spring" is too stiff and above 4000 Hz the mass of the eardrum is too great. The middle ear aids the impedance match by amplifying the pressure by the lever and piston action described above.

The ossicles and their sensory ligaments play an important role in protecting the ear against loud sounds. A loud sound causes the muscles in the middle ear to pull sideways on the ossicles and reduce the sound intensity reaching the inner ear. A decrease of 15 dB is possible by this means. However, it takes about 15 msec or longer for these muscles to react, and damage may be done in this brief period. Persons living or working in an environment of loud sounds permanently lose some of their hearing sensitivity. Noise pollution is not only unpleasant, it can result in permanent physiological damage to the hearing mechanism. The sound levels of some common sounds are given in Table 12.2.

Another structure in the middle ear plays a protective role—the Eustachian tube, which leads down toward the mouth. It is considerably smaller than indicated in Fig. 13.1 and is normally closed rather than open

---

*You can easily feel the vocal cord vibrations with your fingertips on your larynx, or adam's apple. This phenomenon is used to help teach deaf children to speak.

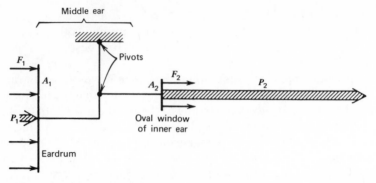

**Figure 13.4. A schematic representation of the lever and piston action of the ear.
The lever action of the ossicles increases the force by 30%. The ratio of the areas of
the large eardrum and small oval window (A_1/A_2) increases the pressure by a factor
of 15. These factors produce a pressure P_2 at the oval window that is about 20 times
higher than the sound pressure P_1 at the eardrum.**

as shown. The middle ear contains air, and it is important for the air
pressures on both sides of the thin eardrum to be essentially the same; the
Eustachian tube serves to equalize the pressures. Air in the middle ear is
gradually absorbed into the tissues, lowering the pressure on the inner
side of the eardrum. The movement of the muscles in the face during
swallowing, yawning, or chewing will usually cause a momentary opening
of the Eustachian tube that equalizes the pressure in the middle ear with
the atmospheric pressure. Pressure differences are usually noticed in
situations in which the outside pressure changes rapidly in a short period
of time, such as when flying, riding in hilly country, or riding in the
elevator of a tall building. You are usually more aware of it upon descent,
when the external pressure is increasing. When for some reason the
Eustachian tube does not open, the resulting pressure difference deflects
the eardrum inward and decreases the sensitivity of the ear; at about 60
mm Hg across the eardrum, the pressure difference causes pain. Common
reasons for the failure of this equalizing system are the blockage of the
Eustachian tube by the viscous fluids from a head cold and the swelling of
tissues around the entrance to the tube.

13.3.   THE INNER EAR

The inner ear, hidden deep within the hard bone of the skull, is man's
best-protected sense organ. The inner ear consists of a small spiral-
shaped, fluid-filled structure called the *cochlea*. The ossicles of the middle

ear communicate with the cochlea via a flexible membrane (the oval window); the stapes transmits its pressure variations of incoming sound waves across this membrane to the cochlea. The cochlea communicates with the brain via the auditory nerve—a bundle of about 8000 conductors that inform the brain via coded electrical pulses which parts of the cochlea are being stimulated by incoming sound waves. The auditory nerve provides information on both the frequency and the intensity of the sounds that we hear. Many aspects of the inner ear are still being studied. The exact mechanism by which the nerve pulses are produced remains a mystery.

The cochlea is about the size of the tip of the little finger. If its spiral were straightened out, the cochlea would be about 3 cm (~1.25 in.) long. It is divided into three small fluid-filled chambers that run its full length. The oval window is on the end of the *vestibular chamber,* the middle chamber is the *cochlear duct,* and the third chamber is the *tympanic chamber* (Fig. 13.5). The vestibular and tympanic chambers are intercon-

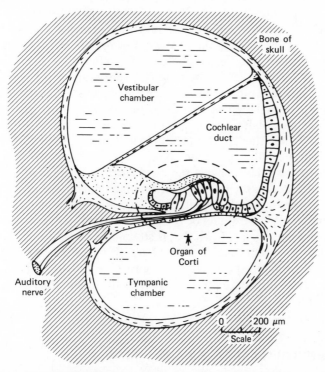

Figure 13.5.   The chambers of the cochlea. The organ of Corti inside the dashed line is shown in more detail in Fig. 13.6.

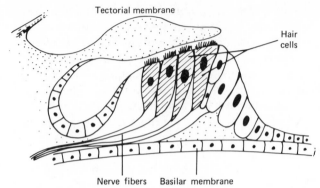

Figure 13.6.   Simplified schematic of the organ of Corti showing the hair cells.

nected at the tip of the spiral. Pressure produced at the oval window by the stapes is transmitted via the vestibular chamber to the end of the spiral and then returns via the tympanic chamber. Since fluid is almost incompressible, the cochlea needs a "relief valve"; the flexible round window at the end of the tympanic chamber serves this purpose (Fig. 13.1).

A sound wave entering at the oval window produces a wave-like ripple in the basilar membrane of the cochlear duct (Fig. 13.6). This duct contains the sensors that convert the sound into nerve signals. The motions of this membrane are about 10 times smaller in amplitude than the motions of the eardrum. Stimulation of nerves in the cochlear duct near the oval window indicates high-frequency sounds. Low-frequency sounds cause "large" motions in the basilar membrane and stimulation of nerves in the cochlear duct near the tip of the spiral.

The transducers that convert the mechanical vibrations into electrical signals are located in the bases of the fine hair cells in the organ of Corti (Fig. 13.6). Apparently the small shear forces on these hair cells induce nerve impulses. When a sound of 10,000 Hz is heard, the nerves located in the portion of the organ of Corti that is stimulated do not send a 10,000 Hz signal to the brain, but rather send a series of pulses that indicates which portion of the audible spectrum is being received. Below about 1000 Hz the frequency of the nerve pulses is synchronized with that of sinusoidal sound waves.

13.4. SENSITIVITY OF THE EARS

The ear is not uniformly sensitive over the entire hearing range. Its best sensitivity is in the region of 2 to 5 kHz (Fig. 13.2). The lower curve in Fig.

13.2 (threshold) shows the average values for a young person with good hearing. The *average* line shows the levels at which half the people tested will detect the sounds. Notice that even a good ear needs about 30 dB more intensity to detect a sound at 100 Hz than to detect one at 1000 Hz.

Sensitivity changes with age. The highest frequency you can hear will decrease as you get older, and the level of sounds will need to be greater for you to detect them. A person 45 years old typically cannot hear frequencies above 12 kHz and needs about 10 dB more intensity than he did at age 20 to be able to hear a 4000 Hz note. A 25 dB loss in sensitivity in the frequencies above 2000 Hz usually has occurred by age 65. This loss is not serious for most activities. Hearing deteriorates more rapidly if the ears are subjected to continuous loud sounds. Some young people who play in rock bands have had serious hearing losses. Factory workers who work under very noisy conditions have also shown measurable losses in hearing.

The property of sound we call *loudness* is a mental response to the physical property called *intensity*. The loudness of a sound is roughly proportional to the logarithm of its intensity and this effectively compresses the huge range of sound intensities to which the ear responds ($\sim 10^{12}$: 1). In addition, the loudness of a sound depends strongly on its frequency. A sound of 30 Hz that is barely audible has the same loudness as a barely audible sound of 4000 Hz, even though their intensities differ by a factor of about 1,000,000, or 60 dB. A special unit has been designed for loudness—the *phon*. One phon is the loudness of a 1 dB, 1000 Hz sound; 10 phons is the loudness of a 10 dB, 1000 Hz sound; and so forth. The loudness of a sound at another frequency is obtained by adjusting the intensity until it appears as loud as the known intensity 1000 Hz sound. Figure 13.7 shows typical curves of equal loudness at the threshold of hearing and at 40 and 60 phons. These curves vary for different individuals but for people with normal hearing the curves have one characteristic in common: as the loudness increases the curves become flat. The threshold of feeling is about 120 dB at all frequencies.

The frequencies of most importance to us are those of the human voice. The shaded area in Fig. 13.7 indicates the general range of frequencies and sound levels of ordinary conversational speech. You can see that the ear is not optimized for the speech frequencies. However, it is possible to have a hearing loss of 40 dB and still hear most conversation. (In Section 13.6 we discuss the physics of deafness and hearing aids.)

If the ear were as sensitive at low frequencies as it is in the 3000 Hz region, we would be aware of many physiological noises, such as blood flow in the arteries in the head, movement of the joints, and possibly the small variations of pressure on the eardrum due to random motion of air

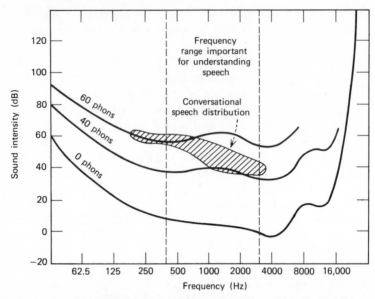

Figure 13.7. Curves of equal loudness at the threshold of hearing and at 40 and 60 phons. (Adapted from *Hearing and Deafness*, Third Edition, by Hallowell Davis and S. Richard Silverman, p. 28. Copyright 1947, © 1960, 1970 by Holt, Rinehart and Winston, Inc. By permission of Holt, Rinehart and Winston, Inc.)

molecules (Brownian motion). If you go into a special soundproof room used for testing hearing, you will be impressed by how many internal body sounds you hear. Most of these sounds are transmitted through bone conduction to the inner ear. These sounds are poorly detected by the ear, which is optimized for sounds coming via the eardrum. In general, a sound must be about 40 dB more intense to be heard by bone conduction than to be heard by air conduction.

13.5. TESTING YOUR HEARING

If you have a hearing problem and consult an "ear doctor"—an otologist or otolaryngologist—he or she may send you to an audiologist to have your hearing tested. If you have a hearing loss, the audiologist will be able to determine whether it is curable; if it is not, your ability to use a hearing aid will be assessed.

The tests are normally done in a specially constructed soundproof testing room. Each ear is tested separately; test sounds can be sent to either ear through a comfortable headset. The subject is asked to give a sign when he hears the test sound. Selected frequencies from 250 to 8000 Hz are used. At each frequency the operator raises and lowers the volume until a consistent hearing threshold is obtained.

The hearing thresholds are then plotted on a chart and can be compared to normal hearing thresholds (Fig. 13.8a). The normal hearing threshold at each frequency is taken to be 0 dB. The chart may show a general loss in one or both ears. Usually a hearing loss is not uniform over all frequencies. Figure 13.8b shows the hearing threshold of a person with imperfect hearing. Notice the sharp hearing loss in both ears at about 4 kHz. In this

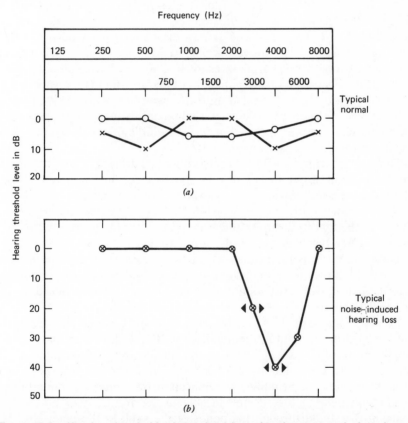

Figure 13.8. Hearing thresholds determined by a hearing test and plotted on a standard hearing chart. The Os represent the threshold for air conduction in the right ear; the Xs are for the left ear. (a) A typical response of a person with normal hearing. (b) A typical noise-induced hearing loss in the region of 4000 Hz. The black triangles indicate thresholds for bone conduction.

case, the loss was due to nerve damage of that frequency portion of the cochlea.

13.6. DEAFNESS AND HEARING AIDS

In 1972 it was estimated that 13 million persons in the United States were either deaf or hard of hearing. The frequency range most important for understanding conversational speech is from about 300 to 3000 Hz (Fig. 13.7). A person who is "deaf" above 4000 Hz but who has normal hearing in the speech frequencies is not considered deaf or even hard of hearing. However, he should not spend a lot of money on good hi-fi equipment. Hearing handicaps are classified according to the average hearing threshold at 500, 1000, and 2000 Hz in the better ear. A person with a hearing threshold 30 dB above normal would probably not have hearing problems. People with hearing thresholds of 90 dB are considered deaf or stone deaf. About 1% of the population have thresholds for speech frequencies greater than 55 dB and must use hearing aids to be able to hear loud speech. About 1.7% have a slight hearing handicap; they have problems with normal speech but have no difficulty with loud speech. Hearing problems increase with age.

The average sound level of speech is about 60 dB (Fig. 13.7). We adjust the sound level of our speech unconsciously according to the noise level of our surroundings. Speech sound levels in a quiet room may be as low as 45 dB; at a noisy party they may be 90 dB. A person with a hearing loss of 45 dB in the 500 to 2000 Hz range may do all right (hearing-wise) at a cocktail party but hear very little of a sermon in church the next day.

There are two common causes of reduced hearing: *conduction hearing loss*, in which the sound vibrations do not reach the inner ear, and *nerve hearing loss*, in which the sound reaches the inner ear but no nerve signals are sent to the brain.

Conduction hearing loss may be temporary due to a plug of wax blocking the eardrum or fluid in the middle ear. It may, however, be due to a solidification of the small bones in the middle ear. This condition can sometimes be corrected by an operation in which the stapes, which pushes on the oval window, is replaced with a piece of plastic. If a conduction hearing loss is not curable, a hearing aid can be used to transmit the sound through the bones of the skull to the inner ear.

A nerve hearing loss may affect only a narrow band of frequencies or it may affect all frequencies. At present there is no known cure or aid for nerve hearing losses.

The hearing threshold that requires a person to use a hearing aid is quite

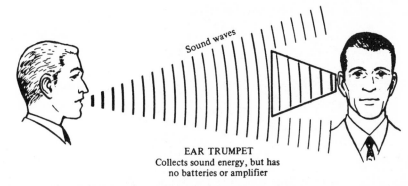

EAR TRUMPET
Collects sound energy, but has
no batteries or amplifier

Figure 13.9. The principle of the ear trumpet. (From *Hearing and Deafness*, Third Edition, by Hallowell Davis and S. Richard Silverman, p. 281. Copyright 1947, © 1960, 1970 by Holt, Rinehart and Winston, Inc. Reprinted by permission of Holt, Rinehart and Winston, Inc.)

variable. Some people lip-read to help them understand speech. The simplest hearing aid, which is quite effective if your hearing loss is not large, is to cup your hand behind your ear. This reflects about 6 to 8 dB of additional sound into your ear canal. In addition, you will usually gain another 10 dB when the speaker notices you and raises his voice.

The earliest artificial hearing aid was the ear trumpet (Fig. 13.9). The large opening catches the sound waves and the funnel concentrates the energy at the ear. The size and shape of the ear trumpet affect its efficiency. The auditory canal has a resonance in the 2000 to 4000 Hz region, at the upper end of the useful speech frequencies, while an ear trumpet has a resonance at the speech frequencies (Fig. 13.10). A good ear trumpet will lower the hearing threshold by 10 to 15 dB. Ear trumpets were never common, probably due to the human tendency to hide handicaps.

Electronic hearing aids are in common use today. Early electronic hearing aids were bulky, and the batteries wore out rapidly. The development of transistor amplifiers and miniaturized electrical components led to the development of hearing aids that can be concealed behind the ear or in the frames of glasses (Fig. 13.11). An electronic hearing aid is like a small public address system. It consists of a microphone to detect sound, an amplifier to increase its energy, and a loudspeaker to deliver the increased energy to the ear (Fig. 13.12). It is possible to obtain an amplification of 90 dB or an increase of 1 billion in sound level. Even though a deaf person may have a hearing threshold of 70 to 80 dB, his discomfort threshold is the same as that for a person with normal hearing, or about

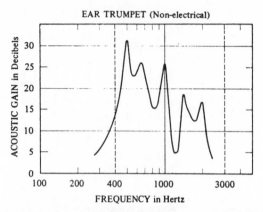

EAR TRUMPET (Non-electrical)

Figure 13.10. The frequency characteristics of the ear trumpet. (From *Hearing and Deafness*, Third Edition, by Hallowell Davis and S. Richard Silverman, p. 281. Copyright 1947, © 1960, 1970 by Holt, Rinehart and Winston, Inc. Reprinted by permission of Holt, Rinehart and Winston, Inc.)

Figure 13.11. Miniaturized electrical components allowed the development of hearing aids to be worn inside the frames of glasses. (Courtesy Zenith Hearing Instruments Corporation.)

308

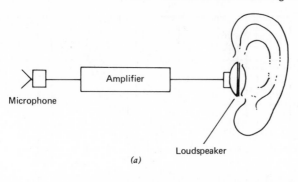

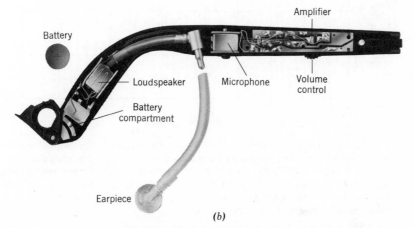

Figure 13.12. (a) A hearing aid is really a small public address system with a microphone to pick up the sounds, an amplifier to increase the energy of the sound, and a loudspeaker to transmit the sounds to the subject's ear. Because the sound is directed into the auditory canal, very little power is needed. Remember that 90 dB is only 10^{-7} W/cm². (b) Components of a hearing aid built into the frames of glasses. (Courtesy Zenith Hearing Instruments Corporation.)

100 to 120 dB. Thus there is a practical upper limit on the sound output from an electronic hearing aid.

Hearing aids cannot return hearing to normal. They can only help compensate for the hearing loss. For example, an abrupt hearing loss above 3000 Hz cannot be completely corrected with a hearing aid. Most hearing aids have a tone control that permits the wearer to adjust the frequency response, but its range of use is very limited (Fig. 13.13).

The frequency response of a hearing aid would be considered terrible by a hi-fi enthusiast. However, the hearing aid does increase the sound

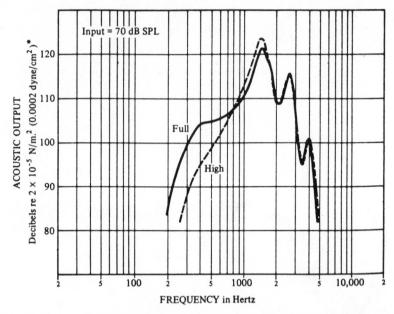

Figure 13.13. The frequency response of a hearing aid. The user can control the frequency response to some extent with the tone control. "Full" indicates maximum bass response and "High" indicates maximum treble response. (From *Hearing and Deafness*, Third Edition, by Hallowell Davis and S. Richard Silverman, p. 300. Copyright 1947, © 1960, 1970 by Holt, Rinehart and Winston, Inc. Reprinted by permission of Holt, Rinehart and Winston, Inc.)

level of the speech frequencies to above the hearing threshold. The good quality of a hi-fi unit is obtained at a considerable cost in space and weight. Shrinking a hi-fi so it would fit into the ear would put a limit on the uniformity of its frequency response.

BIBLIOGRAPHY

Ackerman, E., *Biophysical Science*, Prentice-Hall, Englewood Cliffs, N.J., 1962, Chapters 1 and 6.

Davis, H., and S. R. Silverman, *Hearing and Deafness*, 3rd ed., Holt, Rinehart and Winston, New York, 1970.

Littler, T. S., *The Physics of the Ear*, Macmillan, New York, 1965.

Martin, F. N., *Introduction to Audiology*, Prentice-Hall, Englewood Cliffs, N.J., 1975.

Minifie, F. D., T. J. Hixon, and F. Williams (Eds.), *Normal Aspects of Speech, Hearing, and Language*, Prentice-Hall, Englewood Cliffs, N.J., 1973.

Stevens, S. S., F. Warshofsky, and the Editors of Life, *Sound and Hearing*, Time, New York, 1965.

Towe, A. L., in T. C. Ruch and H. D. Patton (Eds.), *Physiology and Biophysics,* 19th ed., Saunders, Philadelphia, 1965, Chapter 18.

REVIEW QUESTIONS

1. Over what range of sound intensities can the normal ear hear?
2. What are the three main components involved in the sense of hearing?
3. What are the three areas of the ear?
4. What is the difference between an otologist and an audiologist?
5. What is the approximate movement of the eardrum at the threshold of hearing at 3000 Hz?
6. The pressure at the oval window is 20 times greater than that at the eardrum. Calculate this pressure difference in decibels.
7. Over what frequency range is the impedance of the ear fairly well matched to the acoustic impedance of air?
8. What is the main purpose of the round window?
9. What end of the cochlea contains the sensors for high-frequency sounds?
10. From Fig. 13.2, find:
    (a) the frequency at which the ear is most sensitive.
    (b) the minimum intensity (in watts per square centimeter) that a person can hear at 100 Hz.
11. A static pressure of 125 mm Hg across the eardrum can cause it to rupture. How does this compare to the sound pressure from a 160 dB sound that can also cause the eardrum to rupture?
12. What is the average sound level of speech?
13. What are two common causes of reduced hearing?
14. How much gain in decibels results from cupping your hand behind your ear?

CHAPTER 14

Light in Medicine

In this chapter we discuss the medical applications of light in diagnosis and therapy and also the hazards of light. We consider visible light, infrared (IR) light, and ultraviolet (UV) light. We discuss the physics of the eye and vision in Chapter 15.

Even though man is now very efficient at making artificial light, the sun is still the major source of light in the world. The sun is both beneficial and hazardous to our health. The spectrum of light from the sun is shown in Fig. 14.1. The light visible to the human eye is shown shaded. Note that the eye is most efficient in the wavelengths corresponding to the maximum output from the sun.

Light has some interesting properties, many of which are used in medicine:

1. The speed of light changes when it goes from one material into another. The ratio of the speed of light in a vacuum to its speed in a given material is called the *index of refraction*. If a light beam meets a new material at an angle other than perpendicular, it bends, or is refracted. This property permits light to be focused and is the reason we can read and see objects clearly. We discuss this property in Chapter 15.
2. Light behaves both as a wave and as a particle. As a wave it produces interference and diffraction, which are of minor importance in medicine. As a particle it can be absorbed by a single molecule. When a light photon is absorbed its energy is used in various ways. It can cause a chemical change in the molecule that in turn can cause an electrical change. This is basically what happens when a light photon is absorbed in one of the sensitive cells of the retina (the light-sensitive part of the eye). The chemical change in a particular point of the retina

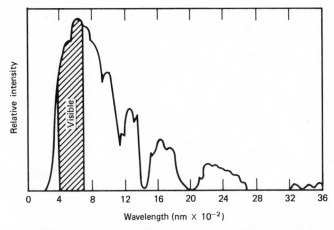

Wavelength (nm × 10^{-2})

Figure 14.1. Relative intensity of solar energy of different wavelengths at the earth's surface. The dips in the spectrum are due to selective absorption by components in the atmosphere.

triggers an electrical signal to the brain to inform it that a light photon has been absorbed at that point.

3. When light is absorbed, its energy generally appears as heat. This property is the basis for the use in medicine of IR light to heat tissues. Also, the heat produced by laser beams is used to "weld" a detached retina to the back of the eyeball and to coagulate small blood vessels in the retina.

4. Sometimes when a light photon is absorbed, a lower energy light photon is emitted. This property is known as *fluorescence;* as you may guess, it is the basis of the fluorescent lightbulb. Certain materials fluoresce in the presence of UV light, sometimes called "black light," and give off visible light. The amount of fluorescence and the color of the emitted light depend on the wavelength of the UV light and on the chemical composition of the material that is fluorescing. One way fluorescence is used in medicine is in the detection of porphyria, a condition in which the teeth fluoresce red when irradiated with UV light. Another important application is in fluorescent microscopes, which are discussed in Section 14.5.

5. Light is reflected to some extent from all surfaces. There are two types of reflection (Fig. 14.2). *Diffuse reflection* occurs when rough surfaces scatter the light in many directions. *Specular reflection* is a more useful type of reflection; it is obtained from very smooth shiny surfaces such as mirrors where the light is reflected at an angle that is equal to the

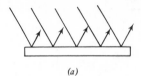

(a)

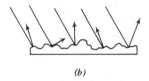

(b)

Figure 14.2. The two types of reflection: (a) specular reflection and (b) diffuse reflection.

angle at which it strikes the surface. Mirrors are used in many medical instruments. One simple instrument is a mirror that is held at the back of a patient's throat to look at his vocal folds.

All medical practitioners use light for some purposes, but a number of specialists use light in unique ways. Pediatricians shine light into the bodies of infants and observe the amount of scattered light (transillumination) produced in order to detect hydrocephalus (water-head) or pneumothorax (collapsed lung). Pediatricians also use visible light for treating jaundice in premature infants (phototherapy). Internists often use tubes with built-in light sources, called *endoscopes,* to see inside the body. Physiatrists (M.D.s in physical medicine) use IR and UV light for therapeutic purposes. Ophthalmologists use lasers to photocoagulate small blood vessels in the eye.

14.1. MEASUREMENT OF LIGHT AND ITS UNITS

The three general categories of light—UV, visible, and IR—are defined in terms of their wavelengths. Wavelengths of light used to be measured in microns (1 μ = 10^{-6} m) or in angstroms (1 Å = 10^{-10} m), but at present the recommended unit is the nanometer (1 nm = 10^{-9} m). Ultraviolet light has wavelengths from about 100 to 400 nm; visible light extends from about 400 to 700 nm; and IR light extends from about 700 to over 10^4 nm. Each of these categories is further subdivided according to wavelength (λ). For example, UV-C has wavelengths from about 100 to 290 nm, UV-B has wavelengths from 290 to 320 nm, and UV-A has wavelengths from 320 to 400 nm.

Visible light is measured in *photometric* units that relate to how light is seen by the average human eye. In photometry the quantity of light striking a surface is called *illuminance* and the intensity of a light source is called its *luminance*. All light radiation, including UV and IR radiation, can be measured in *radiometric* units. In radiometry the quantity of light striking a surface is called *irradiance* and the intensity of a light source is its *radiance*. Photometric and radiometric units are given in Table 14.1.

Table 14.1. Light Quantities and Units

	Quantities		
	On a Surface	From a Source	Units[a]
Photometric (visible light only)	illuminance	—	foot-candles or lumina/m^2 (luces)
	—	luminance	foot-lamberts or watts/m^2 per sr
Radiometric (UV, visible, and IR light)	irradiance	—	watts/m^2
	—	radiance	watts/m^2 per sr

[a]A steradian (sr) is a unit for a solid angle. A typical ice-cream cone is about 0.2 sr.

We have already referred to the wavelengths of light in Fig. 14.1 to describe the spectrum of the sun. Figure 14.3 shows where the wavelengths of light fit into the whole spectrum of electromagnetic radiation. Note that light has wavelengths much shorter than TV and radio waves but much longer than x-rays and gamma rays.

Since light is a form of energy, it is sometimes useful to talk about the energy of individual light photons. Figure 14.3 gives the energies as well as the wavelengths of the different parts of the electromagnetic spectrum. Visible light has energies ranging from about 2 electron volts (eV) up to about 4 eV. For comparison, the kinetic energy of a molecule in air at room temperature is about 0.025 eV and the energy of a typical x-ray photon used in medicine is about 50,000 eV, or 50 keV.

14.2. APPLICATIONS OF VISIBLE LIGHT IN MEDICINE

An obvious use of visible light in medicine is to permit the physician to obtain visual information about the patient regarding, for example, the

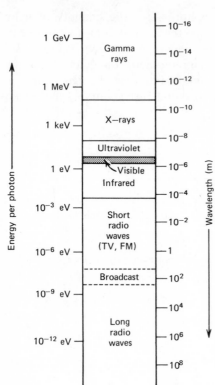

Figure 14.3. The relationship of light wavelengths to the entire spectrum of electromagnetic radiation.

color of his skin and the presence of abnormal structures in or on his body. It is quite easy for a physician to examine the skin under normal lighting conditions, but when she wishes to look into a body opening she is faced with the practical problem of getting light into the opening without obstructing the view. Like a lot of tricks, this one is done with mirrors. Figure 14.4 shows a simple curved mirror with a hole in the middle of it for the physician to look through. The curved surface focuses the light at the region of interest. More sophisticated instruments, such as the ophthalmoscope for looking into the eyes and the otoscope for looking into the ears, use basically the same principle. We discuss the ophthalmoscope in Chapter 15.

A number of instruments, called *endoscopes,* are used for viewing internal body cavities. Special purpose endoscopes are often given names indicating their purpose. For example, cystoscopes are used to examine the bladder, proctoscopes are used for examining the rectum, and bronchoscopes are used for examining the air passages into the lungs. Some

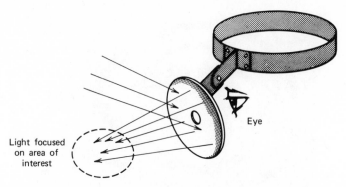

Figure 14.4. **A concave mirror to direct light into the body—the hole is for the physician to look through.**

endoscopes are rigid tubes with a light source to illuminate the area of interest. Many of them are equipped with optical attachments to magnify the tissues being studied.

The development of fiberoptic techniques permitted the construction of flexible endoscopes. Flexible endoscopes can be used to obtain information from regions of the body that cannot be examined with rigid endoscopes, such as the small intestine and much of the large intestine. Some flexible endoscopes are over a meter in length (Fig. 14.5). The image obtained with a flexible endoscope is not as good as that obtained with a rigid endoscope, but often the only alternative to a flexible endoscopic examination is exploratory surgery. Flexible endoscopes usually have an opening or channel that permits the physician to take samples of the tissues (biopsies) for later microscopic examination (Fig. 14.6).

Since light contains energy that largely appears as heat when it is absorbed, there is a limit on the amount of light that can be used in endoscopy. (If you hold your hand close to a 100 W incandescent bulb you will realize the large amount of thermal energy involved.) For endoscopy, the heating can be reduced by reducing the IR light from the source with special IR absorbing glass filters. In this *cold-light endoscopy* the light source contains very little IR radiation and the heating of tissues is minimized.

Transillumination is the transmission of light through the tissues of the body. Most of us have at one time or another shone a flashlight through our fingers to see the red glow that is produced. The glow is primarily red because most of the other colors in the beam are absorbed by the red blood cells; the red light is the only important component that is transmitted.

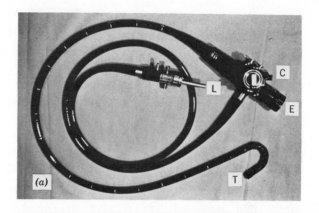

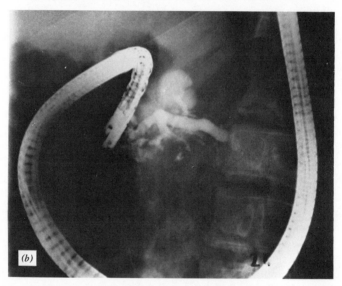

Figure 14.5. Flexible fiberoptic endoscopes are often over a meter in length. (a) The colonoscope has 10 cm markings from the flexible tip T to the controls C. The operator views the image at E. The light source is connected to a fiberoptic bundle at L. (Courtesy of John F. Morrissey, M.D., University of Wisconsin, Madison, Wisc.) (b) An x-ray of a duodenoscope passing through the stomach and into the duodenum. It is used to find the proper location to inject contrast media for an x-ray study.

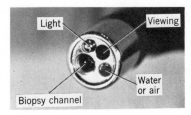

Figure 14.6. The tip of a colonoscope shows the fiberoptic light channel and viewing channel, the water or air channel, and a biopsy channel that permits passage of a device to take tissue samples. (Courtesy of John F. Morrissey, M.D., University of Wisconsin, Madison, Wisc.)

Transillumination is used clinically in the detection of hydrocephalus (water-head) in infants. Since the skull of young infants is not fully calcified, light is able to penetrate to the inside of the skull; if there is an excess of relatively clear cerebrospinal fluid (CSF) in the skull, light is scattered to different parts of the skull producing patterns characteristic of hydrocephalus. The special transilluminating device, which is shown in cross-section in Fig. 14.7, uses a 150 W projection bulb as the light source. This transilluminator was developed at the University of Wisconsin by one of the authors (JRC) and C. Vought, an electronic technician, in cooperation with Dr. Raymond Chun, a pediatric neurologist, and it is called the *Chun Gun* in honor of Dr. Chun. The device has a two-position trigger switch. The infant is taken into a dark room for the study; after a few minutes of dark adaptation the physician pulls the trigger to its first position, which turns on a red light that permits him to find the patient. The physician then points the barrel at the part of the head to be studied and pulls the trigger to the second position, which turns off the red light and turns on the intense white light used for the study. Infrared absorbing glass in the beam removes almost all of the IR radiation so that the light striking the infant is primarily visible light. Figure 14.8*a* shows the light pattern from a normal infant and Fig. 14.8*c* shows advanced hydrocephalus.

Transillumination is also used to detect pneumothorax (collapsed lung) in infants. The bright light penetrates the thin front chest wall and reflects off the back chest wall to indicate the degree of pneumothorax (Fig. 14.9*a*). The physician can then insert a needle attached to a syringe into the area of collapse to remove the air between the lung and chest wall, causing the lung to reinflate. The physics of a pneumothorax is discussed in Section 7.6.

The sinuses, the gums, the breasts, and the testes have also been studied with transillumination.

Visible light has an important therapeutic use. Since light is a form of energy and is selectively absorbed in certain molecules, it should not be

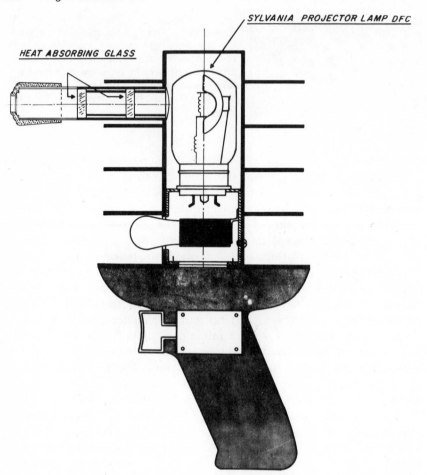

Figure 14.7. Cross-section of the Chun Gun transilluminator. The lower bulb is coated red to allow the operator to dark adapt. (Courtesy of Radiation Measurements, Inc., Middleton, Wisc.)

surprising that it can cause important physiological effects. Many premature infants have jaundice, a condition in which an excess of bilirubin is excreted by the liver into the blood. Relatively recently (1958) it was discovered that most premature infants recover from jaundice if their bodies are exposed to visible light (phototherapy). The exact mechanism is not clear, but blue light (~450 nm) appears to be the most important component.

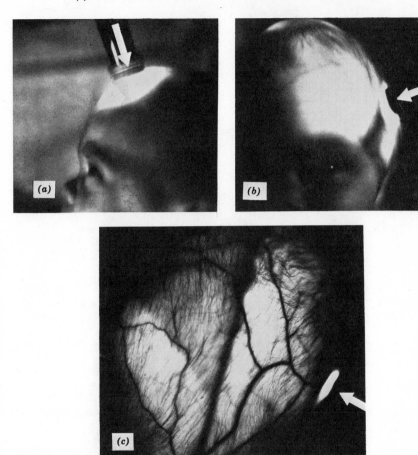

Figure 14.8.   Transillumination can be used to detect certain diseases. (a) A normal infant will demonstrate some transillumination of the skull. (b) Transillumination over half of the skull is indicative of an arachnoid cyst. (c) A large amount of transillumination of the skull, as shown in this top view, indicates severe hydrocephalus. (Courtesy of Raymond Chun, M.D., University of Wisconsin, Madison, Wisc.)

14.3.   APPLICATIONS OF ULTRAVIOLET AND INFRARED LIGHT IN MEDICINE

The wavelengths adjacent to the visible spectrum also have important uses in medicine. Ultraviolet photons have energies greater than visible photons, while IR photons have lower energies. Because of their higher energies, UV photons are more useful than IR photons.

Ultraviolet light with wavelengths below about 290 nm is germicidal—

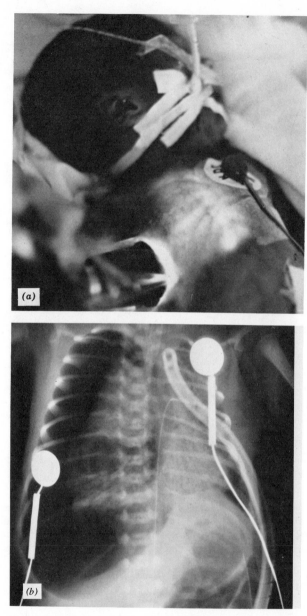

Figure 14.9.   Pneumothorax in infants can be detected by transillumination. (a) The light from a Chun Gun (lower left) enters the chest cavity and lights up the region of the collapsed lung. (b) The x-ray of the chest shows the collapsed lung (dark area on the left). The ECG monitoring electrodes and a catheter are also visible. (From L.R. Kuhns, F.J. Bednarek, M.L. Wyman, D.W. Roloff, and R.C. Borer, *Pediatrics*, Volume 56, Number 3, p. 358. © American Academy of Pediatrics, Inc., 1975.)

that is, it can kill germs—and it is sometimes used to sterilize medical instruments. Ultraviolet light also produces more reactions in the skin than visible light. Some of these reactions are beneficial, and some are harmful. One of the major beneficial effects of UV light from the sun is the conversion of molecular products in the skin into vitamin D. Dermatologists have also found that UV light improves certain skin conditions.

Ultraviolet light from the sun affects the melanin in the skin to cause tanning. However, UV light can produce sunburn as well as tan the skin. The wavelengths that produce sunburn are around 300 nm, just at the edge of the solar spectrum (Fig. 14.1). The amount of 300 nm light in the sun's spectrum depends on the amount of atmosphere that the sunlight must pass through. In winter in northern climates the angle of the sun is such that the atmosphere absorbs nearly all of the wavelengths that produce sunburn (Fig. 14.10). Figure 14.11 shows the UV wavelengths present at noon in winter and summer at a latitude of 45° (e.g., near Chicago). In the early morning and late afternoon of summer days the angle of the sun is again such that the UV wavelengths that produce sunburn are filtered out by the atmosphere. Ordinary window glass permits some near UV to be transmitted but absorbs the sunburn component.

Solar UV light is also the major cause of skin cancer in humans. The high incidence of skin cancer among people who have been exposed to the sun a great deal, such as fishermen and agricultural workers, may be related to the fact that the UV wavelengths that produce sunburn are also very well absorbed by the DNA in the cells. Skin cancer usually appears on those portions of the body that have received the most sunlight, such as the tip of the nose, the tops of the ears, and the back of the neck. Fortunately, skin cancer is easily cured if it is detected in its early stages.

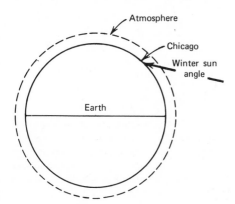

Figure 14.10. Sunlight must travel a much greater distance through the atmosphere in winter than in summer to reach a point far from the equator.

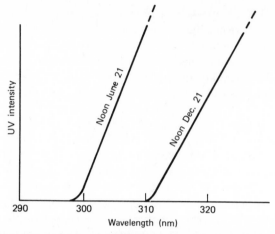

Figure 14.11. The UV component of the solar spectrum at a latitude of 45° at noon in winter and summer. The UV component depends strongly upon the angle of the sun. Because of the winter sun angle, the sunburn UV components (~300 nm) are absorbed in the atmosphere.

You probably know that the sky is blue because light of short (blue) wavelengths is scattered more easily than light of long wavelengths. Ultraviolet light has even shorter wavelengths than blue light and is scattered even more easily. About half of the UV light hitting the skin on a summer day comes directly from the sun and the other half is scattered from the air in other parts of the sky. Thus you can get a sunburn even when you are sitting in the shade under a small tree. Even when the sky is completely covered with clouds about one half of the UV light gets through (Fig. 14.12).

Ultraviolet light cannot be seen by the eye because it is absorbed before

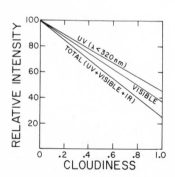

Figure 14.12. Relative intensity of various components of the solar spectrum as a function of the fraction of sky covered by clouds. Note that on a totally overcast day about 50% of the UV light reaches the earth. (Courtesy of Edwin C. McCullough, Mayo Clinic, Rochester, Minn.)

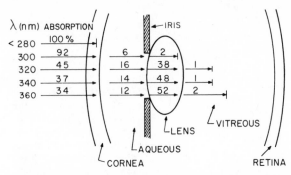

Figure 14.13.   Percentages of UV light of different wavelengths absorbed by the various components of the eye. (Courtesy of Edwin C. McCullough, Mayo Clinic, Rochester, Minn.)

it reaches the retina. Figure 14.13 shows the percentages of UV light of different wavelengths absorbed by the different structures of the eye. The large percentage of near-UV light absorbed by the lens may be the cause of some cataracts (opacities of the lens). Individuals who have had the lens of an eye removed because of a cataract are able to see into the near-UV region because the major absorber is no longer present.

About half of the energy from the sun is in the IR region (Fig. 14.1). The warmth we feel from the sun is mainly due to the IR component. The IR rays are not usually hazardous even though they are focused by the cornea and lens of the eye onto the retina. However, looking at the sun through a filter (e.g., plastic sunglasses) that removes most of the visible light and allows most of the IR wavelengths through can cause a burn on the retina. Some people have damaged their eyes in this way by looking at the sun during a solar eclipse. Dark glasses absorb varying amounts of the IR and UV rays from the sun as shown in Fig. 14.14.

Heat lamps that produce a large percentage of IR light with wavelengths of 1000 to 2000 nm are often used for physical therapy purposes. Infrared light penetrates further into the tissues than visible light and thus is better able to heat deep tissues.

Two types of IR photography are used in medicine: reflective IR photography and emissive IR photography. The latter, which uses the long IR heat waves emitted by the body that give an indication of the body temperature, is usually called *thermography* and is discussed in Chapter 4. We discuss here reflective IR photography, which uses wavelengths of 700 to 900 nm to show the patterns of veins just below the skin. Some of these veins are visible to the eye, but many more can be seen on a near-IR photograph of the skin. Details of the techniques used for IR photography

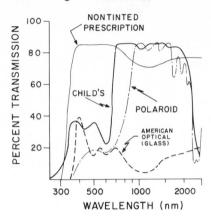

Figure 14.14. Various eyeglass lenses differ greatly in the percent of UV and IR rays they transmit to the eyes. Plastic sunglasses (child's) and Polaroid sunglasses permit more IR to enter the eyes than nontinted glasses or American Optical sunglasses. (From E.C. McCullough and G.D. Fullerton, *Surv. Ophthalmol.*, 16, 1971, 109, © 1971 The Williams and Wilkins Co., Baltimore.)

are described in the Kodak publication *Medical Infrared Photography* (see the bibliography at the end of this chapter). Since the temperature at the skin depends on the local blood flow, a thermogram with good resolution shows the venous pattern much like a near-IR photograph.

There is considerable variation in the venous patterns of normal individuals. Even in the same individual the venous patterns in the two breasts may be quite different. Cancer and other diseases can cause changes in the venous pattern, but these changes can be masked by the normal variations. Also, a layer of fat beneath the skin can reduce the appearance of the venous pattern. Nevertheless, IR photography can be used to follow changes in the venous pattern.

Near IR penetrates about 3 mm below the skin regardless of the color of the skin. Also, differently colored skins reflect about the same amount of IR, so that IR photographs of blacks and whites appear about the same. Figure 14.15 shows photographs of the chest of a young woman with a deep suntan. The IR photograph shows the venous pattern, but the variations in the melanin content of the skin due to the suntan are not apparent.

Infrared can also be used to photograph the pupil of the eye without stimulating the reflex that changes its size. This has no significant clinical use, but it is of interest in some vision research.

Infrared photographs of biological specimens illuminated with blue-green light sometimes show IR luminescence (fluorescence or phos-

Figure 14.15. A young woman with a deep suntan. (*a*) A conventional photograph shows the untanned area of the breasts. (*b*) In the IR photograph, the tan is not obvious but the veins are. (From *Medical Infrared Photography*, Eastman Kodak Company Publication No. N-1, p. 15. © Eastman Kodak Company.)

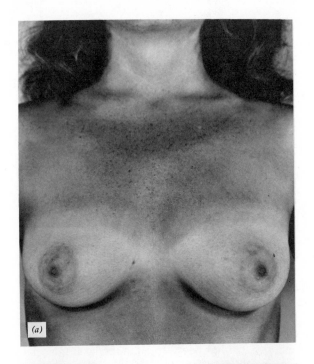

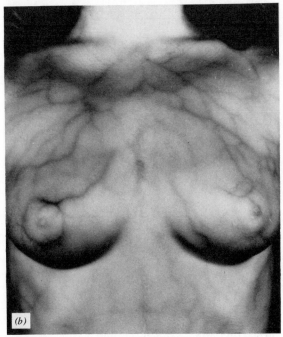

327

phorescence). The usefulness of this phenomenon has not been thoroughly investigated.

Near-IR photography is interesting and occasionally useful, but in 1976 it played a very minor role in clinical medicine.

14.4. LASERS IN MEDICINE

A laser is a unique light source that emits a narrow beam of light of a single wavelength (monochromatic light) in which each wave is in phase with the others near it (coherent light). *Laser* is an acronym for Light Amplification by Stimulated Emission of Radiation. While the basic theory for lasers was proposed by Albert Einstein in 1917, the first successful laser was not made until 1960, when T. H. Maiman produced a laser beam from a ruby crystal. Since 1960 scientists have made many types of lasers using gases and liquids as well as solids as the laser materials. The principles of laser operation are discussed in many physics texts. In this section we discuss some of the physical characteristics of lasers and a few of their applications in medicine.

In a laser, energy that has been stored in the laser material (e.g., ruby) is released as a narrow beam of light—either as a steady beam continuous wave (CW) or as an intense pulse. The beam remains narrow over long distances and can be thought of as an ideal "spot" light. A laser beam can be focused to a spot only a few microns in diameter. When all of the energy of the laser is concentrated in such a small area, the power density (power per unit area) becomes very large. The total energy of a typical laser pulse used in medicine, which is measured in millijoules (mJ), can be delivered in less than a microsecond, and the resultant instantaneous power may be in megawatts. The output of a pulsed laser is usually measured by the heat produced in the detector (calorimetric method). The output of a low-power CW laser is often measured with a photodetector such as a silicon photocell (often called a solar cell).

Since in medicine lasers are used primarily to deliver energy to tissue, the laser wavelength used should be strongly absorbed by tissue. The absorbance and reflectance curve for the skin of a white woman is shown in Fig. 14.16. Note that in the red wavelengths over half of the light is reflected. This curve varies for different individuals, but the short wavelengths (400 to 600 nm) are always absorbed better than the long wavelengths (~700 nm).

Laser energy directed at human tissue causes a rapid rise in temperature and can destroy the tissue. The amount of damage to living tissue depends on how long the tissue is at the increased temperature (Fig.

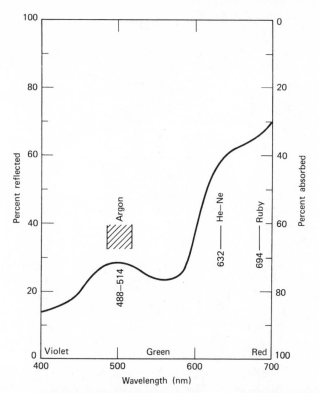

Figure 14.16. The absorbance and reflectance of a white woman's skin. Wavelengths of common lasers are shown. (Adapted from L. Goldman and R.J. Rockwell, Jr., *Lasers in Medicine*, Gordon and Breach, New York, 1971, p. 164.)

14.17). For example, tissue can withstand 70°C for 1 sec. In general, even the briefest exposure to temperatures above 100°C results in tissue destruction. However, not all laser damage is due to heat. Experiments with monkey eyes indicate that laser wavelengths of 1064 nm produce damage to the retina (the light-sensitive part of the eye) from heating, while laser light of 441.6 nm produces primarily photochemical damage. At intermediate wavelengths both types of damage are produced. A 1000 sec exposure of a monkey eye to a 1064 nm laser beam of 24 W/cm^2 on the retina caused a temperature rise of 23°C and produced just noticeable heat damage, while a laser beam of 441.6 nm of just 0.03 W/cm^2 also produced noticeable damage in 1000 sec due to photochemical effects.

The laser is routinely used in clinical medicine only in ophthalmology. Its effectiveness in treating certain types of cancer and its usefulness as a

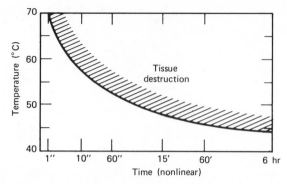

Figure 14.17.   Effects of time and temperature on tissue destruction. (Adapted from L. Goldman and R.J. Rockwell, Jr., *Lasers in Medicine*, Gordon and Breach, New York, 1971, p. 183.)

"bloodless knife" for surgery are under active investigation. Lasers are also being used in medical research for special three-dimensional imaging called *holography*.

In ophthalmology lasers are primarily used for *photocoagulation* of the retina, that is, heating a blood vessel to the point where the blood coagulates and blocks the vessel (Fig. 14.18). Before lasers were available photocoagulation was done with a high intensity xenon arc light source, and for certain applications the xenon source is still preferable. The main disadvantages of the xenon arc for photocoagulation are (1) the retinal spot size is much larger ($\sim$750 μm) than that formed with a laser beam ($\sim$50

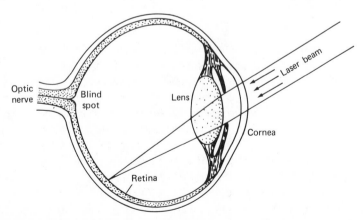

Figure 14.18.   A laser beam is focused by the cornea and lens to a small spot on the retina where it photocoagulates a small blood vessel.

μm), (2) the total amount of energy deposited in the eye is 20 to 50 times greater than that deposited during an equivalent laser treatment, and (3) the xenon arc requires longer exposures (up to 1 sec) than the laser, so local anesthesia must be used during the xenon arc treatment to prevent eye movement.

The amount of laser energy needed for photocoagulation depends on the spot size used. In general, the proper dose is determined visually by the ophthalmologist at the time of the treatment. The minimum amount of laser energy that will do observable damage to the retina is called the minimal reactive dose (MRD). For example, the MRD for a 50 μm spot in the eye is about 2.4 mJ delivered in 0.25 sec. Typical exposures needed for photocoagulation are 10 to 50 times the MRD (i.e., 24 to 120 mJ for a 50 μm spot in 0.25 sec).

Photocoagulation is useful for repairing retinal tears or holes that develop prior to retinal detachment. When the retina is completely detached, the laser is of no help. A complication of diabetes that affects the retina, called diabetic retinopathy, can also be treated with photocoagulation. Because of the small spot sizes available ($\sim$50 μm), it is possible to use the laser even in the small region where our detail vision takes place.

Protective glasses must be worn in medical laser areas to protect the eyes of the patient and the workers. Since the laser energy is concentrated in a narrow beam for long distances, even a reflected beam can be a hazard; thus the walls and other surfaces in a laser installation should have low reflectivity (e.g., flat black paint). The area should have adequate warning signs (Fig. 14.19) and a system that prevents outsiders from entering while lasers are in use. The hazards of laser radiation to the public are controlled by a federal law (Public Law 90-602) that is enforced

Figure 14.19.   A typical laser warning sign. Most are brightly colored.

by the Bureau of Radiological Health—a branch of the Food and Drug Administration.

For more details on lasers and their uses in medicine see the book by Goldman and Rockwell listed in the bibliography at the end of this chapter.

14.5.　APPLICATIONS OF MICROSCOPES IN MEDICINE

There have been few breakthroughs in science that have had as great an impact as the invention of the microscope by Leeuwenhoek ($\sim$ 1670). The use of the microscope in the pathology laboratory is as common as the use of the thermometer in the clinic. The magnification of an object by up to 1000 allows the study of cells (cytology) and of tissues (histology). In addition, electron microscopes and scanning electron microscopes have important research applications in medicine. In this section we describe the major features of the different types of microscopes.

The standard light microscope (Fig. 14.20) usually can be set at any of several magnifications by changing the power of the eyepiece or of the objective lens. The highest magnification that can be obtained is limited by the wavelength of visible light. Since the wavelengths of visible light range from 400 to 700 nm (0.4 to 0.7 μm), the smallest object that can be resolved is about 1 μm in diameter. Since most cells are 5 to 50 μm in diameter, this type of microscope is adequate for resolving all but subcellular objects.

If you put a thin slice of tissue under a microscope you will not see much because most cells are transparent to all wavelengths of visible light—red blood cells are an exception. In order to distinguish different cells it is usually necessary to stain them with a chemical that strongly absorbs certain visible wavelengths. Different chemicals are used to stain the various cell components and aid in the identification of cell structures.

Other techniques in addition to staining are useful in microscopy. One technique takes advantage of the different indexes of refraction of different cell parts. Since light travels at different speeds in the various parts of a cell, the phase relationships of the light waves change in passing through a specimen. The phase-contrast microscope takes advantage of this phenomenon to allow cell structures to be seen without the use of stain. In this type of microscope, a light beam that passes through the tissue is combined with a reference beam directed through an optically uniform zone. The combined beams interfere, producing dark areas where there is destructive interference and light areas where there is constructive interference. The darkness depends on the degree of interference. (This is the

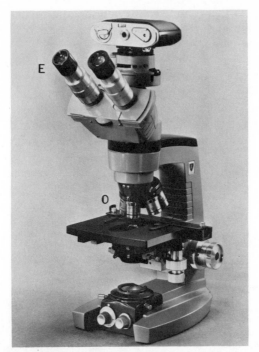

Figure 14.20. The standard light microscope. E indicates the replaceable eyepieces and O indicates the various objective lenses that can be used by rotating the turret. (Courtesy, American Optical Corporation.)

same basic phenomenon that permits you to see colored rings on a pool of water covered with a little oil.)

It is sometimes advantageous to use UV light or x-rays in microscopy. Since our eyes cannot see wavelengths shorter than those of visible light, it is necessary to convert the images produced by UV light or x-ray beams into images that use visible light. There are two common ways to produce visible images from UV light and x-rays. Neither increases the resolution but both offer other advantages.

Ultraviolet light is used in fluorescent microscopy. The tissue sample is stained with a dye that fluoresces when it is irradiated with UV light in a fluorescent microscope. Certain fluorescent stains are useful for identifying cell types. For example, one type of white blood cell is called an eosinophil because it takes up eosin, a fluorescent stain. Sometimes the staining of cells for fluorescent microscopy can be done *in vivo* (in the living body). For example, when a patient is given tetracycline (an antibiotic) some of it is incorporated into new bone growth; a bone sample taken later will fluoresce yellow or orange in the areas of new bone growth when placed in a fluorescent microscope.

Low-energy x-rays are used in a microscopy technique called *historadiography*. X-rays are beamed through a tissue sample that is held in close contact with fine-grain film, and the resultant x-ray image is examined under a conventional microscope. Since low-energy x-rays are strongly absorbed by heavy elements, such as calcium, historadiography is often used to study bone samples that have been cut in thin slices (~ 0.1 mm).

Several microscopy techniques used primarily in research involve electron beams. Electrons can be focused by electric and magnetic lenses, much like light is focused by glass lenses, to form an image. The wavelengths of electrons depend on their energy but are usually very short compared to those of visible light. As a result, electron microscopes can show details much better than light microscopes. Magnifications of up

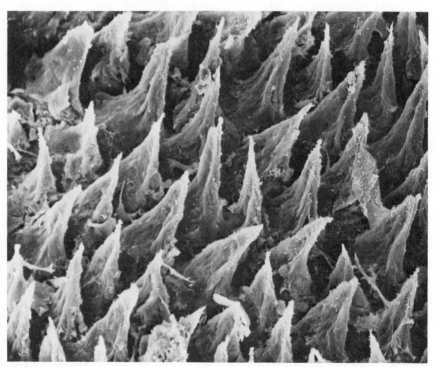

Figure 14.21. Scanning electron micrograph of the surface of a hamster tongue. (Courtesy of Tousimis Research Corporation, Rockville, MD, USA.)

to 250,000 times have been obtained in electron microscopy, while the maximum in conventional light microscopy is about 1000 times. Some electron microscopes use very high accelerating voltages of up to 10^6 V, but 50 kV is more common.

In transmission electron microscopy (TEM) exceedingly thin specimens must be used so that the electrons can pass through them. It is also usually necessary to evaporate a very thin layer of a heavy metal on each sample to act as a stain.

In scanning electron microscopy (SEM) a finely focused beam of electrons scans the surface of the specimen and a detector measures the number of scattered electrons from each point on the surface. This information is used to control the intensity of an electron beam in a TV tube to make an image of the surface of the specimen (Fig. 14.21).

BIBLIOGRAPHY

Daniels, F., *Direct Use of the Sun's Energy*, Yale University Press, New Haven, Conn., 1964.

Goldman, L., and R. J. Rockwell, Jr., *Lasers in Medicine*, Gordon and Breach, New York, 1971.

Hazzard, DeW. G. (Ed.), *Symposium on Biological Effects and Measurement of Light Sources*, Proceedings of a Conference held in Rockville, MD, March 25–26, 1976, HEW Publication (FDA) 77-8002, Bureau of Radiological Health, Rockville, Md., 1976.

Maratka, Z., and J. Setka (Eds.), *Endoscopy of the Digestive System*, Proceedings of the 1st European Congress of Digestive Endoscopy, Prague, July 5–6, 1968, Karger, New York, 1969.

Medical Infrared Photography, 3rd ed., Publication No. N-1, Eastman Kodak, Rochester, N.Y., 1973.

Odell, G. B., R. Schaffer, and A. P. Simopoulos (Eds.), *Phototherapy in the Newborn: An Overview*, National Academy of Sciences, Washington, D.C., 1974.

Urbach, F., *The Biologic Effects of Ultraviolet Radiation (With Emphasis on the Skin)*, Pergamon, Elmsford, N.Y., 1969.

REVIEW QUESTIONS

1. List three ways light can interact with tissue.
2. Give the wavelengths of visible, IR, and UV light.
3. How does the energy of a visible light photon compare to the energy of a typical x-ray photon used in medicine?
4. List three biological effects of UV light.
5. What is an endoscope? Name three special purpose endoscopes and the parts of the body for which they are used.

6. Describe how transillumination is used in the detection of hydrocephalus.
7. What prevents sunburn in northern latitudes in the winter?
8. Why does the eye not see IR light? UV light?
9. List two hazards of UV light to humans.
10. What is the principal medical use of IR light?
11. Why can you get a sunburn in the shade or on a cloudy day in the summer?
12. What is one possible hazard of IR light from the sun?
13. Why are stains used in microscopy of tissue samples?
14. What is fluorescent microscopy?
15. What is the difference between a conventional electron microscope and a scanning electron microscope?

CHAPTER 15

Physics of Eyes and Vision

Most of our knowledge of the world around us comes to us through our eyes. The helplessness we feel when caught in the dark in unfamiliar surroundings is a good indication of our dependence on vision. The sense of vision consists of three major components: (1) the eyes that focus an image from the outside world on the light-sensitive retina (Fig. 15.1), (2) the system of millions of nerves that carries the information deep into the brain, and (3) the *visual cortex*—that part of the brain where "it is all put together." Blindness results if any one of the parts does not function. Physics is obviously involved in all three parts, but we understand the physics of the first part far better than the physics of the other two parts.

In this chapter we discuss the physics of the eye. While the eye has some striking similarities to a camera, a better analogy exists between the eye and a closed circuit color TV system (Fig. 15.2). The lens of the TV camera is analogous to the cornea and lens of the eye; the "signal cable" is the optic nerve, and the "viewing monitor" is the visual cortex. When the light is bright we see things in "living color." In dim light the eye operates like a supersensitive black-and-white TV camera and allows us to see objects with less than 0.1% of the light we need for color vision. This great difference in sensitivity is analogous to the difference between sensitive high-speed black-and-white film and the much less sensitive color film we use in our cameras.

Our optical system has the following special features, most of which are not available on even the most expensive cameras:

1. The eye can observe events over a very large angle while looking intently at an object directly ahead of it (Fig. 15.3).
2. Blinking provides the front lens (cornea) with a built-in lens cleaner

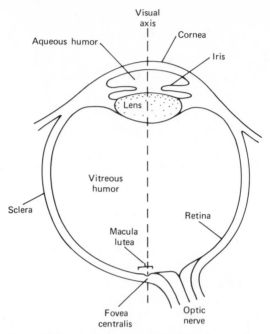

Figure 15.1. Cross-section of the left eye as seen from above.

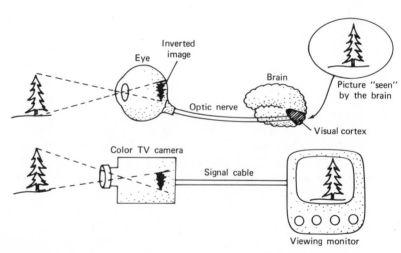

Figure 15.2. The sense of sight is in many ways similar to a closed circuit color TV system. It is superior in all respects except ease of replacement.

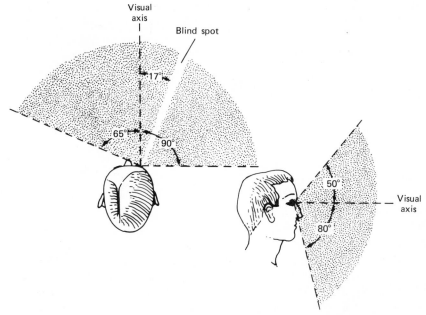

Figure 15.3.   The eye looking straight ahead has a wide angle of vision.

and lubricator. (Each eyelid can be closed independently for communication with the opposite sex.)

3. A rapid automatic focusing system permits viewing objects as close as 20 cm one second and distant objects the next. Under relaxed conditions the focus for normal eyes is set for "infinity" (distant viewing).

4. The eye can operate effectively over a range of light intensity of about 10 billion to one ($10^{10}:1$)—brilliant daylight to very dark night.

5. The eye has an automatic aperture adjustment (the iris).

6. The cornea has a built-in scratch remover; even though it has no blood supply it is made of living cells and can repair local damage.

7. The eye has a self-regulating pressure system that maintains its internal pressure at about 20 mm Hg and thus keeps the eye in shape. If "dented," the eye rapidly returns to its original shape.

8. The eyes are mounted in a well-protected casing almost completely surrounded by bone, and each eye rests on a cushion of fat that reduces sharp shocks.

9. The image appears upside down on the light-sensitive retina at the back of the eyeball (Fig. 15.2), but the brain automatically corrects for this.

10. The brain blends the images from both eyes, giving us good depth perception and true three-dimensional viewing. If vision from one eye is lost, the vision from the remaining eye is adequate for most needs.
11. The muscles of the eye (Fig. 15.4) permit flexible movement up and down, sideways, and diagonally. After a little practice, the eyes can even be made to go in circles.

We discuss most of these features in more detail in this chapter.

Considering the sophistication of the eye mechanism, a surprisingly large percentage of people have good eyesight. These people are called *emmetropes,* but be careful when you use this term—some people may misunderstand! The rest of us have noticeable vision imperfections and are called *ametropes.* If we consider an emmetrope to be anyone who needs correction of less than 0.5 diopters (D), 25% of young adults qualify; 65% fall within the ± 1.0 D range. (We define *diopter* in Section 15.8.)

Various medical specialists deal with problems of the eye. The most highly trained is the *ophthalmologist,* an M.D. who has taken three years of residency training in ophthalmology in addition to his M.D. training. The ophthalmologist is qualified to diagnose and treat any problem of the eye. His treatment may include surgery. The most common eye specialist, the *optometrist,* specializes in prescribing and fitting corrective lenses. An optometrist is now required to have six years of college. He or she is not licensed to treat diseases of the eye. An *ophthalmic technician* is a

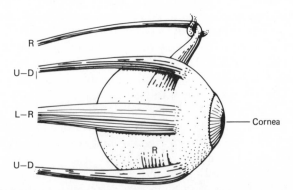

Figure 15.4. The six muscles of the right eye permit a wide variety of motion. The muscles work in pairs: one pair controls up and down movement (U-D), one pair controls left and right movement (L-R), and one pair controls rotation (R). The rotation muscles pass through bony loops. All six muscles are attached to the skull behind the eye.

relatively new type of physician's assistant for ophthalmologists. He usually has had two or more years of college plus two years of on-the-job training. He can perform various eye tests and fit contact lenses (hard and soft); he is also trained in office management. An *optician* is a specialist in making lenses, fitting them in the frames, and fitting the frames to the patient. An optician usually has had two years of on-the-job training after high school. In some states opticians are required to be registered. An *orthopist* is a technician who is concerned with the muscular control of the eyes and problems with this control such as crossed eyes. Orthopists have had two years of on-the-job training and have passed a certification examination.

15.1. FOCUSING ELEMENTS OF THE EYE

The eye has two major focusing components: the *cornea,* which is the clear transparent bump on the front of the eye that does about two-thirds of the focusing, and the *lens,* which does the fine focusing. The cornea is a fixed focus element; the lens is variable in shape and has the ability to focus objects at various distances.

The cornea focuses by bending (refracting) the light rays. The amount of bending depends on the curvatures of its surfaces and the speed of light in the lens compared with that in the surrounding material (relative index of refraction). The indexes of refraction of the cornea and other transparent parts of the eye are given in Table 15.1. When the cornea is under-

Table 15.1. The Indexes of Refraction of the Cornea and Other Optical Parts of the Eye

Part of the Eye	Index of Refraction
Cornea	1.37
Aqueous humor	1.33
Lens cover	1.38
Lens center	1.41
Vitreous humor	1.33

water it loses most of its focusing power because the index of refraction of the water (1.33) is close to that of the cornea (1.37). (Fish have a similar problem out of the water.) Divers keep air around the cornea by wearing a face mask. The index of refraction is nearly constant for all corneas, but the curvature varies considerably from one person to another and is responsible for most of our defective vision. If the cornea is curved too much the eye is near-sighted, not enough curvature results in far-sightedness, and uneven curvature produces astigmatism. We discuss these defects more in Section 15.8. Nearly all of the focusing by the cornea is done at the front surface since the aqueous humor in contact with the back surface has nearly the same index of refraction as the cornea.

Since the living cells in the cornea are not supplied with oxygen by the blood, they must get their oxygen from the air. Having blood vessels in the cornea would not help our vision! The nutrients for the cells in the cornea are supplied by the aqueous humor that is in contact with its back surface. The aqueous humor contains all of the blood components except blood cells. We discuss this fluid in more detail in the next section.

If the cornea is scratched it will heal itself, but some other types of damage are more permanent. Some types of radiation (ultraviolet, neutrons, x-rays, etc.) can cause opacities to develop in the cornea that will block out light. It is now possible to perform cornea transplants using corneas removed from donors shortly after death. Since the cells of the cornea have a low metabolism rate, rejection is not usually as much of a problem as in most organ transplants.

The lens (Fig. 15.1) has focusing properties at both its front surface and its back surface. The lens is more curved in the back than in the front. It changes its focal strength by changing its curvature. The focusing power of the lens is considerably less than that of the cornea because it is surrounded by substances that have indexes of refraction close to its own (Table 15.1). The effective index of refraction is thus only about 1.07. The lens is made up of layers somewhat like an onion, and all layers do not have the same index of refraction. The indexes of refraction in Table 15.1 are average values.

The lens has a flexible cover that is supported under tension by suspension fibers. When the focusing muscle of the eye is relaxed this tension keeps the lens somewhat flattened and adjusted to its lowest power, and the eye is focused on distant objects. The point at which distant objects are focused when the focusing muscle is relaxed is called the *far point*. For a near-sighted person the far point may be quite close to the eye. To focus on closer objects, the circular muscle around the lens contracts into a smaller circle and takes some or all of the tension off the lens. The lens oozes into a more spherical shape, primarily by becoming more curved in front. The lens then has a greater focusing power; objects that are closer to the eye are brought into focus at the retina. The closest point at which objects can be focused when the lens is its thickest is called the *near point*. Young children have very flexible lenses and can focus on very close objects. The ability to change the focal power of the eye is called *accommodation.* As people get older, their lenses lose some accommodation. Presbyopia (old sight) results when the lens has lost nearly all of its accommodation.

Not all animals focus by changing the shape of the lens; fish focus by moving the lens back and forth like we do with a camera. Perhaps they thus avoid presbyopia; you never see old fish wearing bifocals!

The lens, like the cornea, can be damaged by ultraviolet and other forms of radiation. It can develop *cataracts,* which destroy its clarity. It is possible to remove a damaged lens surgically and add extra correction to glasses. Of course no accommodation is then possible, and bifocals must be worn.

15.2. SOME OTHER ELEMENTS OF THE EYE

The elements of the eye discussed in this section either are not directly involved with focusing the image or play a passive role in focusing. The retina plays such a dominant role in the function of the eye that it is discussed separately in Section 15.3. The components we discuss here are the iris and the pupil; the aqueous and vitreous humors; and the "housing" for the eye, the sclera.

The *pupil* is the opening in the center of the *iris* where light enters the lens. It appears black because essentially all of the light that enters is absorbed inside the eye. Under average light conditions, the opening is about 4 mm in diameter. It can change from about 3 mm in diameter in bright light to about 8 mm in diameter in dim light. The physiologic reason for this change in size is not clear. The maximum change of a factor of 7 in the opening area cannot begin to cover the huge range of light intensities the eye can handle—$10^{10}:1$. The iris does not respond instantly to a change of light levels; about 300 sec are needed for it to fully open, and about 5 sec are required for it to close as much as possible (Fig. 15.5).

It is believed that the iris aids the eye by increasing or decreasing incident light on the retina until the retina has adapted to the new lighting conditions. In addition, under bright light conditions it plays an important role in reducing lens defects. Camera bugs will realize that the small aperture increases the *depth of focus,* the range of distance over which objects are in satisfactory focus. You can demonstrate this effect by taking a piece of cardboard, making a hole about 1 mm in diameter in it, and looking through this hole at printed material under bright light. You can move the material very close to your eye and still be able to read it. The small aperture has increased your depth of focus. When you take away the pinhole, you will have to increase the distance to read the printing.

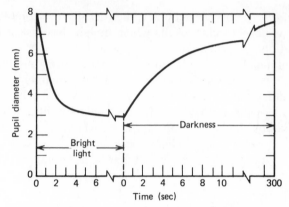

Figure 15.5.   The pupil does not open and close rapidly. Note that the maximum opening is not attained until the eye has been in the dark 5 min.

The *aqueous humor* fills the space between the lens and the cornea. This fluid, mostly water, is continuously being produced, and the surplus escapes through a drain tube, the canal of Schlemm. Blockage of the drain tube results in increased pressure in the eye; this condition is called *glaucoma* (see Chapter 6, p. 108). The aqueous humor contains many of the components of blood and provides nutrients to the nonvascularized cornea and lens. It maintains the internal pressure of the eye at about 20 mm Hg. (See Section 15.10 for a description of how this pressure is measured.) If you press on the eye, you find it is fairly stiff; you cannot indent it much. The reasons are that the fluids in the eye are incompressible at the pressure you use and that the covering of the eyeball does not stretch easily. When you rub your eyes, you greatly increase the internal pressure.

The *vitreous humor* is a clear jelly-like substance that fills the large space between the lens and the retina. It helps keep the shape of the eye fixed and is essentially permanent. It is sometimes called the *vitreous body*.

The *sclera* is the tough, white, light-tight covering over all of the eye except the cornea. The sclera is protected by a transparent coating called the *conjunctiva*.

15.3.   THE RETINA—THE LIGHT DETECTOR OF THE EYE

The retina, the light-sensitive part of the eye, converts the light images into electrical nerve impulses that are sent to the brain. While the role of the retina is similar to that of the film in a camera, a better analogy exists between the retina and the light-sensitive portion of a TV camera tube. Unlike film, the retina does not have to be replaced since a built-in system supplies the light-sensitive chemicals that in some way convert light into electrical nerve impulses. We do not understand completely the mechanisms involved, but we do know many of the characteristics of the photoreceptors in the retina. In this section we discuss the physical aspects of the retina.

The absorption of a light photon in a photoreceptor triggers an electrical signal to the brain—an action potential. The energy of the photon is about 3 eV; the action potential has an energy millions of times greater. The light photon apparently causes a photochemical reaction in the photoreceptor which in some way initiates the action potential. The photon must be above a minimum energy to cause the reaction. Infrared photons have insufficient energy and thus are not seen. Ultraviolet photons have sufficient energy, but they are absorbed before they reach the retina and also are not seen (see Fig. 14.13).

The retina covers the back half of the eyeball. While this large expanse permits useful "warning" vision over a large angle (Fig. 15.3), most vision is restricted to a small area called the *macula lutea,* or yellow spot. All detailed vision takes place in a very small area in the yellow spot (~0.3 mm in diameter) called the *fovea centralis* (Fig. 15.1).

The image on the retina is very small. A convenient equation for determining the size of the image on the retina comes from the ratios of the lengths of the sides of similar triangles. In Fig. 15.6, O is the object size, I the image size, P the object distance, and Q the image distance, usually about 2 cm. Thus we can write $O/P = I/Q$ or $O/I = P/Q$. $I = (Q/P)O$ (see Example 15.1).

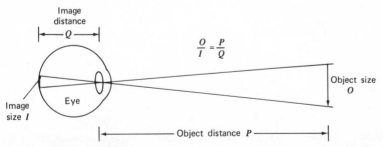

Figure 15.6. There is a simple relationship between the object and image sizes and the object and image distances. The image on the retina is small because of the short image distance of about 2 cm.

Example 15.1

How big is the image on the retina of a fly on a wall 3.0 m away? Assume the fly is 3 mm (0.003 m) in diameter and $Q = 0.02$ m.

$$I = \frac{0.02}{3} \times 0.003 = 2 \times 10^{-5} \text{ m} = 20 \text{ } \mu\text{m}$$

There are two general types of photoreceptors in the retina: the *cones* and the *rods*. Throughout most of the retina the cones and rods are not at the surface of the retina but lie behind several layers of nerve tissue through which the light must pass (Fig. 15.7). However, in the fovea centralis most of this nerve tissue is pushed to the side and there is a slight dip (*fovea* means pit). This decrease in nerve tissue aids our vision in this specialized area. The rods and cones are distributed symmetrically in all directions from the visual axis except in one region—the blind spot (Fig. 15.8).

The cones (~ 6.5 million in each eye) are primarily used for daylight, or *photopic*, vision. With the cones, we can see fine detail and recognize different colors. The cones are primarily found in the fovea centralis, although some are scattered throughout the retina (Fig. 15.8). We see later that the density of the cones in the fovea centralis determines the amount of detail we can resolve in an image. Each of the cones in the fovea has its own "telephone line" to the brain. In the rest of the retina several cones share one nerve fiber. The cones are not uniformly sensitive to all colors but have a maximum sensitivity at about 550 nm in the yellow-green

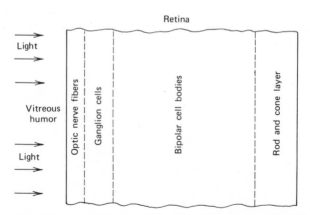

Figure 15.7. The light must pass through various cell layers to reach the rods and cones. In the fovea centralis much of this tissue is pushed to the side, permitting better detail vision in this area. Blood vessels also block the light to some rods and cones.

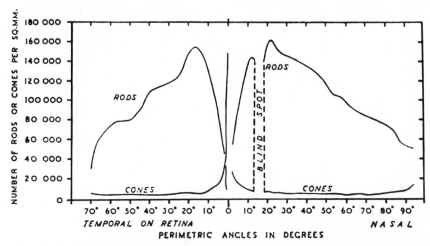

Figure 15.8. The distribution of the rods and cones in the retina of the left eye; notice the blind spot with no sensors. A perimetric angle of 30° indicates the location on the retina of a light source 30° to the left or right when the eye is looking straight ahead. (From M.H. Pirenne, *Vision and the Eye*, 2nd ed., Chapman & Hall Ltd., London, 1967, p. 32, as plotted from data of Osterberg, *Acta Ophthal.*, Suppl. 6, 1935.)

region (Fig. 15.9). This corresponds quite well to the maximum in the solar spectrum at the earth's surface.

The rods are used for night, or *scotopic*, vision and for peripheral vision. They are much more abundant than the cones (~ 120 million in each eye) and cover most of the retina. They are not uniformly distributed over the retina but have a maximum density at an angle of about 20° (Fig. 15.8). That is, if you are looking at the sky at night, the light from a faint star displaced 20° from your line of vision will fall on the most sensitive area of your retina. If you look directly toward the faint star, its image will fall on your fovea which has no rods and you will not see it.

Histological studies have indicated that hundreds of rods send their information to the same nerve fiber. This means that the ability to resolve two close sources of light in peripheral vision is poor. On the other hand, the great sensitivity of the rods and their great expanse permit us to recognize an object approaching from the side when we are looking straight ahead.

The rods are most sensitive to blue-green light (~ 510 nm), which has a wavelength shorter than the optimum for the cones (~ 550 nm). Figure 15.9 indicates that the rods and cones are equally sensitive to red light (650 to 700 nm). The curves in Fig. 15.9 are sometimes plotted so that both

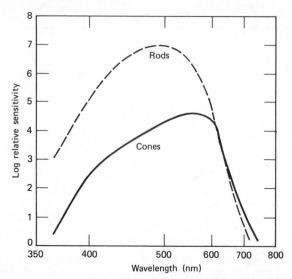

Figure 15.9.   The rods are much more sensitive than the cones. The vertical axis is a log scale; each division represents a factor of 10 in sensitivity. The best sensitivity of cones is at about 550 nm, while the best sensitivity of rods is at about 510 nm.

rods and cones have the same maximum sensitivity. This gives the erroneous impression that cone vision is better in red light than rod vision.

The eyes do not have their greatest sensitivity to light under photopic conditions. If the light level suddenly decreases by a factor of 1000 we are momentarily "in the dark," but after a few minutes we are able to see many of the details that were not visible when it first became dark. This *dark adaptation* is apparently the time needed for the body to increase the supply of photosensitive chemicals to the rods and cones. The rate at which we dark adapt is shown in Fig. 15.10. The cones adapt most rapidly; after about 5 min the fovea centralis has reached its best sensitivity. The rods continue to dark adapt for 30 to 60 min, although most of their adaptation occurs in the first 15 min. It is possible to dark adapt by wearing red goggles that limit the incident light to the red region of the spectrum. You can dark adapt one eye by closing it; this is useful, for example, if you are about to enter a dark theater.

Notice in Fig. 15.8 that there is a region from about 13° to 18° that has neither rods nor cones—the *blind spot*. This is the point at which the optic nerve enters the eye. The blind spot is on the side toward the nose; if an image falls on the blind spot in one eye it misses the blind spot in the other eye. We are normally not aware of the blind spot, but it is easy to

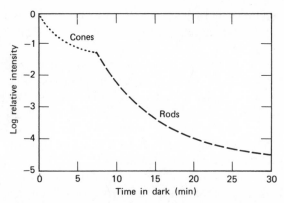

Figure 15.10. **The eye dark adapts in two phases. The phase for the cones is completed in 5 to 10 min; the phase for the rods lasts for over 30 min. The increase in sensitivity is a factor of over 10,000.**

demonstrate. A simple experiment is to place five coins about 10 cm apart in a horizontal row on a table. If you close one eye and look at the central coin with the other, it is possible to adjust your viewing distance to make one of the adjacent coins disappear. If you repeat the experiment with your other eye, the other adjacent coin will disappear. The blind spot is quite large; it covers an angle equal to 11 full moons placed side by side in the sky!

15.4. HOW LITTLE LIGHT CAN YOU SEE?

In 1942, Hecht, Schlaer, and Pirenne published the results of an important experiment on the sensitivity of the rods. The excellent book *Visual Perception* by Cornsweet (see the bibliography at the end of this chapter) gives an extensive discussion of this experiment, and it is only summarized here. The primary question Hecht et al. posed is: What is the minimum number of photons that will produce the sensation of vision at least 60% of the time? To obtain the minimum number, Hecht et al. had to optimize their experimental conditions. They had to determine (1) the optimum color to use in the test flash, (2) the most sensitive location in the eye, (3) the best diameter to use in the flash, and (4) the best length of time to use in the flash. They obtained the following answers: (1) the rods are the most sensitive at 510 nm; (2) the rods are most abundant at about 20° from the visual axis; (3) the detectability is independent of flash diameters up to 10' of arc, while above this size more light is needed for detection;

and (4) for flash times up to about 0.1 sec (100 msec) the length of the flash does not affect the detectability, but for longer times more light is necessary.

The final results of their experiment showed that if about 90 photons enter the eye under these optimum conditions, the flash is seen 60% of the time. When these investigators considered all of the light losses in the eye, they estimated that only 10 photons are actually absorbed in the rods. Since the light is distributed over about 350 rods, they felt that it is unlikely that a single rod receives more than one photon. They thus established that a single photon can activate a single rod. Later experiments indicated that as few as 2 photons actually absorbed in the rods can give a visual signal. For comparison, a typical flashlight emits about 10^{18} photons each second!

You may have been surprised to learn that if 90 photons enter the eye, 10 or less are actually absorbed in photoreceptors. What happens to the others? About 3% are reflected at the surface of the cornea, and about 50% are absorbed in the various structures (cornea, lens, humors). Of those that reach the region of the rods, only about 20% ($\sim$ 10% of the original number) are absorbed in the rods. The photons that miss the rods are absorbed in the "backstop." Some animals, such as cats, have a reflective coating behind the rods that gives the rods another opportunity to absorb the photons. These animals have eyes that "glow in the dark" if a light shines in them.

15.5. DIFFRACTION EFFECTS ON THE EYE

All light waves undergo diffraction as they pass through a small opening. Thus the iris produces a diffraction pattern on the retina (Fig. 15.11). At the normal opening of the pupil ($\sim$ 4 mm) this phenomenon has no practical consequences for our daily vision tasks. However, if the pupil becomes much smaller, for example, 1.0 mm, diffraction produces a measurable effect on visual acuity. You can demonstrate this effect by reading an eye test chart through a 0.75 mm hole; you should notice a decrease in your ability to read the small letters.

All lenses have defects (aberrations). The effect of such aberrations is reduced if the lens opening is made smaller. In the eye, a small pupil improves visual acuity. However, if the pupil is made very small the acuity becomes worse due to diffraction effects. There is an optimum size for the pupil; best acuity for an emmetropic eye is obtained with a pupil size of 3 to 4 mm—its normal size under good illumination.

A point source of light will not be focused on a single cone because of

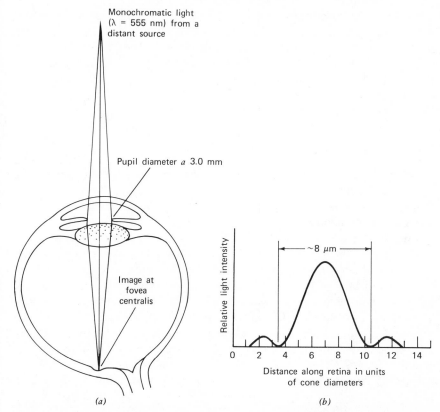

Figure 15.11. Diffraction in the eye. (a) Monochromatic light from a distant point source is brought to a focus at the fovea centralis in the retina. (b) The diffraction pattern on the retina produced by a pupil 3.0 mm in diameter consists of a central bright spot 8 μm in diameter surrounded by a ring of light of reduced intensity.

diffraction effects (Fig. 15.11). The angular spread 2θ of the central bright spot at the retina for $\lambda = 555$ nm and a pupil 3.0 mm in diameter a is given by $2\theta = 2(1.22)\,(\lambda/a) = 2(1.22)\,(555 \times 10^{-9}/3 \times 10^{-3}) = 4.5 \times 10^{-4}$ radians.* The diameter of the central bright spot at the retina is the product of the effective aperture to retina distance (17 mm) times 2θ, or $17(4.5 \times 10^{-4}) = 8 \times 10^{-3}$ mm $= 8\ \mu$m. This spot will include many cones (diameter, ~ 1.1

*Radians are more convenient than degrees for measuring angles. θ (in radians) $= s/r$, where r is the radial distance to an object and s is the dimension perpendicular to the radius. Thus for an angle of 0.3 milliradians and $r = 1$ m, $s = 3 \times 10^{-4}$ m or 0.3 mm. $180° = \pi$ radians, or $1° = 17.45$ milliradians.

μm). If the source is bright, for example, a bright star, the next ring of the diffraction pattern can stimulate even more cones. Thus intense point sources of light appear larger than weak point sources, a fact that early astronomers were not aware of when they assigned "magnitudes" to stars; in astronomy this term is now interpreted as an intensity rather than as a size.

15.6. HOW SHARP ARE YOUR EYES?

The familiar eye charts used to determine whether we need corrective lenses test the property of our eyes called *visual acuity*. A physicist calls visual acuity the *resolution of the eyes*. In this section we discuss several tests for visual acuity.

The optometrist usually uses a Snellen chart (Fig. 15.12) to test visual acuity. If he tells you that your eyes test normal at 20/20, he means that you can read detail from 20 ft that a person with good vision can read from 20 ft. If your eyes test at 20/40, you can just read from 20 ft the line that a person with good vision can read from 40 ft. The Snellen chart tests many things besides acuity. A person who reads a lot recognizes letters more readily than one who reads little. (You can usually read a line of print with the top or bottom half of it covered.) Some letters, such as A and V, are easy to recognize from their general shape; it is not necessary to distinguish their details. Nevertheless, the Snellen chart is a highly useful tool and will probably continue to be used because of its simplicity.

The visual acuity or resolution of the eye is primarily determined by the characteristics of the cones in the fovea. A common way of testing resolution is to use a pattern of alternating black and white lines that become increasingly narrower. A combination of one white line and one black line is called a line pair (lp). Under optimum conditions, the eye can just barely resolve as separate lines a pattern of about 30 lp/mm; when the eye is twice as far away, it can only resolve 15 lp/mm. The resolution is often given in terms of the angle subtended from the eye. This angle is more or less independent of viewing distance. The minimum angle between two black lines that can be seen as separate is about 0.3 milliradians. To be seen as separate, the two lines must fall on alternate rows of cones so that the cones between will see the white strip (Fig. 15.13). The smallest black dot that you can see under optimum conditions is 2.3×10^{-6} radians. The resolution rapidly gets worse as the image moves away from the fovea centralis. At 10° from the fovea the acuity is worse by a factor of 10. If the lighting is not optimum the resolution also deteriorates (Fig. 15.14).

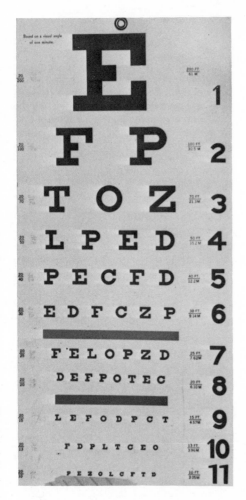

Figure 15.12. A Snellen chart to test vision is usually viewed from 20 ft. The 20/20 line is number 8. The lines that form the letters have an angular width of 1′ of arc at 20 ft. Some letters (e.g., L) are easier to recognize than others (e.g., B and H).

One test of acuity used often by scientists is aligning two lines so that they appear to be a single continuous line (Fig. 15.14). This is involved in using a slide rule and also in using a vernier scale, and it is referred to as *vernier acuity*. It is possible for a trained person to align two fine lines under optimum conditions to within 9×10^{-6} radians, much less than the 3×10^{-4} radians needed to resolve two lines.

The resolution of bright on dark is about 10^{-3} radians of arc while that of dark on bright is 3×10^{-4} radians. This means that a good eye would read only 20/60 on the Snellen chart if the chart were made in the bright-on-dark format! This fact is of practical importance in making projection

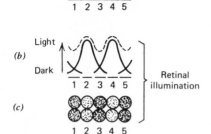

(a) Light pattern of test grating

(b)

(c)

Retinal illumination

Figure 15.13. To see closely spaced lines (*a*) as separate, it is necessary for the images of the lines to fall on alternate rows of cones as shown schematically in *b*. All the cones receive some light (*c*), but more falls on cones corresponding to the white strips. (Adapted from Ruch, T.C., in T.C. Ruch and H.D. Patton (Eds.), *Physiology and Biophysics,* 19th ed., © W.B. Saunders Company, Philadelphia, 1965, p. 428.)

slides for lectures; light letters on a dark background are not as easy to read as the conventional dark letters on a light background.

The ability of the eye to recognize separate lines also depends on the relative "blackness" and "whiteness" of the lines. Resolution is much worse if the lines are two close shades of gray than if one is black and one is white. The contrast C between two areas is defined as $C = (I_1 - I_2)/(I_1 +$

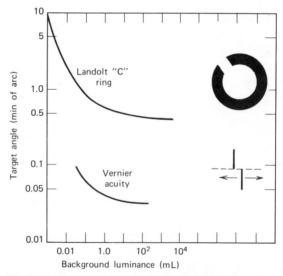

Figure 15.14. Visual acuity improves with better lighting. The top curve shows the acuity for the Landolt C ring, where the gap direction must be recognized. Vernier acuity (the ability to align two lines) is much better than acuity for the Landolt C. The typical brightness level for reading is about 100 millilamberts (mL). (Adapted from J.H. Taylor in J.F. Parker, Jr., and V.R. West (Eds.), *Bioastronautics Data Book*, 2nd ed., National Aeronautics and Space Administration, Washington, D.C., 1973, p. 638.)

I_2), where I_1 and I_2 are the light intensities from the two areas. The low contrast between two areas of interest on an x-ray often severely limits the usefulness of the x-ray.

Before discussing contrast further, we need to define a unit for measuring the "darkness" of an x-ray film or any other optical absorber. The optical density OD is defined as $OD = \log (I_0/I)$ where I_0 is the light intensity without the absorber and I is the intensity with the absorber. For example, a piece of film that transmits 10% of the incident light has an optical density of $\log (1/0.1) = \log 10 = 1.0$. A film that absorbs 99% of the light has an I_0/I of 100, and the $OD = \log 10^2 = 2.0$. An $OD = 3$ means that only 0.1% of the light is transmitted. A light must be bright to be seen through an $OD = 3$ filter (see Example 15.2).

Example 15.2
What is the amount of light transmitted by a film of $OD = 0.5$?

$$0.5 = \log \frac{I_0}{I} \text{ or } \frac{I_0}{I} = \text{antilog } 0.5 \simeq 3.15$$

$I = I_0/3.15$, or about one-third of the light is transmitted.

You can put together two neutral density filters of $OD = 1.0$ (available from most camera stores) to get $OD = 2.0$, you can put three together to get $OD = 3.0$, and so forth. Even a "perfectly clear" piece of film has a small optical density due to the reflection at the surfaces. Typically, about 3% is reflected back from a clean optical surface. The range of optical densities used for most x-ray viewing is 0.3 to 2.0. However, darker areas of as much as 3.0 can be viewed with a spotlight. Taking x-rays of the correct darkness is important for obtaining the maximum amount of medical information.

If two films are placed adjacent to each other, what must be the difference in their optical densities for the eyes to recognize them as different? This difference depends on the light intensity (Fig. 15.15). At very low light levels, when we are using our rods, a factor of 2 difference in light may be required. Under optimum light levels a 1% or 2% difference is detectable. Usually an x-ray does not have a nice sharp border where two areas meet; the optical density gradually changes from one area to another. As much as a 20% change in light intensity may then be required for two films of different optical densities to be recognized as different, even under optimum conditions.

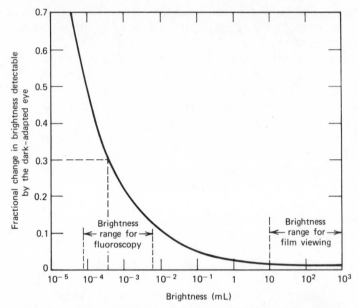

Figure 15.15.   The contrast needed to distinguish different shades of gray at different light levels. At levels used for conventional fluoroscopy, a 30% change in light is needed to be detectable. (Adapted from H.E. Johns, *The Physics of Radiology*, 2nd ed., Charles C. Thomas, Springfield, Ill., 1961, p. 597.)

15.7. OPTICAL ILLUSIONS AND RELATED PHENOMENA

Viewing shades of gray plays tricks on the mind. Figure 15.16*a* shows uniformly colored strips of various shades of gray next to each other. Rather than looking uniform, a given bar looks darker where it borders a lighter bar and lighter on the side near a darker bar. (To help see that a bar is uniform, use white paper to cover the adjacent bars.) This effect is produced by interactions between adjacent groups of light sensors in the eye as illustrated schematically in Fig. 15.16*c*. This "edge enhancement" would appear to help radiologists see borders on x-ray films. Unfortunately, it helps only where they do not need the help—where the contrast is already good. The effect disappears when the contrast is small.

Another optical illusion is that the shade an area of gray appears depends on its surroundings. The circles in Fig. 15.17 are all the same shade of gray, but the circle surrounded by a light area looks darker. This effect has practical significance for viewing x-ray films on an illuminator. If the film does not cover the entire light source, the bright light around the film makes it appear darker.

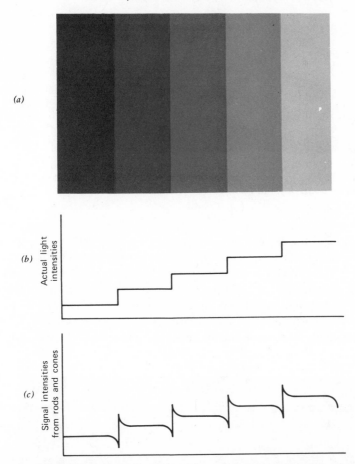

Figure 15.16. Strips of gray (*a*) that are uniformly shaded as shown by their actual light intensities (*b*) are perceived by the rods and cones as darker near a light bar and lighter near a dark bar (*c*).

Like many of the other sensory nerves, the nerve cells in the eye stop sending signals in the presence of a steady stimulus. The eye overcomes this handicap by continuously jiggling during normal use in addition to making gross movements. For a simple demonstration of the fading of the signal from the eye, find a nice large plain ceiling or wall that has a noticeable border between two sections. If you stare steadily at this area, you will see the two sections gradually take on the same shade and the border between them disappear. If you shift your eyes, the border will return!

Figure 15.17.   The perception of the darkness of an image depends on the darkness of its background. All three circles are the same shade of gray.

If you look into the eye with an ophthalmoscope you can see the many blood vessels in the retina (Fig. 15.18). These blood vessels block out light to the rods and cones behind them. The reason we do not normally see these vessels is that the shadows they produce are always over the same rods and cones and this steady signal fades in moments after we open our eyes in the morning. Most people are not prepared to look at the blood vessels in their eyes first thing in the morning anyway! You can, however,

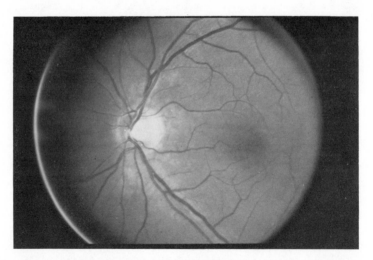

Figure 15.18. The blood vessels in the retina. These blood vessels are not seen during normal vision because their shadows always fall on the same rods and cones. (Courtesy of Thomas Stevens, M.D., University of Wisconsin, Madison.)

see these vessels by transillumination with a penlight. With your eye closed, hold a penlight against your eyelid and move it rapidly back and forth. Some of the light will penetrate the eyelid and sclera and cause the blood vessels to cast shadows on different rods and cones, producing a visible image of the network of blood vessels. This technique is used as a clinical test to ascertain the presence of retinal function in eyes in which the retina cannot be seen.

Everyone has at one time or another seen "lights" when his eyes were closed. These "lights" are called *phosphenes* and can be stimulated by pressing on the eye with the fingers or by closing the eyes very tightly. Phosphenes are produced by the stimulation of some of the normal light sensors. The brain interprets any signals received from the optic nerve as light; it cannot distinguish the various sources of signals from the optic nerve. If you receive a blow on the head, this will stimulate some of the light sensing nerves and you will "see stars." If a small voltage (~ 4 V) is placed across your eye when your eyes are closed and dark adapted, you will see "light" each time the voltage is turned on or off. Since the nerves normally transmit signals of less than 0.1 V, it is not surprising that the rapid changes produced by this technique trigger some nerve pulses.

We will not discuss in detail the important relationships between the brain and the eyes. Everybody has seen illustrations in which straight lines look curved and lines of equal length appear to be quite different in length (Fig. 15.19). When you move your eyes or your head, the image in

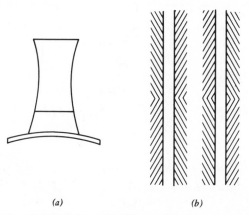

(a)   (b)

Figure 15.19. (a) The width-height optical illusion. The base is as long as the height. (b) The straight-line optical illusion. The surrounding patterns make the straight lines appear curved.

your eyes sweeps rapidly; you might think the brain would get the impression that the room is in motion. Apparently as the brain signals the muscles to move the eyes or head it also informs the visual cortex, and we are not confused by the illusion of a moving room. However, if the eye moves by some external force it can give the impression that the room is moving. Close one eye and half close the other. Push with a finger on the eyelid of the half-closed eye and you will see the image you are viewing move; no signals have been sent to the visual cortex to cancel out your finger pressure.

A characteristic of the eye-brain system that we take for granted is the ability of the brain to fuse the slightly different images from the eyes into a three-dimensional (3-D) image. An artificial 3-D effect can be produced by showing slightly different pictures to the two eyes. About 150 years ago Wheatstone invented the stereoscope, which is still the basic device used to view stereoscopic x-rays. For stereoscopic x-rays, two x-rays are taken of the same part of the body from slightly different angles corresponding to the views normally seen by the two eyes. The two x-rays are then placed in a stereoscope so that each eye sees one image; these images are fused by the brain into a 3-D image (Fig. 15.20). Stereoscopic x-rays are often taken of the skull.

It is also possible to take 3-D pictures in an electron microscope by tilting the specimen slightly before taking the second picture. This technique can be used to see inside a chromosome.

The brain merges the signals from both eyes even if one is badly out of focus or if the image in one eye has a magnification as much as 5% larger than that in the other eye! For a simple demonstration of the brain's ability to merge two quite different images, roll up a tube of paper, hold it to one eye, and look at a plain well-lit wall. Hold your hand alongside the paper tube and look at it with the other eye. You will be surprised (we hope) to see a hole in your hand (Fig. 15.21)!

Sometimes you think you see something because you expect to see it. If a radiologist makes a medical diagnosis on the basis of an imagined image this type of illusion can be dangerous. This situation is summed up in the motto, "If I hadn't believed it with my own mind, I never would have seen it!"

When you view a flash of light, the visual image of the light persists for some time after the flash. That is, after the flash there is a period of many milliseconds when the brain thinks the light is still on. If the frequency of consecutive flashes is increased, at some rate the eye-brain system no longer recognizes the light as flashing—flicker fusion is said to exist. This rate depends on the intensity of the flashes; bright flashes may not fuse

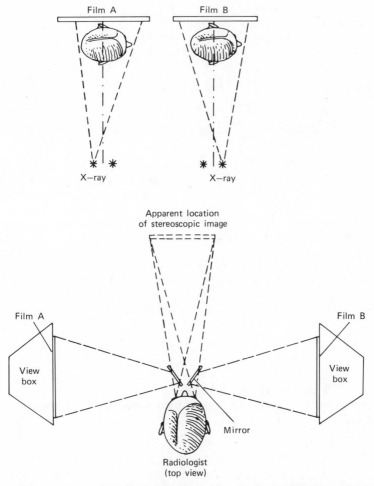

Figure 15.20. The use of a stereoscopic system to obtain a three-dimensional view. The x-rays are taken from slightly different angles to simulate the spacing of the eyes.

into a "steady light" until about 50 Hz, whereas dim flashes will appear as a steady light at as low as 12 Hz. The rods have a higher flicker-fusion rate than the cones. A flicker in your peripheral vision may fuse when you look at it directly.

The ability of the eye to fuse flashing lights into "steady" light is the basis of movies, in which 16 to 32 images per second are flashed on the screen. We are not aware that the screen is blank most of the time.

Figure 15.21. By looking through a tube with both eyes open as shown, you can painlessly put a hole in your hand.

15.8. DEFECTIVE VISION AND ITS CORRECTION

Glasses (corrective lenses) to help defective vision were among the first prosthetic devices invented. It is not clear who gets credit for the invention, but the first reference to the use of glasses (hand held) dates from around 1300. About a century later someone figured out how to attach glasses to the head. However, wearing glasses in public was considered in bad taste even in the 1800s. Until this century, their use was restricted to the well-to-do. The history of corrective lenses includes the invention of the curved lens to reduce aberrations when looking at an angle; bifocals, primarily for old people; and contact lenses, primarily for young people.

In order to discuss the strength of a corrective lens for a defective eye we need to review the basic equations of simple lenses. There is a simple relationship between the focal length F, the object distance P, and the image distance Q of a thin lens (Fig. 15.22)

$$\frac{1}{F} = \frac{1}{P} + \frac{1}{Q}$$

If F is measured in meters, then $1/F$ is the lens strength in *diopters* (D). That is, a positive (converging) lens with a focal length of 0.1 m has a strength of 10 D. The focal length F of a negative (diverging) lens is considered to be negative. A negative lens with a focal length of -0.5 m has a strength of -2 D.

The focal length F of a combination of two lenses with focal lengths F_1 and F_2 is given by $1/F = (1/F_1) + (1/F_2)$. For three lenses the equation is $1/F = (1/F_1) + (1/F_2) + (1/F_3)$. Unless you like working with fractions the solution for F is rather tedious. However, you will note that if each of the focal lengths is given in meters the equation says that the strength of

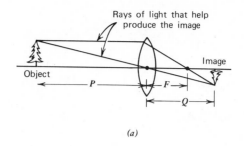

(a)

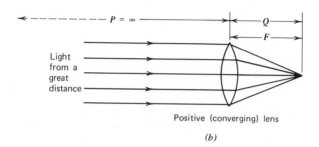

Positive (converging) lens

(b)

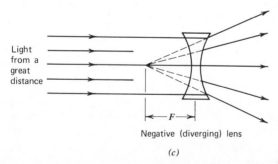

Negative (diverging) lens

(c)

Figure 15.22. **(a) The distance from the lens to the object *P* and the distance from the lens to the place where the image is formed *Q* are related to the focal length *F* of a positive lens by the equation $1/P + 1/Q = 1/F$. (b) Light coming from a great distance to a positive lens converges at the focus of the lens; the image distance *Q* is equal to the focal length *F*. (c) Light from a great distance striking a negative lens diverges. The light appears to diverge from the focal point on the left side of the lens. No image is formed.**

364 *Physics of Eyes and Vision*

the combination in diopters is equal to the sum of the diopters of the various lenses (see Example 15.3).

Example 15.3

Assume lens A with focal length $F_A = 0.33$ m is combined with lens B with focal length $F_B = 0.25$ m. What is the focal length of the combination? What is the dioptric strength of the combination?

$$\frac{1}{F} = \frac{1}{F_A} + \frac{1}{F_B} = \frac{1}{0.33} + \frac{1}{0.25} = \frac{1}{0.143}$$

or $F = 0.143$ m. Note that lens A is 3 D and lens B is 4 D. The combination is simply the sum, or 7 D.

Let us consider the image distance Q of the cornea and lens of the eye to be 2 cm, or 0.02 m (17 mm is a more correct value but the arithmetic is harder). When the normal eye is focused at a great distance (infinity), the focal length F of the eye is the same as Q, or $1/F = 1/Q = 1/0.02$ m, or the eye has a strength of 50 D. If the eye focuses on an object at $P = 0.25$ m, then $1/F = (1/P) + (1/Q)$ gives us $1/F = (1/0.25) + (1/0.02) = 4 + 50 = 54$ D.

Now let us discuss defective eyesight due to focusing (refractive) problems—ametropia. Ametropia affects over half of the population of the United States. It is often possible to correct it completely with glasses. There are four general types of ametropia: *myopia* (near-sightedness), *hyperopia* or *hypermetropia* (far-sightedness), *astigmatism* (asymmetrical focusing), and *presbyopia* (old sight), or lack of accommodation. Figure 15.23 illustrates these conditions schematically and shows the regions where blurring occurs. For each eye we define the near point as the closest distance at which it can see clearly; the far point is the greatest distance at which it has good vision. The various focusing problems and their characteristics are summarized in Table 15.2.

The myopic individual usually has too long an eyeball or too much curvature of the cornea; distant objects come to a focus in front of the retina, and the rays diverge to cause a blurred image at the retina (Fig. 15.24*b*). This condition is easily corrected with a negative lens (see Example 15.4).

Example 15.4

Let us determine the strength of a lens needed to correct a myopic eye with a far point of 1.0 m. We consider the image (lens to retina) distance to be 2 cm (0.02 m).

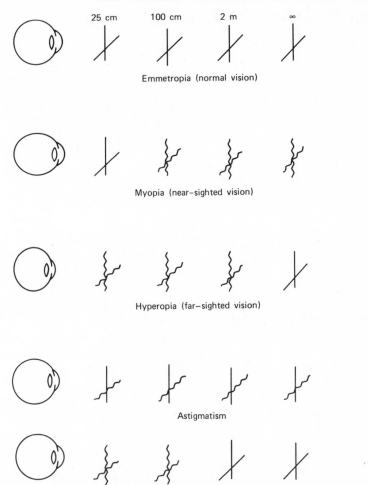

Figure 15.23. Schematic of normal and defective focusing. Wavy lines indicate a blurred image on the retina.

A person who is focusing an object at 1.0 m has a lens strength of $1/F = (1/1.0) + (1/0.02) = 51$ D. An eye able to focus at infinity has a strength of $(1/\infty) + (1/0.02) = 50$ D. Thus a myopic person with a far point at 1.0 m has 1 D too many, and a negative lens of -1.0 D will correct his vision.

A hyperopic eye has a near point further away than normal and uses some of its accommodation to see distant objects clearly. The usual cause

Table 15.2.   A Summary of Various Focusing Problems and Their Characteristics

Focusing Problem	Common Name	Usual Cause	Corrected With
Myopia	Near-sighted vision	Long eyeball or cornea too curved	Negative lens
Hyperopia	Far-sighted vision	Short eyeball or cornea not curved enough	Positive lens
Astigmatism	—	Unequal curvature of cornea	Cylindrical lens
Presbyopia	Old-age vision	Lack of accommodation	Bifocals

of hyperopia is too short an eyeball (Fig. 15.24c). A positive lens is used to correct this condition (see Example 15.5).

Example 15.5

Let us consider a far-sighted eye with a near point of 2.0 m. What power lens will let this person read comfortably at 0.25 m?

The strength of a good eye focused at 0.25 m is given by (1/0.25) + (1/0.02) = 4 + 50 = 54 D. An eye focused at 2 m has a strength of (1/2.0) + (1/0.02)

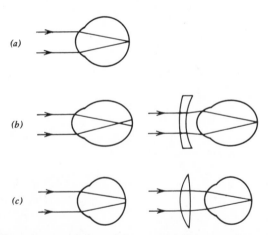

Figure 15.24.   Focusing properties of the eye. (*a*) **The normal, or emmetropic, eye focuses the image on the retina.** (*b*) **The near-sighted, or myopic, eye focuses the image in front of the retina. This problem is corrected with a negative lens.** (*c*) **The far-sighted, or hyperopic, eye focuses the image behind the retina. This problem is corrected with a positive lens.**

= 0.5 + 50 = 50.5 D. A corrective lens of 54 − 50.5 = +3.5 D would be prescribed for this eye.

You can check to see if you are myopic or hyperopic. Look through a pinhole at a well-illuminated distinct object, for example, a streetlight (Fig. 15.25). Move the pinhole up and down in front of your eye. If you are an emmetrope, you will not see any motion; if you are myopic, the image on the retina will move in the direction opposite that of the card and will be interpreted by the brain as moving in the same direction; and if you are hyperopic, the motion of the image on the retina will be in the same direction as the card and will appear to be moving in the opposite direction.

You can also easily check whether glasses have positive or negative lenses by looking at an object through one lens held some distance away. When you move the lens, the object also appears to move. If it moves in the same direction as the motion of the lens, it is a negative lens; if it moves in the opposite direction, it is a positive lens. Another test is to hold the lens over some printing. If it enlarges the printing, the lens is positive; if it makes the printing smaller, the lens is negative.

In astigmatism, the curvature of the cornea is uneven. Astigmatism cannot be corrected by a simple positive or negative lens. A simple test for astigmatism is to look at a pattern of radial lines (Fig. 15.26). An astigmatic eye will see lines going in one direction more clearly than lines going in other directions. Astigmatism is corrected with an asymmetric lens in which the strength is greater in one direction than in the perpen-

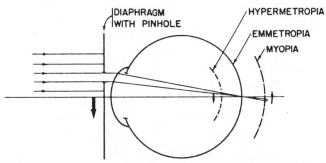

Figure 15.25.   Moving a pinhole in front of the ametropic eye while it is looking at a distant object causes apparent motion of the object. Emmetropes see no motion. (From M.L. Rubin, *Optics for Clinicians*, Triad Scientific Publishers, Gainesville, Fla., 1971, p. 188.)

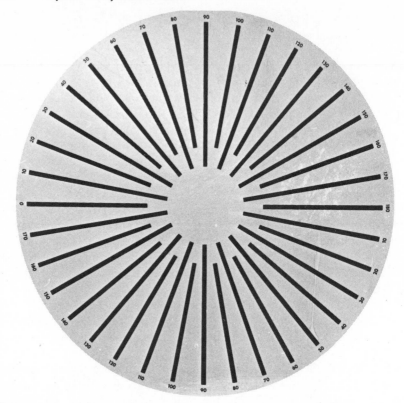

Figure 15.26. A simple test for astigmatism. An eye with astigmatism sees lines going in one direction more clearly than lines going in other directions.

dicular direction (Fig. 15.27). If you wear glasses to correct astigmatism, hold your glasses some distance from your head and rotate them while looking at an object through one lens. You will notice that the object appears to change shape as you rotate the glasses.

Often a person older than 50 notices that he has trouble reading fine print; when he holds the book far enough away to focus clearly, the print is too small for him to distinguish the letters. Although reading in a bright light helps because it narrows the pupil and gives him a better depth of focus, he will need reading glasses. If he already wears glasses to correct a vision defect, he will need bifocals. This problem is due to the loss of accommodation with age (Fig. 15.28). The lens becomes less pliable, and when the tension on it is released, its dimensions change only slightly. As you can see in Fig. 15.28, this loss of accommodation starts at an early

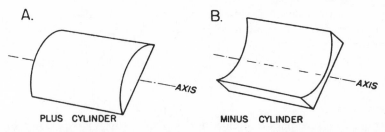

Figure 15.27. Astigmatism is corrected by adding a cylindrical lens to a spherical lens. The cylinder may be either (A) converging (plus cylinder) or (B) diverging (minus cylinder). (From M.L. Rubin, *Optics for Clinicians,* Triad Scientific Publishers, Gainesville, Fla., 1971, p. 94.)

age. (See Examples 15.6 and 15.7 for calculations of the lens strength needed for presbyopic eyes.)

Example 15.6

The optometrist finds that a patient who had good vision now has a near point of 0.5 m and that he likes to read at a distance of 0.25 m. What is his accommodation, and what strength reading glasses should he have? Remember that $1/F = (1/P) + (1/Q)$.

An eye focused at infinity has $1/F = (1/\infty) + (1/0.02) = 50$ D, while an eye focused at 0.5 m has a strength of $(1/0.5) + (1/0.02) = 2 + 50 = 52$ D.

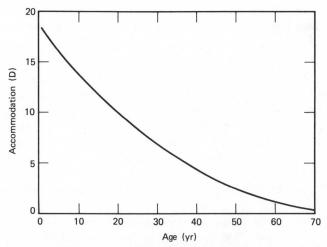

Figure 15.28. Loss of accommodation with age. The decrease in accommodation usually becomes noticeable after age 40.

This patient thus has an accommodation of 2 D. To be able to focus at 0.25 m, the eye needs a strength of $(1/0.25) + (1/0.02) = 4 + 50 = 54$ D. Therefore, his reading glasses should have a strength of $54 - 52$, or $+2$ D. He can buy these in a dime store for a few dollars, but he may have some other eye problem that would not be detected if he did.

Example 15.7
Consider a myopic person already wearing corrective glasses. This person finds that she has to take off her glasses to read fine print, but because she also has astigmatism this solution is not satisfactory. If her near point with glasses is 0.5 m and she wishes to read at 0.25 m, she will need $+2$ D like the subject in Example 15.6. In her case, the optometrist will probably prescribe bifocals in which the lower (reading) part has 2 D more than her regular prescription.

If you wear corrective lenses, you should carry a copy of your prescription. If you lose your glasses far from home, you can have a new pair made without a re-examination. One prescription for glasses reads as follows:

	Sphere	Cylinder		Axis	Add
O.D.	-1.25	-1.25	×	180	$+1.25$
O.S.	-1.75	-1.75	×	163	$+1.25$

This means that the right eye (O.D.) needs a spherical lens of -1.25 D added to a cylindrical lens of -1.25 D in the horizontal plane (180°). In the reading portion of the bifocal, a spherical lens of $+1.25$ D is added to the above prescription. That is, the effective strength in the lower portion of the right lens is a cylindrical lens of -1.25 D to correct the astigmatism. The prescription for the left eye (O.S.) is interpreted in the same way.

Often fractions such as 20/20 or 6/6 are included after the prescription for each eye. This indicates that when the glasses are worn, the eye will be able to read 20/20, that is, to read at 20 ft the line on the Snellen chart that a normal eye can read at 20 ft. The ratio 6/6 is the same statement in meters, since 6 m ≈ 20 ft.

The idea of *contact lenses* dates back to before 1900, but not until the late 1950s were many of the technical and medical problems associated with them satisfactorily solved. Contact lenses are made of either hard or soft plastic. In the United States, the hard plastic type has been the most popular. The contact lens rests on a film of tears on the very front (apex) of the cornea, one of the most sensitive spots on the body. Many people do not wish to go through the necessary and uncomfortable break-in period (that may last up to a year). Contact lenses are used primarily for cosmetic reasons by young people.

From a physics standpoint, the contact lens performs the same function as an ordinary corrective lens. However, the focal power of the combination of two lenses depends on the separation between them. The separation between ordinary glasses and the cornea is fairly well determined by the construction of the frame. Changing this separation does not have much effect unless the glasses have very strong lenses. If you wear glasses, move them a few centimeters from your eyes and see if you can notice much effect. The direct contact between a contact lens and the cornea, however, noticeably affects the prescription. A myopic person switching from glasses to contact lenses would need weaker negative lenses, and a hyperope would need stronger positive lenses. The image size on the retina produced by contact lenses is different from that produced by ordinary glasses; it is larger in myopia and smaller in hyperopia. Wearing contact lenses requires more accommodative effort for myopia and less for hyperopia. Thus a myopic person in the early stages of presbyopia is hindered by contact lenses, and the hyperope in the same situation is helped.

Since a contact lens does not have a fixed orientation, the question of correcting for astigmatism must be resolved. You will recall that astigmatism is caused by nonuniform curvature of the cornea. The hard contact lens is made to fit the curve with the largest radius; when it is worn, the space left in the region of the higher curvature fills with tear fluid. The tears correct for astigmatism by making a symmetrical "new cornea."

Soft plastic contact lenses were developed in Czechoslovakia in the 1950s. They are much more comfortable to wear than hard plastic lenses. New users can often wear them all day long, even on the first day. Also, soft lenses are permeable to gases so that oxygen can reach the cornea directly. With hard plastic lenses, the oxygen is dissolved in the tear layer, and each blink carries a fresh supply under the lens to the cornea. The main disadvantages of soft contact lenses are (1) they cost more; (2) they require a special daily cleaning procedure; and (3) since they conform to the cornea, they do not correct for astigmatism.

Contact lenses are sometimes used for other than cosmetic reasons. In rough sports such as football and soccer, players often wear contact lenses for safety reasons. Soft plastic lenses are also used to administer medication directly to the cornea over a period of hours. The lens is soaked in the medication before it is put in the eye. Readers wishing to learn more about contact lenses should consult the bibliography at the end of the chapter.

15.9. COLOR VISION AND CHROMATIC ABERRATION

One of the remarkable abilities of the eye is its ability to see color. The exact mechanism of color vision is not well understood, but it is fairly well accepted that there are three types of cones that respond to light from three different parts of the spectrum. Images on a color TV are produced by a method similar in some aspects. If you examine a color TV screen with a magnifying glass, you will see a myriad of small red, green, and blue dots. These dots can produce, in different combinations, all of the colors of the spectrum. It is thought that in an analogous way, signals are sent to the brain from the three "colored" cones in different combinations, permitting it to determine color.

If one of the color sets is gone, a person is *color blind*—he confuses certain colors. Approximately 8% of all men and 0.5% of all women are color blind. It is rare that someone is completely color blind, that is, sees only shades of gray.

Chromatic aberration is a common lens defect caused by the change of the index of refraction with wavelength. (This permits a prism to separate white light into a rainbow of colors.) Chromatic aberration causes different colors to come to a focus at different distances. In a simple lens this will produce colored fringes on the image of a white object. Chromatic aberration is greatly reduced in an expensive camera because a lens is used that is made of several types of glass chosen to compensate for the change of the index of refraction with wavelength.

The eyes do not produce colored fringes, or if they do, we are unaware of them. Nevertheless, the acuity of the eye is affected by the different focal lengths for the different colors. For example, the change in focal length from deep blue (390 nm) to deep red (760 nm) is almost 0.7 mm, or more than twice the thickness of the retina! It takes about 2.5 D of lens power to shift the focal length that amount (Fig. 15.29). Chromatic aberration in the eye can be demonstrated by the simple arrangement illustrated in Fig. 15.30. Look at the red filament of a clear lightbulb through a cobalt glass (blue) filter. You will see two images of the filament, one red and one blue, next to each other. If your eye could focus red and blue equally the two images would be superimposed. One of the reasons chromatic aberration is no problem in ordinary vision is that it is rare to encounter such an extreme situation of colors from the ends of the spectrum only. The eyes have their best sensitivity in a rather narrow yellow band in the center of the visible spectrum (Fig. 15.9), and the iris limits the light to the center of the lens, where chromatic aberration is least. The yellowish adult lens acts as a filter to remove some of the reds and blues from the light that strikes the retina, although no observable improvement is made by wear-

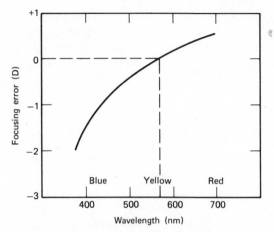

Figure 15.29. **Different colors are focused differently by the eye. There is about 2.5 D difference in focusing power between deep blue and deep red. The graph shows this effect for a normal eye (normalized to yellow light).**

ing yellow-tinted glasses. Monochromatic illumination plus a corrective lens gives higher visual acuity than white light. The best acuity is obtained with yellow light, although the effect is not very dependent on wavelength.

A special color effect that is sometimes noticeable at dusk is called the *Purkinje effect*. Purkinje noticed that at dusk the blue blossoms on his flowers appeared more brilliant than the red blossoms. This effect is

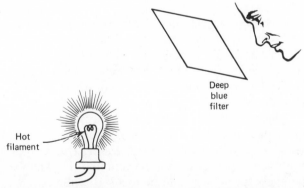

Figure 15.30. **A simple arrangement to demonstrate chromatic aberration. The eye sees separate red and blue filaments due to the different focal properties of the eye for the ends of the spectrum.**

caused by the shift of the best sensitivity of the eyes toward the blue as the rods take over from the cones at low light levels. Since the eyes and corrective lenses are optimized for yellow light, this shift toward the blue produces a refractive error of about 1.0 D. In other words, for night vision you should wear glasses with an additional − 1.0 D!

15.10. INSTRUMENTS USED IN OPHTHALMOLOGY

There are three principal instruments used to examine the eye; the ophthalmoscope, which permits the physician to examine the interior of the eye; the retinoscope, which measures the focusing power of the eye; and the keratometer, which measures the curvature of the cornea. Another instrument, the tonometer, measures the pressure in the eye. The lensometer is not used to study the eye: it determines the prescription of an unknown lens. These instruments are discussed in this section; ultrasound measurements of the structures of the eye are discussed in Section 12.4.

The *ophthalmoscope* is by far the most used, and several versions have been designed. It was invented in 1851 by Helmholtz, an early "medical physicist." The principle of the ophthalmoscope is shown in Fig. 15.31. Bright light is projected into the subject's eye, and the returning light from the subject's retina is positioned so that it can be focused by the examiner. The lens system of the patient's eye acts as a built-in magnifier. A trained individual can detect more than eye problems with an ophthalmoscope since increased pressure inside the skull (for example, due to a brain tumor) can cause a noticeable change in the interior of the eye (papilledema).

Retinoscopy is used to determine the prescription of a corrective lens without the patient's active participation, although the eye has to be open and in a position suitable for examination. The technique can be used, for example, on an anesthetized infant. The retinoscope is also sometimes used to check the prescription determined by the usual "which is better—the first or the second" technique.

A streak of light from the retinoscope is projected into the patient's unaccommodated dilated eye. This streak of light is reflected from the retina and acts as a light source for the operator. The retina's function in retinoscopy is the reverse of its normal function (Fig. 15.32a). Since an object at the eye's far point would be focused at the retina of a relaxed eye, a light from the retina of a relaxed eye will produce a focused image at the far point. The operator views the patient's eye through the retinoscope and adds lenses in front of the patient's eye (positive or negative, as

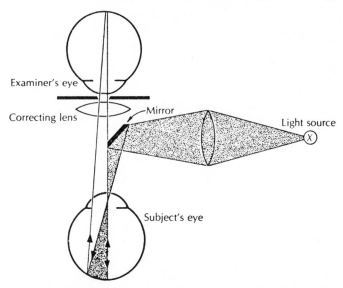

Examiner's eye

Correcting lens

Mirror

Light source

Subject's eye

Figure 15.31. The ophthalmoscope permits the examination of the retina. Light is directed into the patient's eye to permit the examiner to view the retina through a correcting lens. (From T.N. Cornsweet, *Visual Perception*, Academic Press, New York, 1970, p. 62.)

needed) to cause the image from the patient's retina to be focused at the operator's own eye (Fig. 15.32*b*). To determine the prescription needed to correct the patient's eye, the operator must change the lens power of these added lenses by the dioptric power needed to focus the same eye at infinity. If the "operator distance" is 67 cm, -1.5 D must be added.

The *keratometer* is an instrument that measures the curvature of the cornea. This measurement is needed to fit contact lenses. If we illuminate an object of known size placed a known distance from a convex mirror and measure the size of the reflected image, we can determine the curvature of the mirror. In keratometry the cornea acts as a convex mirror. The reflected image is located at the focal plane, a distance $r/2$ behind the surface of the cornea (Fig. 15.33). The keratometer produces a lighted object that is reflected from the cornea while the patient's head is held in a fixed position. The operator adjusts a focus control to place the instrument a known distance from the cornea (Fig. 15.34). Part of the reflected image passes through a prism that causes a second image to be seen by the operator. The operator determines the size of the reflected image by adjusting the angle of the prism to produce a coincidence of marker lines in the two images. The position of the prism after this adjustment is

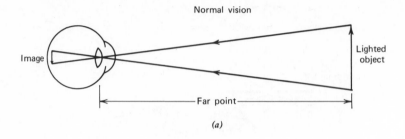

Normal vision

Image

Lighted object

Far point

(a)

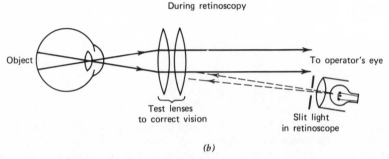

During retinoscopy

Object

To operator's eye

Test lenses
to correct vision

Slit light
in retinoscope

(b)

Figure 15.32. (a) The eye during normal vision. (b) During retinoscopy, reflected light from the patient's retina acts as the object. Lenses are added in front of the eye to focus the image from the retina at the operator's eye.

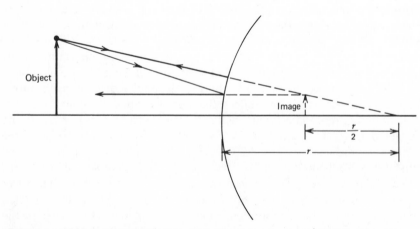

Object

Image

$\frac{r}{2}$

r

Figure 15.33. The reflected image from the cornea is located $r/2$ behind the surface, where r is the radius of curvature of the cornea. This fact is used to measure the curvature of the cornea with the keratometer.

Figure 15.34. The keratometer is an instrument used to measure the curvature of the cornea. This measurement is necessary for prescribing contact lenses. (Courtesy of Bausch & Lomb, Rochester, N.Y.)

indicated on a dial that is calibrated in diopters of focusing power of the cornea. The average value is 44 D, which corresponds to a cornea with a radius of curvature of 7.7 mm. Because astigmatism is common the measurement of the image size is made in both horizontal and vertical directions.

A simple physics experiment involves determining the focal length of a lens. If you have a positive lens (e.g., a magnifying glass), you can produce a focused image of a distant object (e.g., the sun). The image will be at the focal point of the lens. You can measure the distance from the lens to the image to determine the focal length. This technique will not work for a negative lens since no real image will be formed, but a simple modification will permit you to use the same idea. You can combine a negative lens with a strong positive lens of known strength and then produce an image of a distant object to get the focal length of the combination and from this its dioptric strength. You can then determine the diopters of the negative lens from $D_x + D_{known} = D_{measured}$.

These techniques for measuring the power of a lens are suitable for a

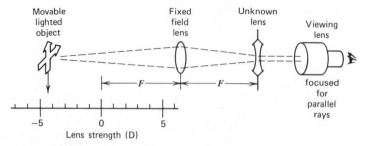

Figure 15.35. A lensometer measures the power of an unknown lens by moving an illuminated object until a sharp image is seen in the viewing lens. The viewing lens is focused for parallel light. In this schematic a −4 D lens is being measured.

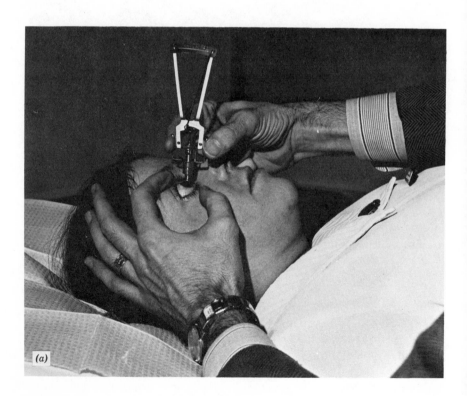

(a)

physics laboratory, but they are clumsy and inconvenient for an ophthalmologist, optometrist, or optician. For routine use, a commercial *lensometer* is more convenient (Fig. 15.35). It moves an illuminated object until it is at the focal point of a lens combination consisting of a fixed field lens and the unknown lens. The parallel rays emerging from the lenses are viewed by a telescope focused at infinity. The fixed field lens is placed a distance equal to its focal length from the unknown lens. This placement conveniently makes the position of the movable lighted object a linear function of the strength of the unknown lens. That is, the scale of diopters (Fig. 15.35) is uniformly divided. When the lighted object is at the focal

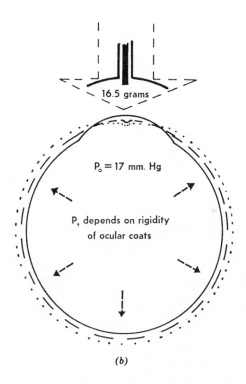

(b)

Figure 15.36. (a) The Schiøtz tonometer measures the pressure in the eye by determining the deflection of the cornea under a given force, usually 16.5 g. (Courtesy of Thomas Stevens, M.D., University of Wisconsin, Madison, Wisc.) (b) Schematic of the Schiøtz tonometer in use. The internal pressure before application of the tonometer is P_0. The pressure after the application of the tonometer P_t depends on the rigidity of the eye. (From Robert C. Drews, *Manual of Tonography*, The C.V. Mosby Co., St. Louis, 1971, p. 10.)

point of the field lens, the lensometer reads 0 D. As it moves further away from the field lens the lensometer reads in negative diopters, and as it moves closer the lensometer reads in positive diopters. For a cylindrical lens (used to correct astigmatism) the strength of each lens axis is measured separately.

It has been known since before 1900 that high eye pressure is related to the condition of glaucoma. This disease narrows the field of view (produces "tunnel vision") and leads to blindness if not treated. The production and outflow rates of the aqueous humor (typically 5 ml/day) determine the pressure in the eye. The fluids in the eyeball are normally under a pressure of 20 to 25 mm Hg; in glaucoma, the pressure may go up to 85 mm Hg (the average arterial blood pressure).

In about 1900 in Germany, Schiøtz invented an instrument for measuring the intraocular pressure—the Schiøtz *tonometer*. The basic technique is to rest the tonometer on the anesthetized cornea with the patient supine (lying face up). The center plunger causes a slight depression in the cornea (Fig. 15.36*a*). The position of the plunger indicates on a scale the internal pressure in the eye. The force of the plunger can be varied by adding various weights. The standard weights have masses of 5.5, 7.5, 10.0, and 15.0 g. The plunger alone has a mass of 11.0 g. With the standard 5.5 g mass, there is 16.5 g resting on a small area of the cornea (Fig. 15.36*b*). This increases the internal pressure about 15 mm Hg depending on the rigidity of the eye.

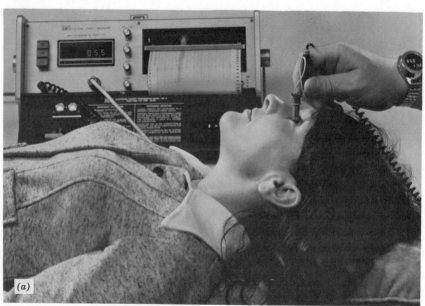

(a)

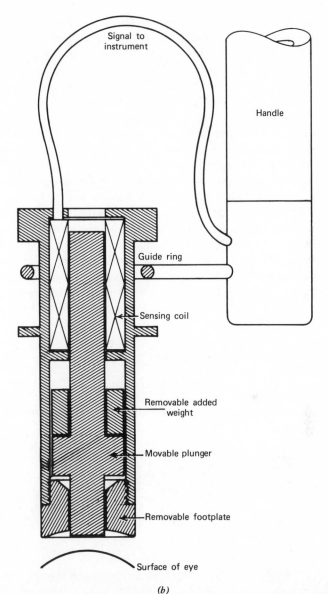

Signal to
instrument

Handle

Guide ring

Sensing coil

Removable added
weight

Movable plunger

Removable footplate

Surface of eye

(b)

**Figure 15.37.   An electronic type Schiøtz tonographer. (a) In use the operator lets
the measuring element rest on the subject's eye. The operator can read the pressure
and the elapsed time on his wrist meter. (b) The weighted movable plunger that rests
on the eye affects the magnetic field of the sensing coil to produce an electrical
signal that indicates the intraocular pressure. (Courtesy of Berkeley Bio-
Engineering, Inc., 600 McCormick Street, Leandro, Calif., 94577.)**

The pressure measured by the tonometer is the original pressure plus the increase due to the instrument. To remove the effect of the rigidity of the eye, another measurement is taken with a heavier weight or with the Goldmann tonometer (described below). The two readings permit the operator to determine, with the help of tables, the original pressure and the rigidity of the eye. The rigidity has no diagnostic value.

The Schiøtz tonometer was modified in about 1950 to give readings electronically. A coil magnetically senses the position of the plunger (Fig. 15.37). One advantage of this model is that it records the change of pressure with time. Figure 15.38 shows such a record, called a *tonograph*. Note the fluctuation in pressure due to the pulse in the arteries. The decrease in pressure indicates that the aqueous fluid is leaving the eye faster than normal under the pressure produced by the tonographer. The outflow can be estimated from the slope of the tonograph The outflow is normally 2 to 6 ml/min with the 15.0 g mass on the plunger. Patients with glaucoma often have an outflow of less than 1 ml/min.

The Goldmann applanation tonometer, developed in about 1955, gives a more accurate measure of the ocular pressure. The measurement is usually taken with the patient in a sitting position (Fig. 15.39). The principle is

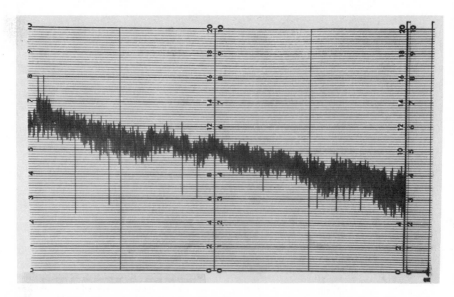

Figure 15.38. Tonograph showing the decrease in pressure with time. The slope permits an estimate of the flow rate. (Courtesy of Thomas Stevens, M.D., University of Wisconsin, Madison, Wisc.)

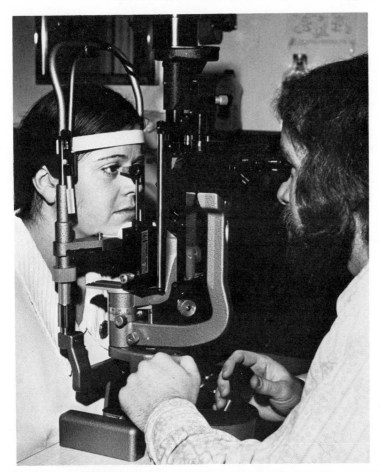

Figure 15.39. The Goldmann applanation tonometer measures the force needed to flatten an area 3.06 mm in diameter on the front of the cornea. (Courtesy of Thomas Stevens, M.D., University of Wisconsin, Madison, Wisc.)

simple; the force needed to flatten an area 3.06 mm in diameter on the front of the cornea is measured. The operator looks through the optical system and adjusts the small force needed to cause the desired flattening. The force needed for a normal eye is equivalent to the weight of a 1.7 g mass. This small force raises the internal pressure only 0.5 mm Hg, while the Schiøtz tonometer increases it 15 mm Hg. The Goldmann tonometer is calibrated directly in millimeters of mercury of internal pressure. The rigidity of the eyeball has little effect on the reading.

A modified applanation tonometer is used somewhat like the Schiøtz with the patient supine. The diameter of the area flattened with a standard weight is measured.

BIBLIOGRAPHY

Begbie, G. H., *Seeing and the Eye: An Introduction to Vision*, Doubleday Anchor, Garden City, N.Y., 1969.
Cornsweet, T. N., *Visual Perception*, Academic, New York, 1970.
Drews, R. C., *Manual of Tonography*, Mosby, St. Louis, 1971.
Gasset, A. R., and H. E. Kaufman (Eds.), *Soft Contact Lens: Symposium and Workshop of the University of Florida, Gainesville*, Mosby, St. Louis, 1972.
Mueller, G., M. Rudolph, and the Editors of Life, *Light and Vision*, Time, New York, 1966.
Pabts, A. M. (Ed.), *The Assessment of Visual Function*, Mosby, St. Louis, 1972.
Rose, A., *Vision: Human and Electronic*, Plenum, New York, 1973.
Rubin, M., *Optics for Clinicians*, Triad, Gainesville, Fla., 1971.
Ruch, T. C., in T. C. Ruch and H. D. Patton (Eds.), *Physiology and Biophysics*, 19th ed., Saunders, Philadelphia, 1965, Chapters 20 and 21.
Southall, J. P. C., *Introduction to Physiological Optics*, Dover, New York, 1961.
Teevan, R. C., and R. C. Birney, *Color Vision*, Van Nostrand, Princeton, N.J., 1961.
Vaughan, D., T. Asbury, and R. Cook, *General Ophthalmology*, 6th ed., Lange, Los Altos, Calif., 1971.
Weymouth, F. W., in T. C. Ruch and H. D. Patton (Eds.), *Physiology and Biophysics*, 19th ed., Saunders, Philadelphia, 1965, Chapter 19.

REVIEW QUESTIONS

1. What is the difference between an optician and an orthopist?
2. What is presbyopia?
3. If you are watching a football game from the end zone, what size will the image of the football on your retina be when the football is at the other end of the field? Assume the football is 30 cm long and 150 m away.
4. At what wavelengths is the eye most sensitive in daylight? At night?
5. What causes the blind spot in the eye? Why are we ordinarily not aware of the blind spot?
6. If 100 light photons enter the eye, about how many are finally absorbed in light sensors of the retina?
7. What is the optimum size of the pupil in an emmetropic eye?
8. What is meant by 20/40 vision?
9. The smallest black dot an emmetrope can see is 2.3×10^{-6} radians. What size does this correspond to at a viewing distance of 25 cm?

10. If the structures in the eye absorb 50% of the photons before they reach the retina, what is the OD of the eye?
11. Why do you not normally see the blood vessels in your retina?
12. If a myope has a near point of 15 cm without glasses and wears a corrective lens of -1.0 D, what is his near point wearing glasses?
13. If an emmetrope has an accommodation of 3 D, what is her near point?
14. If a former emmetrope wears reading glasses of $+2$ D to read at a distance of 25 cm, what is his near point without glasses?
15. How do hard contact lenses correct astigmatism?
16. What is the Purkinje effect?
17. For what is the ophthalmoscope used?
18. What characteristic of the eye does a keratometer measure?
19. What is a tonograph?
20. How does a Goldmann applanation tonometer measure the ocular pressure?

CHAPTER 16

Physics of Diagnostic X-Rays

There is probably not a reader of this book who has not had a number of diagnostic x-rays of her or his teeth or other parts of the body. The use of x-rays in the diagnosis of various medical problems is so common that over half of the people in the United States have at least one x-ray every year. Patients in hospitals have about one x-ray study every three days. In 1976 over 600 million medical and dental x-rays were taken in the United States. This chapter discusses the physical principles involved in the diagnostic use of x-rays in medicine. The therapeutic uses of x-rays are discussed in Chapter 18, and the hazards of x-rays are discussed in Chapter 19.

The x-ray photon is a member of the electromagnetic family that includes light of all types (infrared, visible, and ultraviolet), radio waves, radar and television signals, and gamma rays. Like many important scientific breakthroughs, the discovery of x-rays was accidental. In the fall of 1895, W. C. Roentgen, a physicist at the University of Wurzburg in Germany, was studying cathode rays in his laboratory. He was using a fairly high voltage across a tube covered with black paper that had been evacuated to a low pressure. When he "excited" the tube with high voltage, he noticed that some crystals on a nearby bench glowed and that the rays causing this fluorescence could pass through solid matter. Within a few days Roentgen took the first x-ray film (Fig. 16.1). Perhaps Roentgen suspected the dangers of x-rays: he did not x-ray his own hand, he x-rayed his wife's hand! Within a few weeks the news of his discovery had spread to all of the scientifically advanced countries. It was found that before Roentgen's discovery researchers in a laboratory in the United States had accidentally produced an x-ray image but had not recognized its origin. Within a few months, medical x-rays were being taken in many

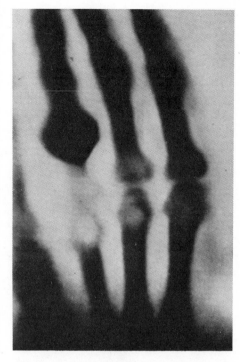

Figure 16.1. The first x-ray taken shows the hand of Roentgen's wife. (Courtesy of the Deutsches Röntgen-Museum, Remscheid-Lennep, West Germany.)

places around the world. In May of 1896, the first medical journal dealing with x-rays was published in Great Britain.

Roentgen very thoroughly analyzed the characteristics of x-rays. In response to an interviewer's question, "What did you think upon discovering the new rays?" he said, "I did not think, I investigated." His investigations established many of the basic facts about the physics of x-rays. We discuss only those aspects related to the use of x-rays in medicine.

The field of radiology has three major branches: diagnostic radiology, radiation therapy, and nuclear medicine. The three areas have relatively little in common except that each uses a part of the electromagnetic spectrum, and the time when a radiologist practiced in all three areas is gone. Today a physician entering radiology can take three or four years of training (residency) in one of these areas of specialization. In 1976 there were about 10,000 diagnostic radiologists (roentgenologists), 1000 radiotherapists, and 3000 nuclear medicine specialists in the United States. While the major radiological societies are made up of members of all three specialties, this situation is changing rapidly as radiation therapy and nuclear medicine grow in importance. Separate organizations for therapists (American Society of Therapeutic Radiologists) and for nuclear medicine specialists (The Society of Nuclear Medicine) have caused the conventional radiological societies (The Radiological Society of North America and the American Roentgen-Ray Society) to become more "diagnostic." In some countries *radiologist* means diagnostic radiologist; perhaps someday this will also be the case in the United States.

Diagnostic radiologists often specialize in particular areas such as pediatric radiology, neuroradiology, and cardiovascular radiology. A physician specializing in one of these areas is usually required to take one year of training beyond the residency requirements for diagnostic radiology.

Working closely with the radiologist is the radiologic technologist, or x-ray technician, who takes most of the x-ray films, or roentgenograms. To become a technologist it is necessary to complete an accredited training program consisting of two or more years after high school and then pass the registry examination. In 1976 there were more than 70,000 registered radiologic technologists in the United States. Unfortunately, many x-rays are taken by untrained operators; in 1976 only three states required operators of x-ray units to be trained. This problem is discussed further in Chapter 19.

16.1. PRODUCTION OF X-RAY BEAMS

A high-speed electron can convert some or all of its energy into an x-ray photon when it strikes an atom, and thus we need to speed up electrons to produce x-rays. Trying to speed up an electron in air is difficult since there are so many electrons on the atoms—about 4×10^{20} in 1 cm^3; before an electron gets going it bumps into another one (Fig. 16.2). It is thus necessary to eliminate most of the electrons, and this is done by using a glass bulb (x-ray tube) from which most of the atoms have been evacuated. For each 1 billion atoms in air only one atom remains in the evacuated x-ray tube, and the electrons can run unimpeded.

The main components of a modern x-ray unit are (1) a source of electrons—a filament, or *cathode*; (2) an evacuated space in which to speed up the electrons; (3) a high positive potential to accelerate the negative electrons; and (4) a target, or *anode*, which the electrons strike to produce x-rays (Fig. 16.3). You may realize that the average home in the United States has a device that contains the same components—a color TV set. In color TVs, quite high voltages (~25 kV) are used to accelerate the electrons. While a few years ago a number of defective color TVs did produce measurable amounts of x-rays, at the present time it is safe to

Figure 16.2.   It is as difficult for an electron to move rapidly through air as for a person to run rapidly through a department store during the biggest sale of the year.

assume that the main danger from color TV is in the programs you watch. X-rays are not emitted from black-and-white TVs because the very weak x-ray photons produced are absorbed in the glass walls of the tube.

Roentgen's x-rays were produced by electrons (cathode rays) from ionized gas in his cathode ray tube. In 1915, Coolidge invented an x-ray tube that produced electrons by "boiling" them off a red-hot filament, and the typical modern x-ray tube is a refinement of this design. In a modern x-ray tube the number of electrons accelerated toward the anode depends on the temperature of the filament, and the maximum energy of the x-ray

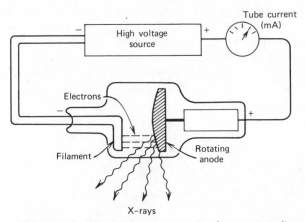

Figure 16.3.   The basic components of an x-ray unit.

photons produced is determined by the accelerating voltage—kilovolt peak (kVp). An x-ray tube operating at 80 kVp will produce x-rays with a spectrum of energies up to a maximum of 80 keV.*

The kilovolt peak used for an x-ray study depends on the thickness of the patient and the type of study being done. X-ray studies of the breast (mammography) are usually done at 25 to 50 kVp, while some hospitals use up to 350 kVp for chest x-rays.

The intensity of the x-ray beam produced when the electrons strike the anode is highly dependent on the anode material. In general, the higher the atomic number (Z) of the target, the more efficiently x-rays are produced. The target material used should also have a high melting point since the heat produced when the electrons are stopped in the surface of the target is substantial. Nearly all x-ray tubes use tungsten targets. The Z of tungsten is 74, and its melting point is about 3400°C.

The electron current that strikes the target is typically 100 to 500 mA—some units even have currents of over 1000 mA. The power put into the surface of the target can be quite large. Recall that the power P is given by $P = IV$, where I is in amperes, V is in volts, and P is in watts. The power at the target of an x-ray tube with a current of 1 A operating at 100 kV (10^5 V) is 1×10^5 W or 100 kW, and over 99% of this power appears as heat.** A power of 100 kW will bring a cup of cold water to the boiling point in less than 1 sec and can heat a typical house on a bitter cold winter day. In an x-ray tube this power is concentrated in an anode area of only a few square millimeters, making overheating a serious problem. If 100 kW is maintained on a target for even 1 sec it can melt it.

Most x-ray tubes have two filaments that can be interchanged to produce either a large or a small focal spot. The small focal spot produces less blurring of the x-ray image than the large focal spot (see Section 16.3), but it concentrates the heat on a smaller area of the target, increasing the chances of overheating and damage. Many years ago radiological engineers found a way to increase the area on the target struck by electrons to avoid overheating without increasing the blurring of the x-ray image. This technique, called the *line-focus principle,* is illustrated in Fig. 16.4. Because of the angle of the target, typically 10 to 20°, the projected focal spot is smaller than the area struck by the electrons; it is analogous to the short shadow cast by a tall man when the sun is almost directly overhead.

---

*One kiloelectron volt (keV) is the energy an electron gains or loses in going across a potential difference of 1000 V. 1 keV $= 1.6 \times 10^{-9}$ erg $= 1.6 \times 10^{-16}$ J. Diagnostic x-rays typically have energies of 15 to 150 keV, while visible light photons have energies of 2 to 4 eV.

**The ratio of the energy that goes into x-ray photons to the energy that goes into heat is approximately $10^{-9} ZV$, where Z is the atomic number of the target. For 100 kV on a tungsten ($Z = 74$) target, about 0.007 or 0.7% of the energy goes into x-ray photons.

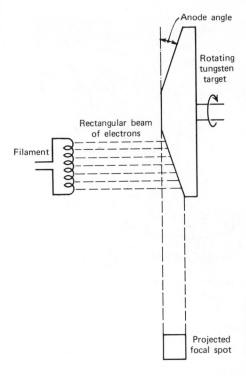

Figure 16.4. Due to the line-focus principle, a relatively large area of the target struck by electrons appears as a much smaller projected focal spot.

The second big breakthrough in designing anodes to avoid overheating was the development in 1930 by Bouwers of the rotating anode x-ray tube (see Fig. 16.3). The normal rotational rate of the anode is 3600 rpm, and the heat is spread over a large area as the anode rotates. Nevertheless, it is still easy to overheat and damage targets (Fig. 16.5), and thus x-ray manufacturers indicate on tube-loading charts how much energy can be safely deposited on the target in a short period of time (Fig. 16.6). When a short exposure is used, the anode does not always make a full rotation at 3600 rpm and thus its full heat capacity is not utilized. For this reason, special high-speed anodes that operate at rates of up to 10^4 rpm were developed.

Since any rotating object can cause vibration if it is not carefully balanced, it is possible for the heavy rotating target to shatter the fragile glass envelope of the x-ray tube. Even a small imbalance can produce large forces at 10^4 rpm. Also, if even a very slight imbalance occurs at a natural frequency of the system, the energy is added to the system in phase with the natural vibration frequency (much like when a child on a swing is pushed at the proper time), and this resonance can shatter the x-ray tube. While it is unlikely that a resonance will occur at exactly the

Figure 16.5. Three damaged rotating tungsten anodes. The upper anode cracked and the surfaces of the lower anodes melted.

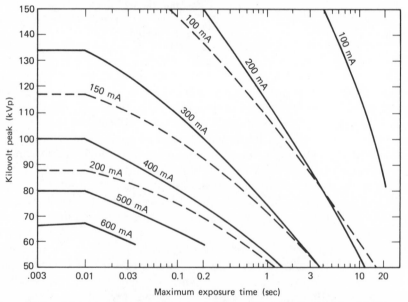

Figure 16.6. A tube loading chart for an x-ray tube with a rotational rate of 3600 rpm. The solid lines are for the large focal spot, and the dashed lines are for the small focal spot. Note that a much shorter exposure time must be used with the small focal spot for a given kilovoltage and tube current.

392

operating frequency of 10^4 rpm, when the anode coasts slowly to a stop after the power is turned off it will pass through one or more resonances of the tube. To avoid shatter the spinning anode is rapidly braked so that it will pass quickly through these resonances.

While the energy of most of the electrons striking the target is dissipated in the form of heat, the remaining few electrons produce useful x-rays. Many times one of these electrons gets close enough to the nucleus of a target atom to be diverted from its path and emits an x-ray photon that has some of its energy (Fig. 16.7a). X-rays produced in this way have a fancy German name, *bremsstrahlung,* which means "braking radiation." Bremsstrahlung is also called *white radiation* since it is analogous to white light and has a range of wavelengths. The amount of bremsstrahlung produced for a given number of electrons striking the anode depends upon two factors: (1) the *Z* of the target—the more protons in the nucleus, the greater the acceleration of the electrons—and (2) the kilovolt peak—the faster the electrons, the more likely they will penetrate into the region of the nucleus.

Sometimes a fast electron strikes a K electron in a target atom and knocks it out of its orbit and free of the atom. The vacancy in the K shell is filled almost immediately when an electron from an outer shell of the atom falls into it, as indicated schematically in Fig. 16.7b, and in the process, a characteristic K x-ray photon is emitted. An x-ray photon emitted when an electron falls from the L level to the K level is called a K_α characteristic x-ray, and that emitted when an electron falls from the M shell to the K shell is called a K_β x-ray. Since the energies of the electrons in the various shells of an atom are precisely determined by nature, an electron falling from an outer shell to an inner shell will always produce an x-ray with an energy characteristic of that atom. Table 16.1 gives the energies of the K_α x-rays of several elements. (The K-edge is explained in the next section.) Characteristic x-rays are of little use at present except in mammography,

Table 16.1. Approximate Energies of the K_α X-Rays and K-Edge for Several Elements

	K_α (keV)	K-Edge (keV)
Aluminum	1.5	1.6
Calcium	5	6
Copper	8	9
Molybdenum	17.5	20
Iodine	28	33
Tungsten	59	70
Lead	75	88

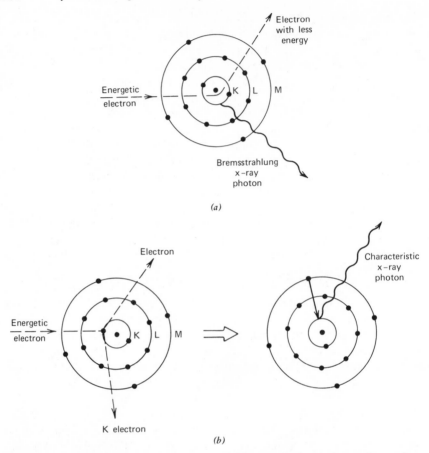

Figure 16.7. Generation of x-rays. (a) A fast electron is diverted near the nucleus
and loses some of its energy as a bremsstrahlung x-ray photon. (b) A fast electron
knocks a K electron free of the atom; when an outer electron falls into the vacancy
an x-ray characteristic of the target atom is emitted.

where a molybdenum target with K_α x-rays of about 18 keV is sometimes
used.

The spectrum of x-rays produced by a modern x-ray generator is shown
in Fig. 16.8. The broad smooth curve is due to the bremsstrahlung, and
the spikes represent the characteristic x-rays. Many of the low-energy
("soft") x-ray photons produced are absorbed in the glass walls of the
x-ray tube.

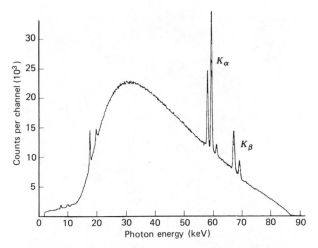

Figure 16.8. The spectrum from a tungsten target x-ray tube operated at 87 kVp. All of the photons below 12 keV and some of those in the 12 to 30 keV region are absorbed in the glass wall of the x-ray tube; the counts shown below 12 keV are due to electrical noise in the detector system. (Courtesy of Robert Jennings, University of Wisconsin, Madison, Wisc.)

16.2. HOW X-RAYS ARE ABSORBED

X-rays are not absorbed equally well by all materials; if they were, they would not be very useful in diagnosis. Heavy elements such as calcium are much better absorbers of x-rays than light elements such as carbon, oxygen, and hydrogen, and as a result, structures containing heavy elements, like the bones, stand out clearly. The soft tissues—fat, muscles, and tumors—all absorb about equally well and are thus difficult to distinguish from each other on an x-ray image. Of course, air is a poor absorber of x-rays (Fig. 16.9).

The *attenuation* of an x-ray beam is its reduction due to the absorption and scattering of some of the photons out of the beam. A simple method of measuring the attenuation of an x-ray beam is shown in Fig. 16.10. A narrow beam of x-rays is produced with a collimator—a lead plate with a hole in it—and an x-ray detector measures the beam intensity. The unattenuated beam intensity is I_0. As sheets of aluminum are introduced into the beam, the intensity I decreases approximately exponentially as shown in Fig. 16.11. (See Appendix A for a general discussion of exponential behavior.) The lower energy (soft) x-rays are absorbed more readily than

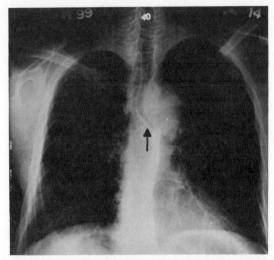

Figure 16.9. An x-ray of the chest showing the different amounts of absorption of x-rays by the various structures. The dark areas (e.g., the lungs) represent little absorption and the white areas (e.g., below the lungs) represent much absorption. The arrow indicates a curve in the esophagus caused by an enlarged artery, or aneurysm, that cannot be seen directly on the x-ray.

the higher energy (hard) x-rays; the greater penetration of the hard x-rays is shown by the flattening of the curve in Fig. 16.11.

The intensity of a monoenergetic x-ray beam would decrease exponentially as shown by the dashed line in Fig. 16.11. The exponential equation describing the attenuation curve for a monoenergetic x-ray beam is

$$I = I_0 e^{-\mu x} \tag{16.1}$$

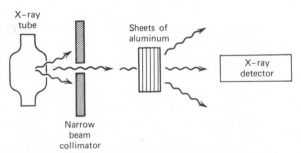

Figure 16.10. Arrangement for measuring the attenuation of an x-ray beam. The x-ray photons are both absorbed and scattered out of the beam.

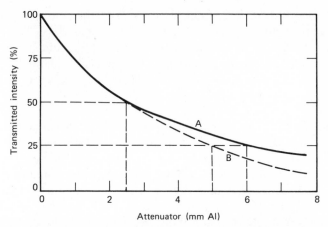

Figure 16.11. Graphs of the transmitted intensity versus the thickness of aluminum attenuator for (A) an x-ray beam and (B) a monoenergetic x-ray beam.

where $e = 2.718$, x is the thickness of the attenuator, and μ is the linear attenuation coefficient of the attenuator. The linear attenuation coefficient is dependent on the energy of the x-ray photons; as the beam becomes harder, it decreases.

The half-value layer (HVL) for an x-ray beam is the thickness of a given material that will reduce the beam intensity by one-half. The half-value layer for the x-ray beam in Fig. 16.11 is 2.5 mm Al. Note that another 3.5 mm Al is needed to reduce the intensity in half again; this value is the second half-value layer. For a monoenergetic x-ray beam, the second half-value layer equals the first half-value layer. The half-value layer is related to the linear attenuation coefficient by

$$\text{HVL} = \frac{0.693}{\mu}$$

In lead the half-value layer of the x-ray beam in Fig. 16.11 would be about 0.1 mm. You can see why lead is used for shielding material. A lead sheet 1.5 mm thick (1/16 in.) would be about 15 HVLs and would reduce the beam intensity by a factor of 2^{15} or about 30,000!

The equivalent energy of an x-ray beam is determined by its half-value layer; it is the energy of a monoenergetic x-ray beam with the same half-value layer. For example, a typical x-ray set operating at 80 kVp with a filter of 3 mm Al would have a half-value layer of about 3 mm Al. Since a beam of monoenergetic 28 keV x-rays also has a half-value layer of 3 mm Al, the equivalent energy of the x-ray beam would be 28 keV.

The mass attenuation coefficient μ_m is used to remove the effect of density when comparing attenuation in several materials. The mass attenuation coefficient of a material is equal to the linear attenuation coefficient μ divided by the density ρ of the material. Equation 16.1 can be rewritten as

$$I = I_0 e^{-(\mu/\rho)(\rho x)} = I_0 e^{-\mu_m(\rho x)} \qquad (16.2)$$

The quantity ρx is in grams per square centimeter and is sometimes called the area density; μ_m is in square centimeters per gram. The mass attenuation coefficient emphasizes that the mass is primarily responsible for attenuating the x-rays. That is, 1.0 g of lead covering an area of 1 cm² will absorb the same amount of x-rays whether its density is 11 g/cm³ or whether it is mixed with plastic to reduce its density to 2 g/cm³. The half-value layer in area density units (g/cm²) is given by $0.693/\mu_m$.

Figure 16.12 shows the mass attenuation coefficients of fat, muscle, bone, iodine, and lead as a function of x-ray energy. Note that on a gram-for-gram basis, iodine is a better absorber than lead from about 30 to about 90 keV. This phenomenon is due to the *photoelectric effect*.

The photoelectric effect is one way x-rays lose energy in the body. It

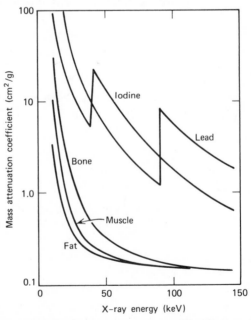

Figure 16.12. Mass attenuation coefficients for various tissues, lead, and iodine. Note that on a mass basis, all tissues attenuate about the same above 100 keV.

occurs when the incoming x-ray photon transfers all of its energy to an electron which then escapes from the atom (Fig. 16.13a). The photoelectron uses some of its energy (the binding energy) to get away from the positive nucleus and spends the remainder ripping electrons off (ionizing) surrounding atoms.

The photoelectric effect is more apt to occur in the intense electric field near the nucleus than in the outer levels of the atom, and it is more common in elements with high Z than in those with low Z. Of course, for a given electron to be liberated its binding energy must be lower than the energy of the x-ray. The binding energy of a K electron in iodine is 33 keV, while that in lead is 88 keV, and from 33 to 88 keV an x-ray photon can release a K electron from iodine but not from lead. When the energy of the x-ray is just slightly greater than the binding energy, the probability that the photoelectric effect will occur increases greatly, and this accounts

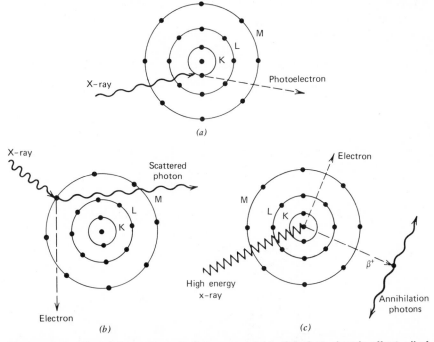

(a)

(b)   *(c)*

Figure 16.13. X-rays lose energy in three ways: (a) In the photoelectric effect, all of the photon energy is given to the photoelectron. (b) In the Compton effect, some energy is given to an electron and some goes into a scattered photon. (c) In pair production, a high-energy photon is converted into an electron and a positron (β^+). The β^+ annihilates to form two photons of 511 keV each that go in opposite directions.

for the sharp rises in the curve for iodine at 33 keV and in the curve for lead at 88 keV in Fig. 16.12. These sharp rises are called *K-edges*. The elements in bone, muscle, and fat have K-edges, but they are at such low energies (~6 keV for calcium) that they do not appear in Fig. 16.12. The K-edge energies for some common elements are given in Table 16.1.

Another important way x-rays lose energy in the body is by the Compton effect. In 1922 A. H. Compton suggested that an x-ray photon can collide with a loosely bound outer electron much like a billiard ball collides with another billiard ball. At the collision, the electron receives part of the energy and the remainder is given to a Compton (scattered) photon, which then travels in a direction different from that of the original x-ray (Fig. 16.13*b*).

The energy transferred to the electron can be calculated in the same way as the energy transferred during a billiard ball collision by using the laws of conservation of energy and momentum. The x-ray has an effective mass m of E/c^2 (from Einstein's famous equation $E = mc^2$), and its momentum is E/c. We can also calculate the energy equivalent of the electron mass to be 511 keV, and the Compton effect is most likely to occur when the x-ray has this energy (Fig. 16.14).

The number of Compton collisions depends only on the number of

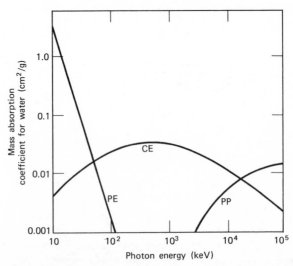

Figure 16.14. The mass absorption coefficient for water in the 10 to 10⁵ keV range. The photoelectric effect (PE) and the Compton effect (CE) are about equally probable at about 30 keV. Pair production (PP) occurs only at high energies and is of no importance in diagnostic radiology.

electrons per cubic centimeter, which is proportional to the density. A gram of bone has about the same number of electrons as 1 g of water, and thus the number of Compton collisions will be about the same. Note in Fig. 16.12 that the mass attenuation coefficients for fat, muscle, and bone are essentially identical at 150 keV where the Compton effect is dominant. However, since the photoelectric effect is more apt to occur in high Z materials than in low Z materials, the fraction of x-rays that lose energy by the Compton effect is greatest in low Z elements. For example, in water or soft tissue the Compton effect is more probable than the photoelectric effect at energies above about 30 keV. Even in bone the Compton effect is more probable than the photoelectric effect at energies above 100 keV.

Pair production is the third major way x-rays give up energy. When a very energetic photon enters the intense electric field of the nucleus, it may be converted into two particles: an electron and a positron (β^+), or positive electron. Providing the mass for the two particles requires a photon with an energy of at least 1.02 MeV, and the remainder of the energy over 1.02 MeV is given to the particles as kinetic energy. The positron is a piece of antimatter. After it has spent its kinetic energy in ionization it does a death dance with an electron. Both then vanish, and their mass energy usually appears as two photons of 511 keV each called *annihilation radiation* (Fig. 16.13*c*).

Since a minimum of 1.02 MeV is necessary for pair production, this type of interaction is only important at very high energies (Fig. 16.14). Because the intense electric field of the nucleus is involved, pair production is more apt to occur in high Z elements than in low Z elements.

How are these interactions related to diagnostic radiology? You can see that pair production is of no use in diagnostic radiology because of the high energies needed and that the photoelectric effect is more useful than the Compton effect because it permits us to see bones and other heavy materials such as bullets in the body. At 30 keV bone absorbs x-rays about 8 times better than tissue due to the photoelectric effect. To make further use of the photoelectric effect radiologists often inject high Z materials, or *contrast media,* into different parts of the body. Compounds containing iodine are often injected into the bloodstream to show the arteries (Fig. 16.15*a*), and an oily mist containing iodine is sometimes sprayed into the lungs to make the airways visible (Fig. 16.15*b*). Radiologists give barium compounds orally to see parts of the upper gastrointestinal tract (upper GI) and barium enemas to view the other end of the digestive system (lower GI) (Fig. 16.15c).

Since gases are poorer absorbers of x-rays than liquids and solids, it is possible to use air as a contrast medium. When a pneumoencephalogram

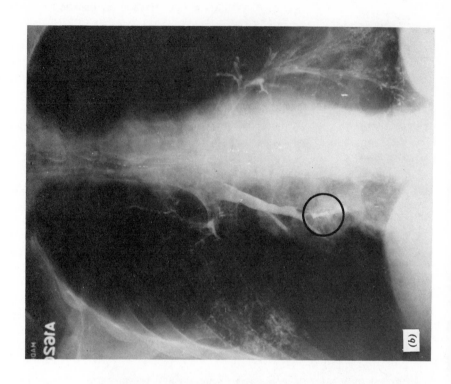

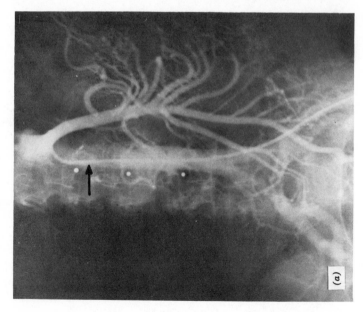

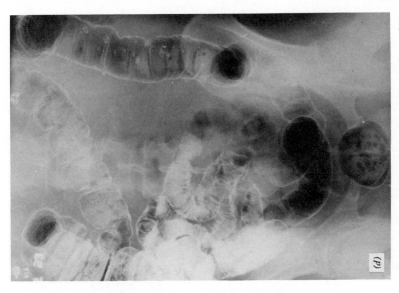

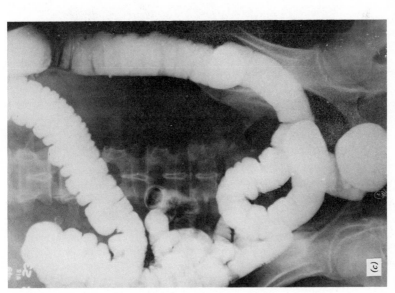

Figure 16.15. Use of contrast media. (a) A compound containing iodine is injected into the arteries through a catheter (arrow) to make them visible. (b) An oily mist containing iodine is sprayed into the lungs to show the airways. (The object inside the circle is a piece of metal from an old wound.) The large bowel can be studied by giving a barium enema (c) or by injecting air into the bowel (d). The two studies of the bowel shown here were done on the same patient.

403

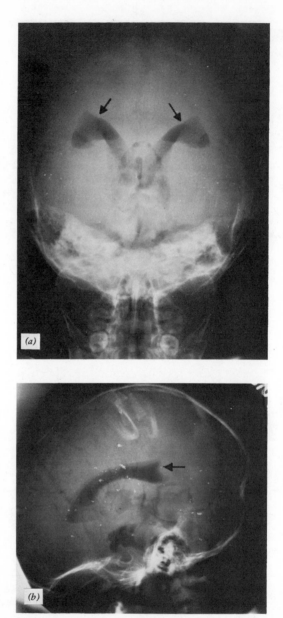

Figure 16.16. Front view (a) and side view (b) pneumoencephalograms taken after some of the cerebrospinal fluid was removed from the ventricles (arrows) and replaced with air.

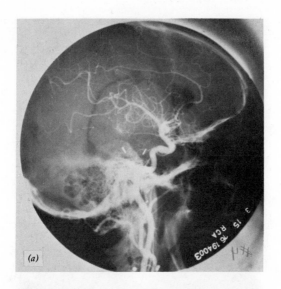

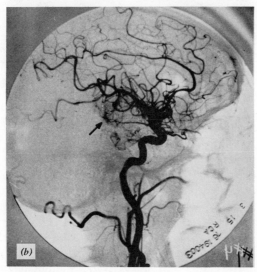

Figure 16.17.   X-ray images of the brain. (a) A lateral x-ray of the skull taken after a contrast medium was injected into the arteries. (b) The image obtained by subtracting a from an x-ray taken before the contrast medium was injected. Note the visualization of a tumor (arrow) that cannot be seen clearly in a.

405

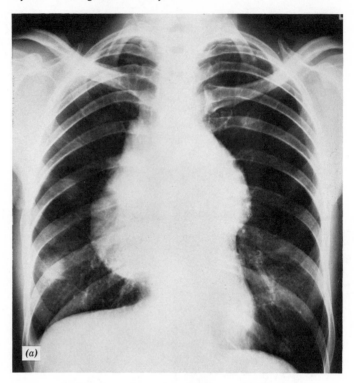

(a)

is taken, air is used to replace some of the fluid in the ventricles of the brain (Fig. 16.16). In a double-contrast study, barium and air are used separately to show the same organ (Fig. 16.15c and d).

To obtain additional information from a contrast study of the arteries it is possible to use a subtraction technique in which an x-ray taken after the injection of a contrast medium is photographically subtracted from an x-ray of the same body part taken before the contrast medium was injected. A subtraction x-ray often contains information that cannot be seen on either of the two conventional x-rays (Fig. 16.17).

If the photoelectric effect did not exist and radiologists had to rely on the Compton effect, x-rays would be much less useful because the Compton effect depends only on the density of the materials. Bone is about twice as dense as soft tissue and will still be seen on an x-ray film taken with high-energy x-rays, but the contrast will be low. This low contrast is not always undesirable, however; for example, on a chest x-ray the ribs are often of no medical interest and hide some of the lungs, and some

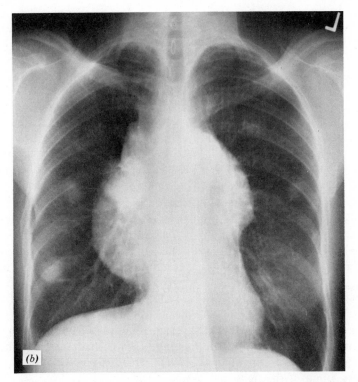

(b)

Figure 16.18. Radiographs taken at 80 kVp (*a*) and at 350 kVp (*b*) of a patient with a large mass in the center of the chest as well as other abnormalities. (Courtesy of George S. Hallenbeck, M.D., Oak Park Hospital, Oak Park, Ill., and Hewlett-Packard, manufacturer of the 350 kV x-ray system.)

medical centers use high potentials (~350 kVp) to make the ribs less obvious (Fig. 16.18).

The Compton effect seriously degrades x-ray images of thick body parts since the scattered radiation that gets through the patient and strikes the film reduces the useful information by reducing the contrast in the image. Methods used to reduce the amount of scattered radiation striking the film are discussed in Section 16.3.

16.3. MAKING AN X-RAY IMAGE

It is relatively easy to make an x-ray image, or roentgenogram—all that is needed is an x-ray source and a film wrapped in black paper on which to record the image. However, making a good x-ray image while keeping the x-ray exposure at a minimum requires considerable knowledge and the use of modern technology. In this section we discuss how a modern x-ray image is produced and the factors that influence the quality, or the detail, of the image. Some of the same factors affect the quality of fluoroscopic images (see Section 16.5).

Unfortunately, x-rays cannot be focused to make a picture as with a camera. X-ray images are basically images of the shadows cast on film by the various structures in the body; they were once called *skiagraphs,* which is Greek for shadow graphs. To better understand the physical problems of recording sharp x-ray shadows, let us consider the problems of casting sharp shadows with visible light.

You can try the experiments illustrated in Fig. 16.19 with your hand as the object. A large lightbulb produces a blurred shadow because the light from different parts of the bulb casts shadows in different places. The blurred edge of the shadow is called the penumbra, which means "next to the shadow." The width of the penumbra can be calculated from the dimensions of the lightbulb and the distances to the object and the paper (Fig. 16.20). The penumbra can be reduced by using a smaller diameter lightbulb or by moving the object closer to the paper (Fig. 16.19*b* and *c*). (Moving the lightbulb further away also reduces the penumbra.)

Another problem involved in casting a sharp shadow is illustrated in Fig. 16.19*d*. The sediment in the water absorbs some of the light and scatters much of the light that is not absorbed.

The problems involved in obtaining good x-ray shadows are analogous, and blurring can be reduced by using a small focal spot, positioning the patient as close to the film as possible (and increasing the distance between the x-ray tube and the film as much as possible), and reducing the amount of scattered radiation striking the film as much as possible. It is also necessary to avoid motion during the exposure, since motion causes blurring.

The nominal sizes of the focal spots on many x-ray units are 1 mm (small focal spot) and 2 mm (large focal spot). However, the focal spots are nearly always larger than the nominal sizes. In addition, a focal spot is generally not uniform and sometimes appears to be two spots close together. The actual size of the focal spot can be determined by several techniques. The physics approach is to make a pinhole image of the focal spot and calculate the size of the focal spot from the size of the image and the distances involved (Fig. 16.21). In another method of measuring the focal spot, a metal plate with patterns of openings of different sizes is placed 20 cm above an x-ray film and an exposure is made (Fig. 16.22). The penumbra prevents the smaller patterns from being resolved, and the

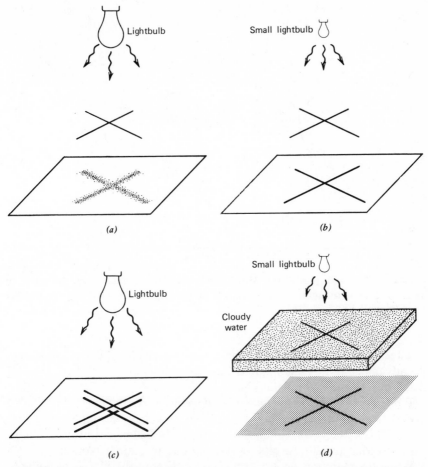

Figure 16.19. The principles involved in casting shadows with visible light. (a) The shadow of an object some distance from a piece of paper is blurred when a large lightbulb is used. This shadow can be made much sharper (b) by using a smaller diameter lightbulb or (c) by moving the object closer to the paper. (d) Cloudy water between the lightbulb and paper absorbs some light and scatters much of the rest, reducing the contrast of the shadow.

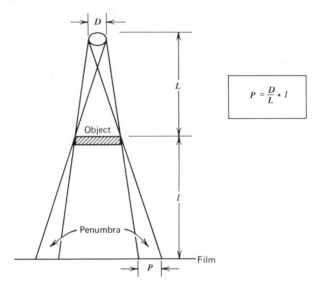

$$P = \frac{D}{L} \cdot l$$

Figure 16.20. The width of the penumbra *P* can be calculated from the ratios of sides of similar triangles if we know the diameter *D* of the light source and the dimensions *L* and *l*.

smallest of the patterns that can be resolved indicates the focal spot size. An advantage of this technique is that it visually demonstrates the loss of detail due to the size of the focal spot.

While decreasing the size of the focal spot reduces the penumbra, it also necessitates lowering the power to avoid damaging the target (see Fig. 16.6). This reduces the intensity of the x-ray beam, requiring a longer exposure that usually results in blurring due to patient motion.

While the patient is generally placed as close to the film as possible in order to reduce the penumbra, sometimes it is also possible to further reduce the penumbra by increasing the distance from the x-ray tube to the film. Chest films are usually taken from a distance of 180 cm (72 in.) for this reason. Unfortunately, increasing the distance reduces the beam intensity according to the inverse square law, making it impractical to take many x-rays from a large distance; at 90 cm (36 in.) the beam intensity is four times that at 180 cm.

To obtain a satisfactory x-ray image of thick body parts such as the abdomen and hips, it is necessary to reduce the scattered radiation at the film. The amount of scattered radiation at the film depends somewhat on the energy of the x-rays, but the thickness of the tissue that the x-ray beam passes through is the most important factor—the thicker the tissue,

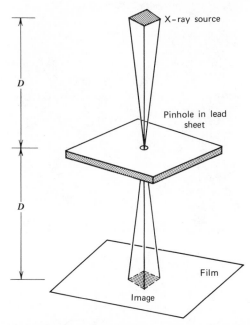

Figure 16.21. The focal spot can be measured by making a pinhole image of it on a piece of film. If the pinhole is midway between the x-ray source and the film, the image will be the same size as the source.

the greater the scatter. Also, the larger the beam, the greater the scatter, and thus one simple way of reducing scattered radiation is by keeping the x-ray beam as small as possible.

The most significant way of reducing the amount of scattered radiation striking the film is by using a grid consisting of a series of lead and plastic strips. The strips are aligned so that unscattered x-rays from the source will go through the plastic strips and strike the film while most of the scattered radiation will strike the lead strips and be absorbed (Fig. 16.23). The grid was invented by G. Bucky in 1915; H. E. Potter improved it in 1919 by making it move during the exposure so that the lead strips do not produce visible shadows on the image. Some modern grids are stationary but have lead strips so fine (4/mm) that their shadows do not interfere with the image.

The grid shown in Fig. 16.23 is called a *focused grid*—it does not focus the x-rays, but its strips are slanted toward the edges so that only unscattered x-rays from an optimum distance (e.g., 1 m) can go through unimpeded. When the x-ray source is very far from this optimum distance, many of the unscattered rays will be absorbed in the lead strips. If a

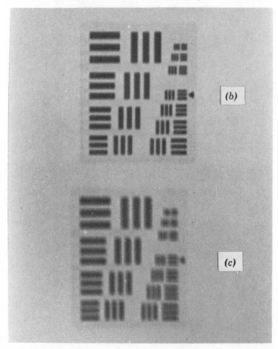

Figure 16.22.   The Wisconsin focal spot test tool uses a pattern of rectangular holes (top of *a*) to cast an image on the film. The image *b* was made with a 1 mm focal spot, and the image *c* was made with a 2 mm focal spot. Notice the greater amount of blurring from the larger focal spot. (Courtesy of Radiation Measurements, Inc., Middleton, Wisc.)

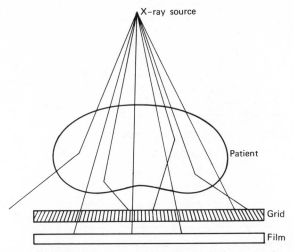

Figure 16.23. **The grid consists of alternating thin lead strips and wide plastic strips. The unscattered x-rays pass through the plastic strips, while most of the scattered x-rays are absorbed by the lead strips.**

focused grid is placed upside down, only the center of the field will be seen on the image.

If two equivalent x-rays are taken, one with and one without a grid, the one taken with the grid will be clearer because of the reduced scatter. However, it will also require a larger exposure to the patient. Since reducing the scatter reduces the darkening of the film, it is necessary to increase the exposure in order to obtain an optimum darkness (optical density) of the film. In addition, a higher exposure must be given because the lead strips absorb some of the unscattered radiation.

When you have a chest x-ray taken, the technologist tells you to hold your breath since reducing motion reduces blurring. However, it is not possible to "hold" your heart motion, and x-rays of the heart are somewhat blurred. This blurring can be reduced by making the exposure as short as possible. The desire for short exposures has led to the development of x-ray tubes with large current capacities that can produce intense x-ray beams.

Most x-ray images are made on a special film sandwiched tightly between two intensifying screens—cardboards covered with a thin coating of crystals (e.g., $CaWO_4$) that absorb x-rays well and give off visible or UV light (fluoresce) when struck by x-rays (Fig. 16.24). The film is coated on both sides with a light-sensitive emulsion, and each side takes a "picture" of the light from the intensifying screen with which it is in

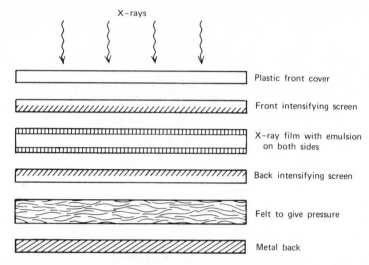

X-rays

Plastic front cover

Front intensifying screen

X-ray film with emulsion on both sides

Back intensifying screen

Felt to give pressure

Metal back

Figure 16.24.   An expanded cross-section of part of an x-ray cassette. The intensifying screens absorb most of the x-rays and give off light that exposes the film. The felt holds the screens in close contact with the film.

contact. Intensifying screens are much more efficient for making x-ray images than film alone. In Section 16.4 we discuss x-ray exposures, and you will see that a dental x-ray taken with film alone (nonscreen technique) requires almost 30 times the x-ray exposure of a chest x-ray taken with intensifying screens. However, since the film records the light emitted by the screen rather than the x-rays striking it, the image is more blurred than when film alone is used (Fig. 16.25).

The screens are mounted in a cassette with a compressible felt backing that holds the film and the screens in close contact (Fig. 16.24). To get a better contact, vacuum cassettes have been developed; outside air pressure holds a single screen in close contact with the film. Vacuum cassettes are often used in mammography, where blurring must be minimum to allow the fine detail indicative of cancer to be seen.

Just as we can use film of different speeds in photography, we can use x-ray film of different speeds in diagnostic radiology. The speed is the inverse of the amount of exposure in roentgens (R) needed to darken the film so that it transmits only 10% of the incident light (optical density = 1.0); that is, if 0.1 R is needed, the speed is 10 R^{-1}. (The roentgen is defined in the next section.) In general, high-speed film requires less exposure, that is, is more sensitive, but shows less detail than lower speed film. Intensifying screens also come in different sensitivities; the most

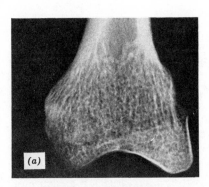 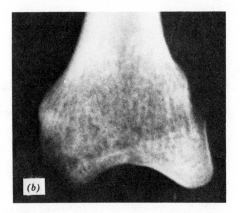

Figure 16.25. Two x-rays of the same bone taken with (*a*) nonscreen film and (*b*) parspeed screen film. Note that more detail is visible in *a*. However, the exposure for *a* was about 20 times greater than that for *b*.

sensitive, called fast screens, are more efficient but show less detail in the image since a thicker layer of crystals is used in them. Special screens with thin layers of crystals, called detail screens, have good resolution but require more exposure. For most medical work screens of intermediate sensitivity, or parspeed screens, are used.

It is important to use the proper exposure when making a radiograph; a good x-ray should be neither too dark nor too light. Some x-ray units have an automatic exposure meter, or a phototimer, that measures the x-rays striking the film and stops the exposure at the best time. The first phototimer was developed by R. H. Morgan in 1942.

On an x-ray of the chest, you can see that if the exposure in the lungs is good, some of the other parts are too light (underexposed). While this problem cannot be solved, the amount of underexposure can be reduced in several ways. Film that has greater latitude, or produces a larger range of useful shades of gray in the image, can be used, or the effective latitude can be increased by using a higher kilovoltage. Some overexposure can be avoided by using tapered absorbers in the x-ray unit to reduce the beam intensity over the thin portions of the body.

After the x-ray exposure has been made the film must be developed (processed). Just as many photographs are ruined by careless processing, many properly taken radiographs are spoiled by poor processing. When the chemicals are old or at the wrong temperature, the image can be seriously degraded. Much of the variability of hand developing has been reduced by the use of automatic film processors. The operation of automatic processors should be checked daily by developing a film given a standard exposure; if the processor is not working properly, the darkness of the image will not be correct.

16.4. RADIATION TO PATIENTS FROM X-RAYS

Within weeks after Roentgen discovered x-rays it was found that x-rays could damage the skin. Other hazards of x-rays and ways of reducing unnecessary exposure through the proper education of all x-ray operators and the application of good techniques are discussed in Chapter 19. In this section we discuss the amount of radiation received from typical x-ray studies and factors that affect the radiation received by the patient.

The unit used for radiation exposure is the *roentgen* (R), a measure of the amount of electric charge produced by ionization in air; $1 \text{ R} = 2.58 \times 10^{-4}$ C/kg of air. Typical exposures received by an adult for various x-ray studies are given in Table 16.2; however, there is a large variation in the

Table 16.2. Typical Amounts of Radiation Received by Adults in the United States in 1974

X-Ray Study	Exposure (mR)	Beam Area/Film Area	Exposure-Area Product (raps)
Chest	23	2	0.5
Skull	270	1.1	1.3
Abdomen	560	1.1	4.7
Upper (cervical) spine	230	1.9	1.5
Lower (lumbar sacral) spine	790	1.1	6.6
Dental bitewing	650	2.9	0.2

exposure given at different medical facilities. For example, in a study of exposures from over 500 chest x-ray units across the United States, it was found that the radiation received by a "standard" patient ranged from 3 mR to 2300 mR.

Since an exposure to a large area is more hazardous than the same exposure to a small area, a useful quantity for describing radiation to the patient is the exposure-area product (EAP), obtained by multiplying the exposure in roentgens by the area in square centimeters. A new unit called the *rap* (Roentgen-Area Product) has been proposed for this quantity; 1 rap = 100 R cm², and thus if you receive an exposure of 0.6 R to an area of 33 cm² (a typical dental exposure) you receive 20 R cm² or 0.2 rap. Table 16.2 gives typical EAP values for various x-ray studies.

The thickness of tissue the x-rays must pass through to reach the film affects the amount of radiation to the patient. The x-ray beam is reduced by a factor of about 2 for each 2.5 cm of tissue thickness, and thus a 25 cm thick patient reduces the intensity by a factor of about 2^{10} or 1000. When x-raying some body parts, such as the breast and the abdomen, it is possible to compress the tissue and thus reduce the amount of radiation needed (Fig. 16.26). Compressing the tissue also results in an image of more uniform density, which is easier for the radiologist to interpret.

The image detector used greatly affects the radiation exposure. A sensitive screen-film combination requires much less exposure than a low sensitivity nonscreen film system, but shows less detail (see Fig. 16.25).

In general, the radiation to the patient is reduced when the kilovoltage is raised, since high-energy x-rays are more penetrating than low-energy x-rays and thus are less likely to be absorbed in the body. However, high-energy x-rays also produce more scatter. Using a grid greatly reduces the scattered radiation but requires a larger exposure as explained

Figure 16.26. When x-raying the breast, the radiation to the patient can be reduced by compressing the tissues with an air-filled balloon. The arrow shows a thermoluminescent dosimeter (TLD) used to measure the exposure for the lateral view.

in Section 16.3. The use of high kilovoltage with a grid sometimes results in a greater exposure than the use of low kilovoltage without a grid, but the image produced generally contains more information.

One of the most important determinants of the radiation to the patient is the amount of filtration in the beam. As explained in Section 16.2, adding filtration reduces the intensity of the beam (Fig. 16.27). Filters selectively remove many more low-energy x-rays than high-energy x-rays (Fig. 16.28), and since most of these low-energy x-rays would be absorbed in the body and would not reach the film, the primary effect of filtration is to reduce patient exposure. The first few millimeters of aluminum filtration

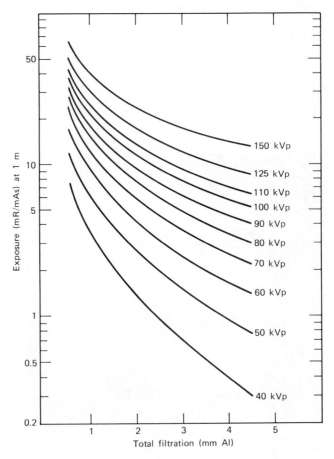

Figure 16.27.   The output of an x-ray set as a function of kilovoltage and amount of filtration.

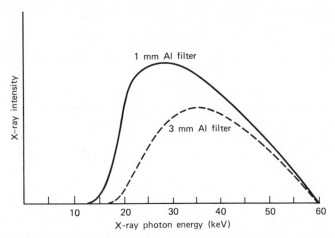

Figure 16.28. Filters primarily absorb low-energy x-ray photons. The amount of more penetrating high-energy photons is only slightly reduced.

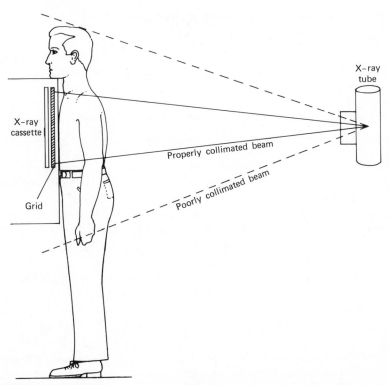

Figure 16.29. Excess radiation to the patient results from a poorly collimated x-ray beam. This is a major source of unnecessary radiation to the gonads.

419

are the most effective in reducing the exposure, and most states require that the filtration for conventional x-ray equipment be at least the equivalent of 2.5 mm Al.

Causes of unnecessary radiation to patients are covered in Chapter 19, but it is appropriate to mention here one common cause—the use of much larger x-ray beams than necessary. Any portion of an x-ray beam that falls outside of the area of the film is obviously wasted and actually reduces the clarity of the image by producing more scattered radiation. Yet in 1974 the beam area used for chest x-rays in the United States was on the average twice as large as the film area used (Table 16.2). Figure 16.29 shows a poorly collimated x-ray beam. New medical x-ray equipment purchased in the United States after August 1, 1974, must have a device to collimate the x-ray beam to the size of the film being used, but no such regulations apply to the many older units still in use.

16.5. PRODUCING LIVE X-RAY IMAGES—FLUOROSCOPY

As mentioned previously, Roentgen discovered x-rays by noticing the weak fluorescent light given off by crystals next to his cathode ray tube. Although early researchers used film to make their first x-ray images, it was soon found that x-ray images could be viewed directly on a sheet coated with a fluorescent material, or a fluorescent screen. Fluoroscopic techniques are useful where motion, such as the movement of contrast media in the digestive tract, must be studied. Conventional (old style) fluoroscopes are occasionally used today in hospitals and clinics because of the high cost of replacing them with modern image intensified fluoroscopes. In this section we discuss both conventional fluoroscopy and image intensified fluoroscopy.

Figure 16.30 shows a conventional fluoroscope. The coating of the fluoroscopic screen gives off yellow fluorescence when struck by x-rays, and the screen is covered with a sheet of lead glass that absorbs nearly all of the transmitted radiation.

The light produced in conventional fluoroscopy is very weak, and the radiologist views the image with his night vision sensors, the rods, which are about 1000 times more sensitive than the cones used for normal daylight vision but which cannot see as much detail (see Chapter 15, p. 346). Since the rods respond poorly to red light, the radiologist wears red goggles for about 30 min before viewing the screen to dark-adapt his rods while continuing other work using his cones (Fig. 16.30). Some early fluoroscopic operators did not take the time to dark-adapt but instead

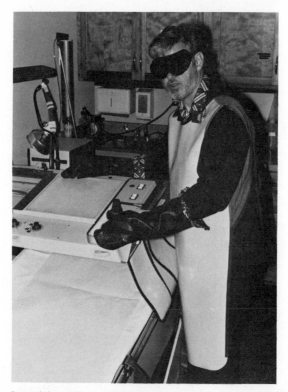

Figure 16.30. One of the authors (JRC) standing next to a conventional fluoroscope. The x-ray tube, which is mechanically coupled to the viewing screen, is below the couch on which the patient would lie, and the controls for adjusting the size of the x-ray beam are at the operator's left hand. The red goggles, used to dark-adapt the eyes, are removed after the patient is in place and the room is darkened, and the lead lined apron and gloves reduce the radiation to the operator.

increased the x-ray output to make the image brighter. This gave both the patient and the operator large amounts of unnecessary radiation.

Fluoroscopes were used in shoe stores as gimmicks to help sell shoes up until about 1960, when they were outlawed throughout the United States (Fig. 16.31). They gave large amounts of unnecessary radiation to the population. The amount of scattered radiation striking the gonads of little boys, eager users of the shoe fluoroscopes, was particularly large—it greatly increased the genetically significant dose of the population (see Chapter 19, p. 524).

Because the light emitted from conventional fluoroscopic screens is so

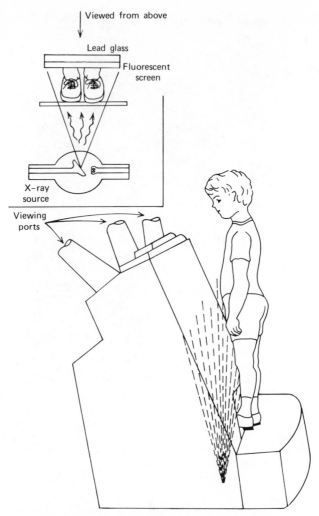

Figure 16.31. A shoe fluoroscope. Note that a considerable amount of scattered radiation (dashed lines) strikes the gonads of little boys. These units have been outlawed throughout the United States.

weak, physicists and engineers sought ways to improve the brightness of the image without increasing the radiation to the patient. In 1948 Coltman developed the *image amplifier*, or *image intensifier tube*. The basic components of this tube are shown in Fig. 16.32. The x-rays strike a fluorescent input screen inside a large evacuated bulb, and the light photons produced strike the adjacent photocathode. Upon striking the photocathode, some of the photons release electrons (by means of the photoelectric effect) that are then accelerated toward a small fluorescent output screen. The electrons are given energies of up to 25,000 eV, and each one produces many light photons when it strikes the output screen, which is viewed by the radiologist, a movie camera, or a television camera.

Even though the light output of a modern image intensifier system may be 1000 times greater than that of a conventional fluoroscope, it does not result in a drastic reduction in patient exposure. Most of the gain in output is used to permit the radiologist to view the image with his cones, which

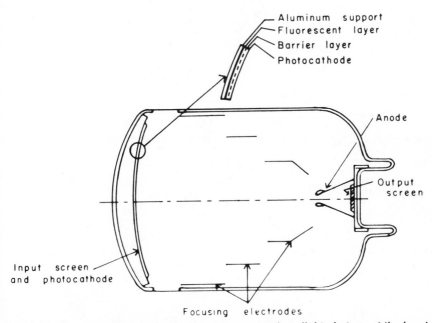

Figure 16.32. The image intensifier tube. X-rays produce light photons at the input screen which release electrons from the photocathode. These electrons are accelerated to the output screen and produce a bright image. (From *Medical Radiation Physics* by Hendee, W.R., p. 466. Copyright © 1970 by Year Book Medical Publishers, Inc., Chicago. Used by permission.)

have better resolving power than his rods. The increased brightness also makes it possible to take movies (cines) of the fluoroscopic image (Fig. 16.33); these movies can be studied at a later time or used for teaching purposes.

The movie camera is sometimes replaced with a TV camera and the image viewed on one or more TV monitors. This arrangement permits radiologists to use many of the techniques of the TV industry, such as video tape recording for later study and video disc recording for "instant replay." In 1973 the *image storage tube* was developed. This tube can "freeze" a TV image as a distribution of electric charge on about 1 million microscopic islands to provide a steady signal to a conventional TV monitor without continuing the radiation to the patient. The image on the TV monitor can be electronically changed to make the different shades of gray easier to see.

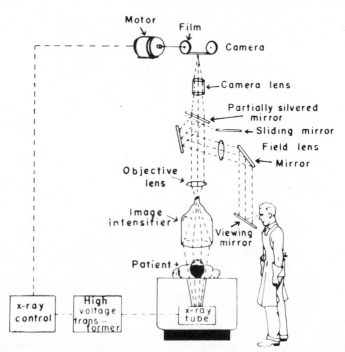

Figure 16.33. A modern fluoroscopy system with movie camera. (From *Medical Radiation Physics* by Hendee, W.R., p. 489. Copyright © 1970 by Year Book Medical Publishers, Inc., Chicago. Used by permission.)

16.6. X-RAY SLICES OF THE BODY

On an ordinary x-ray image the shadows of all the objects in the path of the x-ray beam are superimposed, and thus the shadows of normal structures may mask or interfere with the shadows that indicate disease. In order to distinguish the shadows indicating disease, the radiologist often takes x-ray images from different directions, such as from the back, the side, and an intermediate (oblique) angle. Taking x-ray images of slices of the body, or body section radiography, better known as *tomography,* was first proposed in about 1930 as a better way to distinguish these shadows. Both conventional tomography and computerized tomography, which was developed in 1972, are discussed in this section.

The basic action of a conventional linear tomographic unit is shown in Fig. 16.34. The x-ray tube and film are mechanically linked and move so that the shadows of structures at a chosen level in the patient—the plane of the cut—are cast at the same points on the film and thus are imaged clearly. The shadows of structures above or below the plane of the cut are blurred on the image because they strike different points on the film and

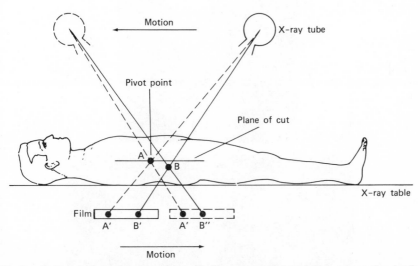

Figure 16.34. A linear tomographic unit. The x-ray tube and film are mechanically linked and pivot about an axis in the plane of the cut. An object in the plane of the cut such as A will cast its shadow at the same place on the film as the system moves and appear clearly on the image, while an object above or below the cut such as B will cast its shadow at various places on the film and appear blurred on the image.

serve as a more or less uniform background for the shadows of structures in the plane of the cut.

We can illustrate the selective blurring of objects by using a simple device for testing tomographic equipment. The device, shown in Fig. 16.35a, is a plastic disc in which lead numbers from 1 to 12 are embedded in a spiral, with each number 1 mm above the previous one. Figure 16.35b is an ordinary x-ray of this disc, and Fig. 16.35c is a tomograph of it; note that in the tomograph only the number 5 is imaged clearly, while the numbers adjacent to 5 are blurred and distant numbers (1 and 12) are not visible.

The simple linear motion of the tomographic unit shown in Fig. 16.34 is useful for many situations, but it can introduce confusing artifacts; for example, the shadows of structures either above or below the cut that lie along the direction of motion will not be adequately blurred. In order to eliminate many of the artifacts, the motion of tomographic equipment was

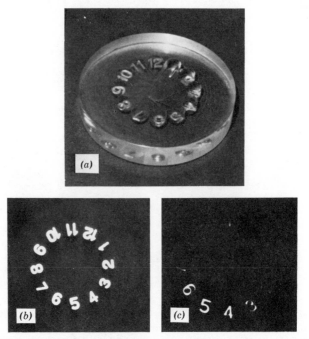

Figure 16.35. The Wisconsin tomographic test tool consists of a spiral of lead numbers embedded in plastic (a). An ordinary x-ray shows all the numbers (b), while a tomograph at the level of number 5 shows only a few numbers clearly (c). (Courtesy of Radiation Measurements, Inc., Middleton, Wisc.)

made more complex; in many modern units the x-ray beam and film follow either a cycloidal or an elliptical pattern (Fig. 16.36).

An *axial tomograph* is an image of a slice across the body and is taken by rotating the x-ray tube and film around the patient (see Fig. 18.6). Axial tomographs are useful in planning the treatment of cancer with radiation since they often show both the normal structures and the tumor.

Axial tomography was dramatically improved in 1972 when Hounsfield developed *computerized axial tomography* (CAT), sometimes called *computerized tomography* (CT), for EMI Limited in England. Hounsfield made use of a technique for analyzing data by computer that was originally developed for use in astronomy.

The original CAT unit, designed to be used on the head, is shown in Fig. 16.37. In this unit an x-ray tube with a relatively high potential ($\sim$140 kVp) produces two narrow beams of x-rays that scan linearly across the patient's head, and the intensities of the transmitted x-ray beams are recorded by two detectors moving with the beams on the other side of the patient (Fig. 16.38a). The data from the linear scan are stored in the

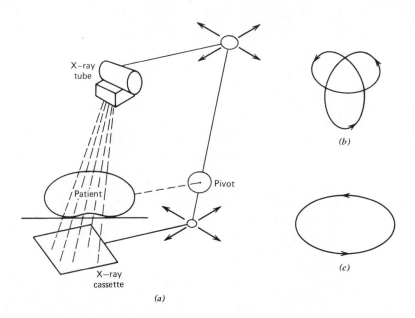

Figure 16.36. **(a) A tomographic unit permitting motion in a complex path produces better blurring of structures above and below the cut than a linear tomographic unit. The most common patterns for the x-ray tube and film are cycloidal (b) and elliptical (c).**

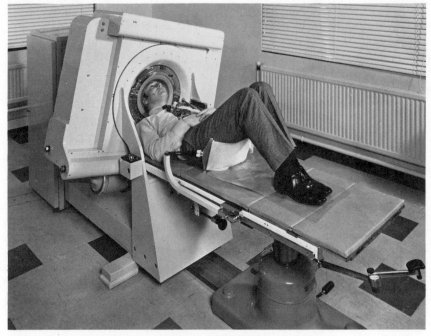

Figure 16.37.   The first commercial CAT scanner. The head is surrounded by a rubber bag filled with water. (Courtesy of EMI Limited, England.)

memory of the computer, and the tube and detectors are then rotated 1° and the process repeated (Fig. 16.38*b*). After 180 scans, which require about 4 min, the computer analyzes the data to determine the distribution of densities in the slice. The operator can choose to have these densities printed out as numbers or represented by shades of gray to produce an image. The numbers assigned for the printout range from about −500 for the density of air to about +500 for the density of compact bone, with 0 as the density of water (Fig. 16.39). When used to produce an image (Fig. 16.40), the instrument is usually adjusted so that the range of black to white covers a small enough range of densities for a 1% difference in density to be easily seen. The computer program can introduce artifacts in

Figure 16.38.   Principle of operation of the CAT head scanner shown in Fig. 16.37. (*a*) An x-ray tube is collimated to produce 2 narrow x-ray beams that scan across the head, and the intensities of the transmitted beams are measured by 2 detectors moving with the beams. (*b*) The apparatus is then rotated 1° and another scan is made; this process is repeated until 180 scans have been obtained. (Courtesy of EMI Limited, England.)

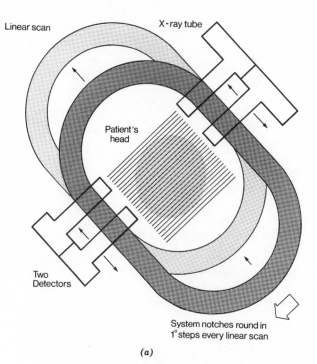

Linear scan

X-ray tube

Patient's head

Two Detectors

System notches round in 1° steps every linear scan

(a)

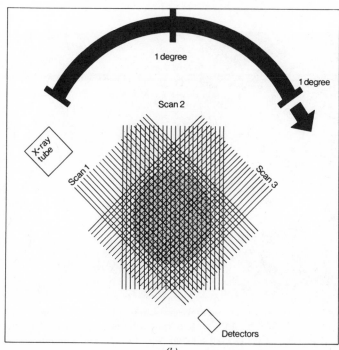

1 degree

1 degree

Scan 2

X-ray tube

Scan 1

Scan 3

Detectors

(b)

429

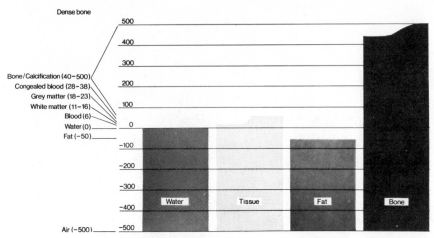

Dense bone

500

400

300

Bone/Calcification (40−500)
Congealed blood (28−38)
Grey matter (18−23)
White matter (11−16)
Blood (6)
Water (0)
Fat (−50)

200

100

0

−100

−200

−300

−400

Air (−500)

−500

Water

Tissue

Fat

Bone

Figure 16.39. The typical density values of materials in the head assigned by the CAT scanner. (Courtesy of EMI Limited, England.)

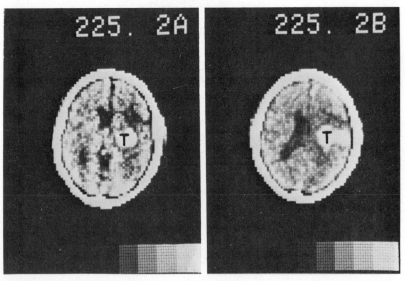

Figure 16.40. Two scans obtained simultaneously with a CAT scanner similar to the one shown in Fig. 16.37. The dark area in the center is caused by the normal fluid in the ventricles. A tumor (T) is also shown. (Courtesy of EMI Limited, England.)

430

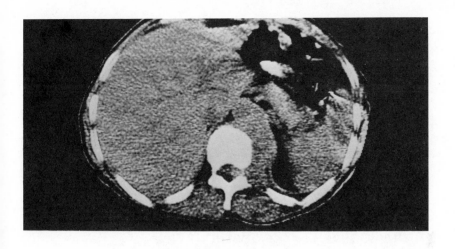

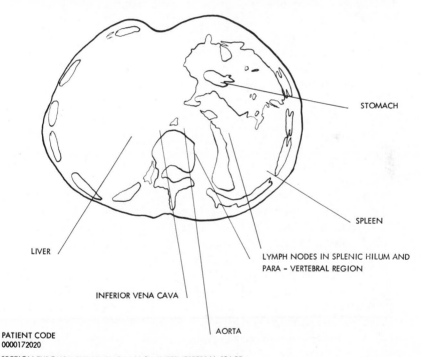

STOMACH

SPLEEN

LIVER

LYMPH NODES IN SPLENIC HILUM AND
PARA - VERTEBRAL REGION

INFERIOR VENA CAVA

AORTA

PATIENT CODE
0000172020

A SECTION THROUGH THE LEVEL OF D11/12 INTERVERTEBRAL SPACE

Figure 16.41. A CAT scan of the thorax just below the lungs shows the spine and ribs (white areas) as well as the various soft tissues. (Courtesy of Dr. L. Kreel, Head of Radiology, Clinical Research Centre, Northwick Park Hospital, Harrow, and EMI Limited, England.)

431

regions where there are large abrupt changes in density; the dark line on the inside edge of the skull in Fig. 16.40 is such an artifact.

Within a few years of the invention of the CAT scanner shown in Fig. 16.37, CAT scanners that could be used on any portion of the body were developed. It was necessary to shorten the original scan time of 4 min to eliminate blurring on scans of the thorax (Fig. 16.41) since a patient has difficulty holding the breath for more than 30 sec. The scan time was reduced by using an x-ray beam collimated in a fan shape and many detectors to measure the transmitted segments of the beam. The CAT scanner used for Fig. 16.41 has a scan time of 20 sec.

16.7. RADIOGRAPHS TAKEN WITHOUT FILM

The Xerox Corporation has used an electrostatic technique to make copies of printed material for so long that the procedure is often called xeroxing even when it is done on a competitor's machine. The electrostatic technique used to make a xerox copy can also be used to make an x-ray image called a *xeroradiograph*.

In xeroradiography, a special selenium-coated plate is used instead of film to record the image. The plate is given a uniform positive charge and placed in a light-tight cassette, which is then used in the same manner as a film cassette (Fig. 16.42a). X-rays coming through the patient strike the xerox plate and release electrons that neutralize part of the positive charge. The areas of the plate under thick body parts will retain most of the original charge, while on the areas of the plate under thin parts of the body, most of the charge will be neutralized. The plate is then placed in a processor and sprayed with a negatively charged fine dark blue powder (Fig. 16.42b); the areas with remaining charge hold the powder, producing a positive image of the x-rayed object. Some radiologists prefer a negative image, which can be obtained by changing the charge on the powder. The powder pattern is transferred by heat to a sheet of paper coated with plastic for viewing and storage. The processing takes only 90 sec, and the plate is reusable provided the old charge is thoroughly removed before the plate is recharged (Fig. 16.42a).

During processing, the powder near an area that has little remaining charge is attracted to the edge of the nearest area with more remaining charge, producing a dark and well-defined image of that edge. This *edge enhancement effect* is also produced along the edges of underexposed areas that have retained most of their original charge. Due to this effect, a

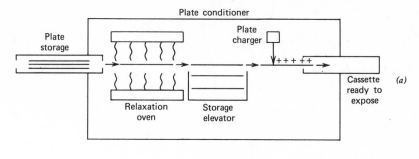

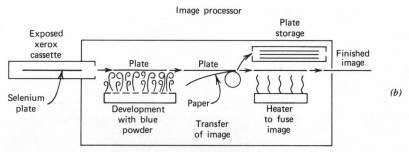

Figure 16.42. Basic steps in xeroradiography. (a) Before use the plate is electrically charged in a conditioner and placed in a light-tight cassette. An exposure is made, and the cassette is placed in the processor (b) where the image is developed by spraying dust on the plate. This dust image is transferred to a paper that exits after it is heated to fuse the dust in the plastic coating. The plate is then cleaned and returned to storage for later use.

xeroradiograph shows detail in thick body parts better than a conventional radiograph (Fig. 16.43).

The main disadvantage of xeroradiography is that it is less sensitive than conventional radiography, sometimes requiring exposures that are 10 times greater. Further research may improve the sensitivity of the technique.

Ionography is an alternative imaging system that is similar to xeroradiography in the way the image is processed. In ionography the x-rays transmitted through the body produce ionization in a layer of solid, liquid, or gas—the image-detecting material. The ions are attracted to a plastic sheet by an electric field, typically 1000 V/cm or larger, and the plastic sheet is then processed by the powder technique to show the charge pattern. In 1976 one commercial ionography system with xenon gas as the detecting material was on the market. Research on liquid

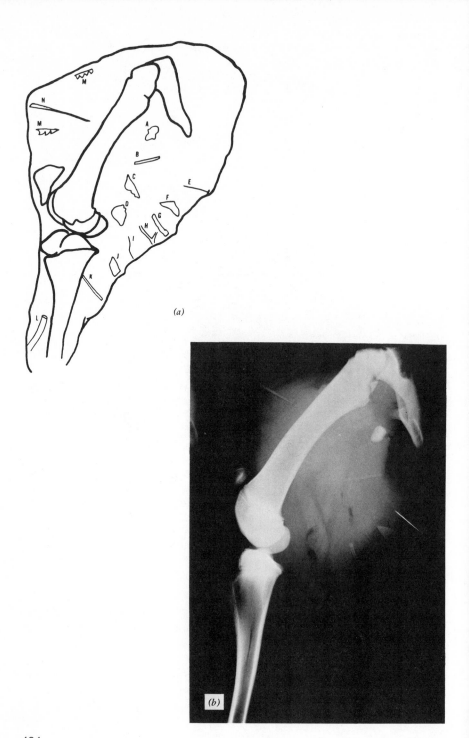

(a)

(b)

434

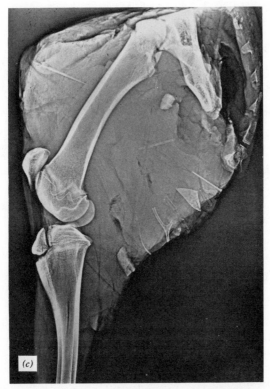

Figure 16.43. **Conventional x-ray and xeroradiographic images.** (*a*) **Various foreign bodies were embedded in a lamb shank: (A) gravel, (B) graphite (lead pencil), (C) gauze, (D) cotton ball, (E) metallic pin, (F) glass, (G) wood, (H) chicken bone, (I) fish bone, (J) beef bone, (K) polyethylene tubing, (L) rubber, (M) palm thorn, and (N) plastic pick. A conventional x-ray (*b*) shows some of the objects, while a xeroradiograph (c) shows most of the objects. (From M.E. Woesner and I. Sanders,** *Amer. J. Roentgen. Rad. Ther. and Nuc. Med.,* **115, 1972, pp. 637–642.)**

ionography was underway at the University of Toronto, and solid ionography was being investigated at the University of Wisconsin.

BIBLIOGRAPHY

Christensen, E. E., T. S. Curry, and J. Nunnally, *An Introduction to the Physics of Diagnostic Radiology,* Lea and Febiger, Philadelphia, 1972.

Howl, B., *Simplified Physics for Radiology Students,* Thomas, Springfield, Ill., 1971.

Jaundrell-Thompson, F., and W. J. Ashworth, *X-Ray Physics and Equipment,* 2nd ed., Davis, Philadelphia, 1970.

Meredith, W. J., and J. B. Massey, *Fundamental Physics of Radiology,* Williams and Wilkins, Baltimore, 1968.

Ridgway, A., and W. Thumm, *The Physics of Medical Radiography,* Addison-Wesley, Reading, Mass., 1968.

Selman, J., *The Fundamentals of X-Ray and Radium Physics,* 5th ed., Thomas, Springfield, Ill., 1975.

Wright, D. J. (Ed.), *Physics of Diagnostic Radiology,* Proceedings of a summer school held at Trinity University, San Antonio, Texas, July 12–17, 1971, Bureau of Radiological Health, Rockville, Md., 1973.

REVIEW QUESTIONS

1. What fraction of the United States population has a medical or dental x-ray each year?
2. What are the three major branches of radiology?
3. What are the four main components of a modern x-ray unit?
4. How does the source of electrons in a modern x-ray tube differ from the source of electrons in the x-ray tube used by Roentgen?
5. A fluoroscope can operate continuously at a potential of 80 kV and a current of 3 mA. What is the power into the target?
6. If the x-ray tube in Question 5 has a tungsten target, what percentage of the energy goes into x-ray photons?
7. What is the purpose of a tube-loading chart?
8. Why is a rotating anode used?
9. If an exposure of 0.01 sec is used, what angular rotation of the anode occurs at a rotational rate of 3600 rpm?
10. Why is an anode rapidly braked when it is used at 10^4 rpm?
11. What is the energy of the K_α x-rays from a tungsten target?
12. If the half-value layer of an x-ray beam is 3 mm Al, what is the linear attenuation coefficient of the beam?
13. How is the mass attenuation coefficient of a material related to its linear attenuation coefficient?
14. Which interaction is most useful for producing a diagnostic x-ray image—photoelectric, Compton, or pair production? Why?
15. When a beam of 50 keV monoenergetic x-rays enters a water absorber, which is more probable—the photoelectric effect or the Compton effect?
16. List three contrast materials commonly used in diagnostic radiology.
17. List three causes of blurring in a radiograph.
18. How is the amount of scattered radiation striking the film reduced?
19. What is the purpose of intensifying screens?
20. What is the unit used to measure radiation exposure?

21. What vision sensors in the eye are used for viewing conventional fluoroscopic images?
22. What is tomography? How does computerized tomography differ from conventional tomography?
23. Give one advantage and one disadvantage of xeroradiography over conventional radiography.

CHAPTER 17

Physics of Nuclear Medicine (Radioisotopes in Medicine)

With the assistance of Jerome Wagner, Rochester Institute of Technology, Rochester, New York

Many scientists started investigating x-rays after their discovery by Roentgen in 1895. In France in 1896, Becquerel investigated the hypothesis that sunlight would cause certain minerals to give off x-rays. He placed minerals on a light-tight envelope containing film and exposed them to sunlight. One of those coincidences of science occurred—the minerals he was using were naturally radioactive. When he developed the film he found darkening due to "x-rays." As part of his continuing studies, he accidentally did the "control" experiment—on a rainy day he left his minerals and film in a drawer and afterward decided to develop the film even though he assumed it would be blank. He was surprised to find that "x-rays" came from the minerals without exposure to sunlight. After some confusion it became clear that he had discovered natural radioactivity; he was on his way toward his 1905 Nobel Prize.

Natural radioactivity was more amazing to physicists than x-rays since to produce x-rays it was necessary to put in electrical energy, while natural radioactivity provided a steady source of high-energy particles for an indefinite period of time with no input of energy. This seemed to violate the conservation of energy law, which was already well established.

The story of natural radioactivity involved many more scientists than did the story of x-rays, which was almost a "one-man show." Madame Curie stands out as one of the pioneers. Working with her husband Pierre from 1898 to 1904, she managed to separate from tons of uranium ore a very powerful radioactive element—radium. It was to play an important role in medicine in the treatment of cancer, and it is still used today (see Chapter 18).

Investigators of radioactivity discovered that certain natural elements, primarily the very heavy ones, have unstable nuclei that disintegrate to emit various rays—alpha (α), beta (β), and gamma (γ) rays.* The alpha, beta, and gamma rays were found to have quite different characteristics. Alpha and beta particles bend in opposite directions in magnetic and electric fields; alpha particles are positively charged and beta particles are negatively charged. Alpha particles, which stop in a few centimeters of air, are the nuclei of helium atoms. Beta rays are more penetrating but can be stopped in a few meters of air or a few millimeters of tissue—they are high-speed electrons. Gamma rays are very penetrating and are physically identical to x-rays. They usually have much higher energies than the x-rays used in diagnostic radiology. Alpha and gamma rays from a given source have fixed energies, but beta rays have a spread of energies up to a maximum characteristic of the source.** The properties of alpha, beta, and gamma rays are summarized in Table 17.1.

Each element has a specific number of protons in the nucleus; for example, carbon has six protons, nitrogen has seven protons, and oxygen has eight protons. However, for each element, the number of neutrons in the nucleus can vary. Nuclei of a given element with different numbers of neutrons are called *isotopes* of the element. If they are not radioactive they are called *stable isotopes,* and if they are radioactive they are called *radioisotopes*; for example, carbon has two stable isotopes (^{12}C and ^{13}C) and several radioisotopes (e.g., ^{11}C, ^{14}C, and ^{15}C). Most elements do not have naturally occurring radioisotopes, but radioisotopes of all elements can now be produced artificially. *Isotope* means "in the same place" and should be used when referring to a single element. The word *radionuclides* is appropriate when several radioactive elements are involved.

Natural radioactivity played an important role in developing our understanding of the nucleus, and some of the early "tracer" work in medicine was done using naturally radioactive elements. However, the dramatic turning point in the use of radioactivity in medicine was the development of the nuclear reactor during World War II in connection with the atomic bomb project. The first one, called an "atomic pile," was built by Fermi and his co-workers in 1942. The nuclear reactor made possible the pro-

*Alpha and beta rays are often called alpha and beta particles.
**The neutrino, a massless, chargeless "particle," takes up the difference in energy between the actual beta energy and the maximum beta energy. All physicists believe the neutrino exists even though it is extremely difficult to detect. It is of no practical interest in medical physics.

Table 17.1. The Characteristics of Alpha, Beta, and Gamma Rays

	Symbol	Rest Mass (kg)	Charge	Nature	Penetration	Energies
Alpha	α	6.6×10^{-27}	$+2$	Helium nucleus	Stopped in a sheet of paper or 7 cm of air	Fixed
Beta	β	9×10^{-31}	-1	Electron	Stopped in a few mm of tissue	Continuous up to a fixed maximum
Gamma	γ	—	0	Identical to an x-ray	Absorbed exponentially in many cm of tissue	Fixed

440

duction of many artificial radionuclides in huge quantities. These radioactive elements have been used in medicine for research, diagnosis, and therapy. In the late 1940s radioactivity from reactors was used primarily for medical research rather than for diagnosis. For example, radioactivity has been used to study the function of many organs and the chemical changes that take place within the body. In this chapter we do not discuss these research uses of radioactivity; we concentrate on the clinical uses of radioactivity for the diagnosis of disease—*nuclear medicine*.

The field of nuclear medicine has been growing at the rate of 15 to 25% per year for the last two decades. This expansion has been greatly aided by the technical developments in nuclear physics research. Many of the instruments described in Section 17.4 were developed for basic research in nuclear physics.

There are now about 30 different routine nuclear medicine studies performed on patients in a modern medical center. In a modern hospital about one-third of the patients have a nuclear medicine study of some type. Although many nuclear medicine tests are concerned with the detection of cancer, others are used to detect problems of the heart, lungs, kidneys, and joints.

The most useful radionuclides for nuclear medicine are those that emit gamma rays. Since gamma rays are very penetrating, a gamma emitting radioactive element inside the body can be detected outside the body (see Sections 17.4 and 17.5). Extremely minute quantities of radioactive substances are used in nuclear medicine, and these small quantities (typically less than 1 μg) do not affect the normal physiological functioning of the body. A molecule labeled with a radioactive element will behave physiologically like its stable counterpart.

In 1972 most of the practitioners of nuclear medicine were trained in the fields of radiology, internal medicine, or pathology. Since then many medical centers have developed training programs in nuclear medicine. With the official recognition of nuclear medicine as a separate specialty, many of these training programs have become three-year residencies. In the future, specialists in nuclear medicine will go through these training programs rather than gather their knowledge in the haphazard manner of the early workers in the field.

There are several other professionals in nuclear medicine in addition to the nuclear medicine physician. A large medical center usually has one or more medical physicists assisting in the operation of the nuclear medicine department. The physicists are often involved with computer applications and the design of new instrumentation. Most large centers also have a radiopharmacist who prepares the radiopharmaceuticals for injection; he or she normally has a pharmacy degree plus special training in radiopharmacy. The operators of the nuclear medicine equipment are technologists with one to four years of special training. Many nuclear medicine technologists take two years of training as radiological technologists and one year of additional training in nuclear medicine. These technologists usually take the registry examination of the American Registry of Radiologic Technologists (ARRT) to demonstrate their competence. Persons trained as medical technologists often take additional training in nuclear medicine, and they are examined by the American Society of Clinical Pathology (ASCP) to establish their credentials. A few colleges offer B.S. degrees in nuclear medicine technology.

17.1. REVIEW OF BASIC CHARACTERISTICS AND UNITS OF RADIOACTIVITY

There are whole families of natural radioactive elements. When one element decays its daughter, which is also radioactive, is formed; each daughter decays until the final daughter, a stable isotope of lead, is formed (Table 17.2).

The abbreviations used for radionuclides have varied over the years,

Table 17.2. Successive Steps Through Which ^{238}U Decays to Become Stable ^{206}Pb

Element	Symbol[a]	Half-Life	Type of Radiation
Uranium	$^{238}_{92}$U	4.55×10^9 years	α
Thorium	$^{234}_{90}$Th	24.1 days	β, γ
Protactinium	$^{234}_{91}$Pa	1.14 min	β, γ
Uranium	$^{234}_{92}$U	2.69×10^5 years	α
Thorium	$^{230}_{90}$Th	8.22×10^4 years	α, γ
Radium	$^{226}_{88}$Ra	1600 years	α, γ
Radon	$^{222}_{86}$Rn	3.8 days	α
Polonium	$^{218}_{84}$Po	3.05 min	α
Lead	$^{214}_{82}$Pb	26.8 min	β, γ
Bismuth	$^{214}_{83}$Bi	19.7 min	α, β, γ
Polonium	$^{214}_{84}$Po	1.5×10^{-4} sec	α
Lead	$^{210}_{82}$Pb	22.2 years	β, γ
Bismuth	$^{210}_{83}$Bi	4.97 days	β
Polonium	$^{210}_{84}$Po	139 days	α, γ
Lead	$^{206}_{82}$Pb	Stable	—

[a]The subscript indicates the number of protons in the nucleus, or the atomic number. This subscript is usually omitted.

but the changes are minor. For example, the abbreviation ^{131}I used to be written I^{131}. Both indicate an atom of iodine with a nucleus containing 131 protons and neutrons. Stable iodine, ^{127}I, has four fewer neutrons. Sometimes a subscript indicates the atomic number, for example, $^{131}_{53}$I.

There are over 1000 known radionuclides, most man-made. Heavy elements tend to have many more radioisotopes than light elements; for example, iodine has 15 known radioisotopes, while hydrogen has 1, tritium (^{3}H). A particular radionuclide can be identified by its radioactivity alone just as humans can be identified by their fingerprints. Characteristics that help identify the radionuclide are the type and energy of its emitted particles or rays.

Table 17.3 lists some of the important characteristics of many of the radionuclides used in the daily practice of nuclear medicine. Note that all except ^{32}P emit photons.

The most common emissions from radioactive elements are beta particles and gamma rays. Since beta particles are not very penetrating, they are easily absorbed in the body and are generally of little use for diagnosis. However, some beta-emitting radionuclides such as ^{3}H and ^{14}C play an important role in medical research. ^{32}P is used for diagnosis of tumors in the eye because some of its beta particles have enough energy to emerge from the eye. Most clinical diagnostic procedures use photons of some type—usually gamma rays. Gamma rays with energies above 100 keV can penetrate many centimeters of tissue, and a gamma emitter in the body can be located and mapped by a detector outside the body. All of the gamma-emitting radionuclides of the common organic elements—carbon, nitrogen, and oxygen—are short lived, which makes their use in clinical medicine difficult without an accelerator. A few medical centers have installed cyclotrons for producing short-lived radionuclides (see Section 17.2).

Some man-made radionuclides emit types of radiation not emitted by natural radioactive substances. The *m* in ^{99m}Tc in Table 17.3 stands for *metastable*, which means "half stable." A metastable radionuclide decays by emitting gamma rays only, and the daughter nucleus differs from its parent only in having less energy. For example, ^{99m}Tc decays to form ^{99}Tc by emitting a gamma ray of 140 keV. This is a very useful energy for nuclear medicine applications since it is penetrating enough to get out of the body easily and is easy to shield with a few millimeters of lead. When a metastable radionuclide is used internally, the absence of beta rays greatly reduces the radiation dose to the patient.

Nuclear reactors produce radioactivity by adding neutrons to stable nuclei; the resulting nuclei thus have too many neutrons and usually decay by emitting beta particles, which make the nuclei more positive.

Table 17.3. Characteristics of the Most Common Radionuclides Used in Clinical Nuclear Medicine

Element	Radionuclide	Emission or Decay Mode[a]	Main Photon Energy (MeV)	Half-Life
Carbon	^{11}C	β^+	0.511	20 min
Nitrogen	^{13}N	β^+	0.511	10 min
Oxygen	^{14}O	β^+,γ	0.511, 2.312	71 sec
	^{15}O	β^+	0.511	2 min
	^{19}O	β^-,γ	0.197	29 sec
Fluorine	^{18}F	β^+,ec	0.511	110 min
Phosphorus	^{32}P	β^-	none	14.5 days
Chromium	^{51}Cr	ec,γ	0.320	28 days
Iron	^{52}Fe	β^+,ec,γ	0.165, 0.511	8 hr
Cobalt	^{57}Co	ec,γ	0.122, 0.136	270 days
Gallium	^{67}Ga	ec,γ	0.093, 0.184, 0.296, 0.388	78 hr
Krypton	^{68}Ga	β^+,ec	0.511	68 min
	^{81m}Kr	IT,γ	0.190	13 sec
Rubidium	^{81}Rb	β^+,ec,γ	0.253, 0.450, 0.511	4.7 hr
Technetium	^{99m}Tc	IT,γ	0.140	6 hr
Indium	^{113m}In	IT,γ	0.393	102 min
Iodine	^{123}I	ec,γ	0.159	13 hr
	^{125}I	ec,γ,IT	0.028, 0.035	60 days
	^{131}I	β^-,γ	0.364	8 days
Xenon	^{133}Xe	β^-,γ	0.081	5.3 days
Ytterbium	^{169}Yb	ec,γ	0.057, 0.110, 0.131, 0.177, 0.198, 0.308	31 days
Gold	^{198}Au	β^-,γ	0.412	2.7 days
Mercury	^{197}Hg	ec,γ	0.069	65 hr
	^{203}Hg	β^-,γ	0.279	47 days
Thallium	^{201}Tl	ec,γ	0.081, 0.135, 0.167	73 hr

[a]ec signifies electron capture; IT, isomeric transition.

444

A type of disintegration that occurs only in man-made radionuclides is the emission of a positive beta ray or *positron* (β^+). (The ordinary negative beta ray is sometimes called a negatron—β^-.) Most positron emitters are produced by cyclotrons, which accelerate positive particles to bombard the target material. A positron is physically identical to an electron except that it has a positive charge. Radionuclei that are "too positive," such as ^{18}F, often decay by positron emission.

Associated with positron emission is *annihilation radiation*. After the positron has come to a stop it does a brief "death dance" with an electron, and they annihilate each other (see Fig. 16.13c). The energy equivalent of their masses (511 keV each) is usually emitted as two 511 keV photons, the annihilation radiation, which go in opposite directions. You will see in Section 17.5 that this characteristic of annihilation radiation is useful for mapping the distribution of positron emitters in the body.

A nucleus that is too positive can decay in another way. The nucleus can capture one of its own electrons (usually a K electron), which then combines with a proton in the nucleus to form another neutron and thus reduce the charge on the nucleus. Electron capture (ec) occurs in the decay of ^{125}I (Table 17.3). A characteristic (K_α or K_β) x-ray is always emitted after the K-capture as an electron falls into the open spot in the K shell.

Iodine 125 decay illustrates another phenomenon of nuclear physics. After ^{125}I has "swallowed" one of its own electrons to become ^{125}Xe, the nucleus has 35 keV of energy, which is sometimes emitted as a 35 keV gamma ray. When it is not, the nucleus transfers this energy directly to the remaining K electron, which then uses the energy to escape, leaving another vacancy in the K shell. This process is called *internal conversion* or isomeric transition (IT). In this case two K x-rays are emitted. For every 100 ^{125}I nuclei that decay, there are emitted 140 K_α x-rays plus some K_β x-rays and a few 35 keV gamma rays.

Each radionuclide decays at a fixed rate commonly indicated by the *half-life* ($T_{1/2}$), the time needed for half of the radioactive nuclei to decay. Figure 17.1 is the graph of radioactivity versus time for the common radionuclide ^{99m}Tc. Figure 17.1a shows the activity on a linear scale, while Fig. 17.1b shows the activity on a log scale. The straight line on the semi-log graph indicates that the decay is exponential (see Appendix A). To measure the half-life of relatively short-lived radioactivity in the laboratory, it is usually convenient to plot the activity on a semi-log graph. If the graph shows a straight line you can be reasonably sure only one radionuclide is present. A curved line indicates the presence of more than one radionuclide.

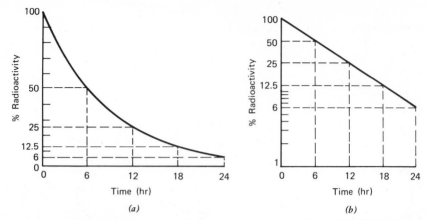

Figure 17.1.   Radioactivity versus time for the common radionuclide ⁹⁹ᵐTc plotted on (*a*) a linear scale and (*b*) a log scale.

The basic equation describing radioactive decay is

$$A = A_0 e^{-\lambda t} \qquad (17.1)$$

where A is the activity in disintegrations per second, A_0 is the initial activity, λ is the decay constant, and t is the time since the activity was A_0. If t is measured in hours, λ must be in hours⁻¹. If λ is small, that is, less than 0.1, it is very nearly the fraction of the radionuclide that decays per unit time. For ¹⁹⁸Au, λ is 0.01 hr⁻¹, which means that 0.01 (or 1%) decays per hour.

We can express Equation 17.1 as

$$A = \lambda N \qquad (17.2)$$

where N is the number of radioactive atoms. Equation 17.2 can be used to determine the half-life of a long-lived radionuclide (see Example 17.1*a*).

The decay constant is related to the half-life by the simple equation

$$T_{1/2} = \frac{0.693}{\lambda} \qquad (17.3)$$

The constant 0.693 is the natural logarithm of 2. The relationship between the decay constant and the half-life is illustrated in Example 17.1*b*.

Example 17.1
a. If you have 1 g of pure potassium 40 (⁴⁰K) that is experimentally determined to emit about 10⁵ beta rays per second, what is the decay constant λ?

40 g of ^{40}K contains 6.23×10^{23} (Avogadro's number) potassium atoms. Therefore, 1 g contains $(6.23/40) \times 10^{23}$ atoms. Using Equation 17.2,

$$\lambda = \frac{A}{N} = \frac{10^5}{1.55 \times 10^{22}} \text{ sec}^{-1} = 6.5 \times 10^{-18} \text{ sec}^{-1}$$

b. Estimate the half-life of ^{40}K from the decay constant.
From Equation 17.3,

$$T_{1/2} = \frac{0.693}{\lambda} = \frac{0.693}{6.5 \times 10^{-18} \text{ sec}^{-1}}$$

$$T_{1/2} \simeq 10^{17} \text{ sec}$$

Since there are 3.15×10^7 sec/year,

$$T_{1/2} \simeq \frac{10^{17} \text{ sec}}{3.15 \times 10^7 \text{ sec/yr}} \simeq 3 \times 10^9 \text{ years}$$

(The accepted half-life is 4×10^9 years.)

The value $1/\lambda$ is the average or mean life τ of the radionuclide. For example, since $\lambda = 0.01$ hr^{-1} for ^{198}Au, the average life of ^{198}Au is 100 hr. Since $1/\lambda = 1.44 \, T_{1/2}$ (from Equation 17.3), the mean life can be calculated from $\tau = 1.44 \, T_{1/2}$.

The unit of radioactivity, the *curie* (Ci, formerly abbreviated c), is equal to 3.7×10^{10} disintegrations per second. This number originally represented the radioactivity of 1 g of radium, but it is now independent of the disintegration rate of radium, which kept "changing" when different people measured it. The curie is a rather large quantity for nuclear medicine: the millicurie (mCi)—10^{-3} Ci—and the microcurie (μCi)—10^{-6} Ci—are usually more convenient. For radiation protection purposes even smaller units are used—the nanocurie (nCi), which is 10^{-9} Ci, and the picocurie (pCi), which is 10^{-12} Ci.

The International Commission on Radiological Units (ICRU) recommended in 1975 that the International System (SI) unit for radioactivity be the *becquerel* (Bq), which is defined as 1 disintegration/sec. Thus 1 Ci = 3.7×10^{10} Bq. Because the becquerel is so small, it is necessary to use multiples such as the kilobecquerel (kBq), 10^3 disintegrations/sec; the megabecquerel (MBq), 10^6 disintegrations/sec; and the gigabecquerel (GBq), 10^9 disintegrations/sec. Table 17.4 can be used to convert from curies to becquerels.

Table 17.4.   Conversion Factors: Curies to Becquerels

Curies (Ci)		Becquerels (Bq)^a	
1000 Ci	1 kCi	3.7×10^{13} Bq	37 TBq
100	—	3.7×10^{12}	3.7 TBq
10	—	3.7×10^{11}	370 GBq
1	—	3.7×10^{10}	37 GBq
10^{-1}	100 mCi	3.7×10^{9}	3.7 GBq
10^{-2}	10 mCi	3.7×10^{8}	370 MBq
10^{-3}	1 mCi	3.7×10^{7}	37 MBq
10^{-4}	100 μCi	3.7×10^{6}	3.7 MBq
10^{-5}	10 μCi	3.7×10^{5}	370 kBq
10^{-6}	1 μCi	3.7×10^{4}	37 kBq

^{a}T (tera) = 10^{12}, G (giga) = 10^9, M (mega) = 10^6, and k (kilo) = 10^3.

17.2.   SOURCES OF RADIOACTIVITY FOR NUCLEAR MEDICINE

Radioactive drugs, or radiopharmaceuticals, must meet rigid standards of purity from both the drug aspect and the radioactivity aspect. Most radiopharmaceuticals used in nuclear medicine are bought from commercial drug companies, who check on the pharmaceutical quality as well as on the radioactive quality of the radiopharmaceuticals. They sell prepackaged fixed amounts of commonly used radiopharmaceuticals so that the physician can use them much like other drugs. The only additional work the physician has to do is use a table to determine how much these radiopharmaceuticals have decayed since they were calibrated.

One such drug that has had a long and productive career in nuclear medicine is ^{131}I. Its 8-day half-life permits delivery across the country, and weekly shipments are adequate for most medical facilities. However, it is no longer used extensively because its beta rays give a relatively large radiation dose to the patient. For both beta and gamma ray emitters, a radionuclide with a long half-life gives a larger radiation exposure than a radionuclide with a shorter half-life and emissions of similar energy.

The delivery system used for long-lived radionuclides is not suitable for short-lived radionuclides, and modern medical centers must have safe and reliable methods of producing their own radiopharmaceuticals with short-lived radionuclides. A new specialty, radiopharmacy, has developed to meet this need. Radiopharmacists cannot take their products off the shelf but must compound prescriptions like their brothers of old.

One commonly used radionuclide is ^{99m}Tc, which has a 6-hour half-life and decreases by four half-lives or a factor of 16 each day. The delivery of ^{99m}Tc is accomplished in an interesting way. ^{99m}Tc is the daughter of ^{99}Mo, which has a 2.5-day half-life, and by buying 3 GBq (~ 100 mCi) of ^{99}Mo it is possible to obtain useful amounts of ^{99m}Tc. About 3 GBq of ^{99m}Tc can be rinsed off the "parent" on the first day. The parent continues to produce ^{99m}Tc, and the next day about 2 GBq of ^{99m}Tc can be washed off; after another day about 1.5 GBq can be obtained; and so on for a week. A system that produces a radionuclide in this way is called a "cow," or generator, and it can be "milked" or eluted daily or twice daily to obtain a supply of the daughter product. Figure 17.2 is a schematic of a ^{99m}Tc generator.

The radioactivity obtained from the ^{99m}Tc generator will not exceed the radioactivity of the ^{99}Mo source, but it may be considerably less if all of the daughter product is not rinsed off or if the generator is milked too soon

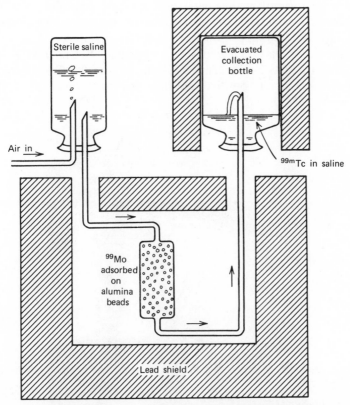

Figure 17.2.   Milking a ^{99m}Tc generator. A bottle of sterile saline solution is placed on the inlet and an evacuated bottle inside a lead shield is placed on the outlet. Air pressure forces the saline solution over the alumina beads, which contain ^{99}Mo. The daughter product, ^{99m}Tc, is dissolved in the saline solution and carried to the collection bottle.

after the last milking. Therefore, an ionization calibration device must be used in the laboratory to determine the ^{99m}Tc activity in the elutant (Fig. 17.3). Since the calibration device uses electronics to amplify the current it can easily change with time and should be checked with a radioactive source with a long half-life that has been compared to standard sources at the National Bureau of Standards.

^{123}I, which has a 13-hr half-life and no beta emission, is suitable for nuclear medicine studies. While it cannot be produced from a generator like ^{99m}Tc it can be easily produced in a cyclotron (Fig. 17.4), and some large medical centers have bought cyclotrons for the production of ^{123}I and other short-lived radionuclides such as ^{11}C, ^{13}N, ^{15}O, ^{19}O, ^{18}F, ^{52}Fe, ^{68}Ga, and ^{81}Rb (Table 17.3). In the future, nuclear medicine will probably involve more dynamic studies with short-lived nuclei produced by cyclotrons at medical centers. The same cyclotrons may be used to produce fast neutrons for the treatment of cancer (see Chapter 18, p. 515).

It might seem that ^{15}O would be impractical for nuclear medicine studies because of its 2-min half-life, but this is not the case. This radioactive gas can be rapidly purified and delivered to the patient, and since

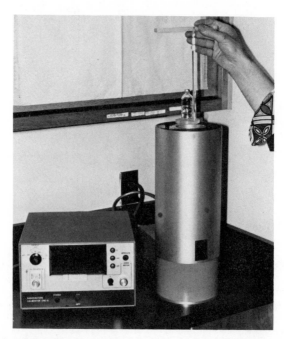

Figure 17.3.   A small bottle of a radiopharmaceutical being lowered into a detector to determine its radioactivity. The results appear on the instrument on the left.

Figure 17.4. A small cyclotron that can be used to produce short-lived radionuclides. The accelerated positive particles are taken from the machine to the left in the beam tube (which is about waist high) and used to bombard a target. (From J.S. Laughlin in W.H. Blahd (Ed.), *Nuclear Medicine*, 2nd ed., McGraw-Hill, New York, 1971, pp. 677–692, Fig. 24-2.)

oxygen is such an active element, it is used by the body within seconds after it enters the lungs. However, much research remains to be done before ^{15}O can be used routinely as a diagnostic tool in nuclear medicine.

17.3. STATISTICAL ASPECTS OF NUCLEAR MEDICINE

The statistical laws used in nuclear medicine are based on the random nature of radioactive disintegrations.* If you have 10^6 atoms of ^{99m}Tc that you can individually identify (a good trick!) and you are asked when the one identified as "5236" is going to disintegrate, you could not say. You could say that after 6 hr you will have about 5×10^5 left, after 12 hr you will have about 2.5×10^5 left, and so on. The one identified as "5236" may die in the first second, but it has a 1 out of 2^4 chance of being alive

---

*The mathematics of random events was one of the first applied fields of mathematics; it was used to predict probabilities in honest games of chance. (The probabilities involved in dishonest games do not require much mathematics.)

after four half-lives (one day) and a 1 out of 2^8 chance of being alive after two days.

A common procedure in nuclear medicine is to count the number of gamma rays detected from a patient in 1 min. To illustrate the random nature of radioactivity let us look at what happened when this measurement was repeated many times. We assumed that the patient did not move and that the equipment worked correctly throughout the experiment, and we ignored the decay of the radionuclide. The results are summarized in Fig. 17.5. The average number of counts was 400, and about two-thirds of the measurements fell between 380 and 420. The values 380 and 420 are not arbitrary; they correspond to $400 \pm \sqrt{400}$. The quantity $\sqrt{400}$ is the standard deviation σ of the measurement. If the disintegration rate is random and the detection equipment is functioning correctly, about two-thirds of the measurements will fall within $\pm 1\ \sigma$ of the average value, about 95% will fall within $\pm 2\ \sigma$ of the average, and about 99.7% will fall within $\pm 3\ \sigma$. Thus if we repeated our measurement 100 times about 67 values would be between 380 and 420, about 95 would be between 360 and 440, and perhaps none would be below 340 ($400 - 3\ \sigma$) or above 460 ($400 + 3\ \sigma$). However, getting two values in this last range would not necessarily mean that our equipment was working incorrectly or that the laws of statistics were failing. It is possible that it could happen by accident. Anyone who gambles realizes that the laws of statistics do not exclude rare events, like drawing a card to fill an inside straight in a poker hand.

What is the standard deviation and what does it mean when only one

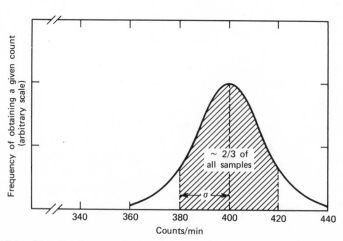

Figure 17.5.   Results of measuring gamma rays from a patient many times at 1 min intervals.

measurement is taken, not 100? If one measurement resulted in a reading of 441, σ would equal $\sqrt{441}$ or 21, and there would be a 67% probability of the "true" answer being within 441 ± 21 and a 95% probability of it being within 441 ± 42. For a longer count time of 2.5 min resulting in a reading of 1000, the standard deviation would be $\sqrt{1000}$ or 31, and there would be a 67% chance of the true answer being within 1000 ± 31. Note that in this case $\sigma \simeq 3\%$ of the counts, while at 400 ± $\sqrt{400}$, $\sigma = 5\%$ of the counts. As the count time increases the uncertainty of the answer decreases, but it decreases slowly. Table 17.5 shows the standard deviations and the percent standard deviations for various counts N. Note that to improve the percent standard deviation by a factor of 10, it is necessary to increase the number of counts by a factor of 100.

Table 17.5. The Standard Deviation σ and the Percent Standard Deviation %σ for Various Counts N

N	σ	%σ
100	10	10
400	20	5
1,000	31	3
4,000	62	1.5
10,000	100	1
40,000	200	0.5
10^6	10^3	0.1

The reading obtained when a source is counted is the gross count N_g. If you repeated the measurement with the radioactive source absent you normally would not get a reading of zero. The natural radioactivity in all materials, cosmic rays, and sometimes electrical noise in the electronic circuits contribute to the background count N_b. If N_g and N_b are each the counts for 1 min, then the rules for obtaining the net count N_{net} and the standard deviation of the net count σ_{net} are:

$$N_{net} = N_g - N_b \qquad (17.4a)$$

$$\sigma_{net} = \sqrt{N_g + N_b} \qquad (17.4b)$$

If N_g is counted for t_g minutes and N_b is counted for t_b minutes, then the gross count rate is N_g/t_g and the background count rate is N_b/t_b. The standard deviation of the gross count rate is

$$\sigma_g = \frac{\sqrt{N_g}}{t_g} \qquad (17.5a)$$

and the standard deviation of the background count rate is

$$\sigma_b = \frac{\sqrt{N_b}}{t_b} \qquad (17.5b)$$

The net count rate is

$$\frac{N_{net}}{min} = \frac{N_g}{t_g} - \frac{N_b}{t_b} \qquad (17.6)$$

The standard deviation of the net count rate is given by

$$\sigma_{net} = \sqrt{\sigma_g^2 + \sigma_b^2} \qquad (17.7)$$

These equations are applied in Example 17.2. Note that in Example 17.2b the standard deviation of the net count rate comes primarily from the standard deviation of the gross count rate. Counting the background for 10 min was a waste of time.

Example 17.2

a. If the gross count for 1 min is 400 and the background count for 1 min is 100, what is the net count and its standard deviation?

Using Equations 17.4a and 17.4b,

$$N_{net} \pm \sigma = (400 - 100) \pm \sqrt{400 + 100} = 300 \pm 23$$

b. If the gross count for 4 min is 1600 and the background count for 10 min is 1000, what is the net count rate and its standard deviation?

Using Equations 17.5a and 17.5b,

$$\sigma_g = \frac{\sqrt{1600}}{4} = 10 \text{ counts/min}$$

$$\sigma_b = \frac{\sqrt{1000}}{10} = 3.16 \text{ counts/min}$$

Using Equation 17.6,

$$\frac{N_{net}}{min} = \frac{1600}{4} - \frac{1000}{10} = 300 \text{ counts/min}$$

Finally, to get the standard deviation of the net count rate we use Equation 17.7.

$$\sigma_{net} = \sqrt{(10)^2 + (3.2)^2} \simeq 10.5 \text{ counts/min}$$

or

$$\frac{N_{net}}{min} \pm \sigma = 300 \pm 10.5 \text{ counts/min}$$

17.4. BASIC INSTRUMENTATION AND ITS CLINICAL APPLICATIONS

Two quite different types of counting are done in nuclear medicine: determining the amount of radioactivity in a given sample or volume, and determining the distribution of the radioactivity in the body, or imaging. In this section we discuss basic instrumentation and how it is used to count radioactivity, and in Section 17.5 we discuss the various imaging devices used to map the distribution of radioactivity in the body. In Section 17.6 we discuss some of the most common nuclear medicine procedures.

As previously mentioned, natural radioactivity was discovered through the use of film. In general, film is not very convenient for detecting radioactivity, although in autoradiography, a research area, it is placed in close contact with a radioactive specimen to obtain an image from the emitted beta rays.

Early workers in nuclear physics (~1900 to 1910) used scintillation screens to detect alpha particles. When an alpha particle strikes a crystal of zinc sulfide it gives off a weak flash of light, or a scintillation, and these researchers viewed and counted the flashes. Since this was a very tedious process, they welcomed more convenient techniques. One breakthrough was the invention in 1906 of the Geiger–Mueller counter (often called the Geiger counter or GM counter). The principle of the GM counter, an extremely simple instrument, is shown in Fig. 17.6. Even the small amount of ionization produced by a single beta ray entering the tube can trigger a discharge, producing a large pulse of electricity that can be heard on a loudspeaker or counted electrically. The GM counter does not distinguish between large and small amounts of ionization; when it is triggered, it discharges completely and then resets to await the next event. The GM counter is now seldom used in research, but it is convenient for use in radiation protection (see Chapter 19, p. 533). Since it is inefficient for detecting gamma rays, the most important rays in nuclear medicine, it is of little use in clinical nuclear medicine.

In about 1950 several new developments in nuclear physics detectors dramatically advanced nuclear medicine. One of these was the invention of the photomultiplier tube (PMT), which can both detect a weak flash of light and estimate the amount of light. The principle of operation of the

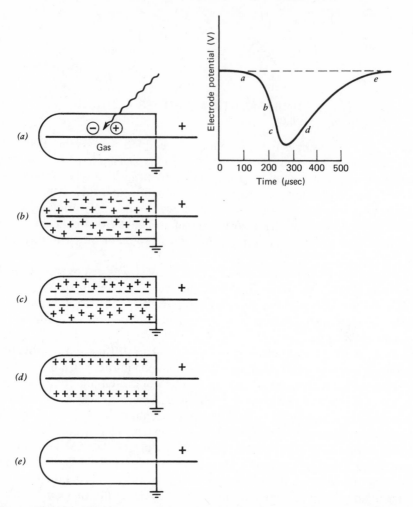

Figure 17.6. The GM tube contains a gas that is ionized by a gamma ray (*a*). The center electrode, which is at a high positive potential (e.g., 1000 V), attracts the electrons and gives them energy to produce further ionization until the whole volume contains ion pairs (*b*). As the electrons are rapidly collected (*c*), the voltage on the center electrode drops as shown in the graph. The slow heavy + ions go to the outer wall (*d*). After about 400 μsec, the tube is ready to repeat the process (*e*).

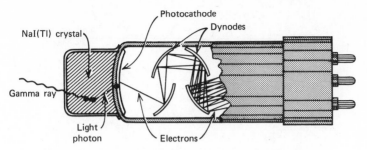

Figure 17.7. Cross-section of a photomultiplier tube.

PMT is shown in Fig. 17.7. A light photon releases an electron at the photocathode that is accelerated to the first dynode, where it causes several more electrons to be emitted; these electrons are accelerated to the second dynode, which is more positive than the first and where further multiplication takes place. Most PMTs have 10 dynodes, so that an electron multiplication of 10^5 to 10^6 times occurs from the photocathode to the anode. In order for each succeeding dynode to be more positive, a high voltage supply (~1000 V) is necessary.

A related development was the production of large clear crystals of sodium iodide (NaI) that would scintillate with good efficiency when a gamma ray is absorbed. These crystals were attached directly to the PMT to detect the weak flashes of light. They were improved by the addition of a trace amount of thallium (Tl) and are often referred to as NaI(Tl) crystals (Fig. 17.7). Since these crystal detectors are about 2000 times more dense than the gas used in the GM detector, they are quite efficient for detecting gamma rays. The 140 keV gamma rays emitted by ^{99m}Tc are almost completely absorbed in a NaI(Tl) crystal 1 cm thick. NaI(Tl) detectors, which come in a wide variety of sizes and shapes for special purposes, are the most widely used detectors in nuclear medicine.

A nuclear medicine scintillation detector system is shown in Fig. 17.8. Since the NaI(Tl) detector is very sensitive, it has to be well shielded from background radiation. It is often surrounded by 5 or more cm of lead except at the collimator opening. The intensity of the scintillation produced when a gamma ray deposits its energy in the crystal is proportional to the energy of the gamma ray. The electrons emitted at the photocathode of the PMT produce an electrical pulse at the output. This pulse is electronically amplified in the *amplifier*. The pulse can be counted directly by the *scaler,* but it is more common to first pass the pulse to a single channel *pulse height analyzer* (PHA) to determine the energy of the gamma ray that caused it. The upper discriminator of the PHA rejects

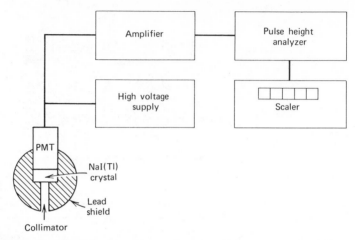

Figure 17.8. A scintillation detector system. One stage of amplification is often mounted in the base of the PMT (the preamplifier). The four components shown in boxes are often mounted in a single unit.

pulses larger than a chosen size and the lower discriminator rejects pulses smaller than a chosen size. The energy difference between the upper and lower limits is called the *window* of the PHA (Fig. 17.9). All pulses in the window are passed on to the scaler to be counted.

For certain purposes it is convenient to determine the pulse height distribution of all the pulses from the detector. This is conveniently done with a *multichannel analyzer* (MCA), which sorts the pulses according to size into 256 or 512 groups. If a NaI(Tl) detector is exposed to a source of monoenergetic gamma rays, such as the 140 keV gamma rays from ^{99m}Tc, the pulse height distribution from an MCA would look similar to that shown in Fig. 17.10. The total energy peak is centered at 140 keV. Pulses corresponding to low energy are usually due to electrons from Compton interactions (see Chapter 16, p. 399) where the scattered photons escape from the crystal. The spread in the total energy peak is due to several factors, but it is primarily statistical in origin. When gamma rays of 140 keV are absorbed in the crystal, they will not all produce the same number of electrons at the photocathode. Nonuniformities in the crystal detector and the photocathode surface cause the total energy peak to broaden still further. The ability of a scintillation detector system to distinguish two gamma ray sources with similar energies is limited by this lack of resolution. The resolution of a NaI(Tl) detector system is usually expressed as the energy spread of the peak (ΔE) divided by the energy of the photons that produced the peak. In Fig. 17.10, the resolution is 30

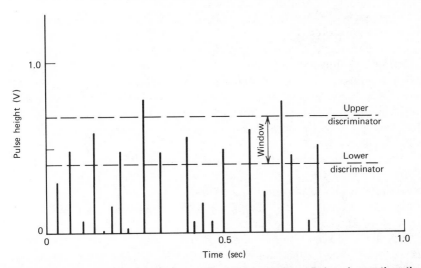

Figure 17.9. The pulses from a scintillation detector system. Pulses larger than the upper discriminator and pulses smaller than the lower discriminator are not counted. The discriminators can be adjusted for gamma rays of different energies. For example, if a pulse of 0.5 V corresponds to the average signal produced by a 140 keV gamma ray, the lower discriminator would correspond to about 100 keV and the upper discriminator to about 180 keV.

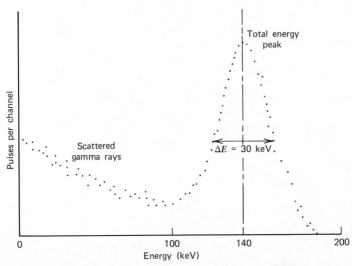

Figure 17.10   A typical MCA pulse height spectrum from a scintillation detector. The source is ^{99m}Tc, which emits gamma rays of 140 keV. The width of the total energy peak at half its height is about 30 keV.

459

keV/140 keV, or 21%. In general, the resolution gets worse as the energy of the gamma rays decreases.

Other types of gamma ray detectors are being tested as alternatives to scintillation detectors. One alternative detector that is widely used in nuclear physics research is the solid-state semiconductor detector. Its main advantage is that it has much better resolution than the NaI(Tl) scintillation detector. Its main disadvantages are that it is not available in large sizes and that it is much more expensive than a scintillation detector of equivalent size.

The principle of operation of the solid-state detector is quite simple. The semiconductor acts as a solid ionization chamber; that is, it acts as an insulator and does not allow current to flow until ionization has taken place in its volume. It is usually kept very cold in order to minimize current produced by the thermal activation of electrons. When a gamma ray is absorbed in the semiconductor it produces a large number of ion pairs, about one ion pair for each 3 eV of energy absorbed. When a 140 keV gamma ray from ^{99m}Tc is absorbed it produces about 47,000 ion pairs. Since only about 1000 electrons are produced from the photocathode of the scintillation detector and PMT system, the statistical variation of the pulse sizes from a solid-state detector is much smaller than that from the scintillation detector, greatly increasing the resolution (Fig. 17.11).

In most clinical nuclear medicine studies, it is important to detect the gamma rays from a limited part of the body only. A specially designed lead shield with one or more holes in it, the collimator, is used on the scintillation detector to limit its counting volume. Cross-sections and

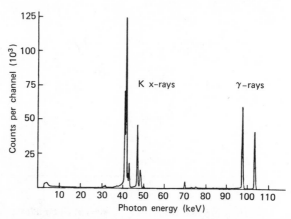

Figure 17.11.   The spectrum of gadolinium 153 produced with a solid-state semiconductor detector. (Courtesy of R. Jennings, University of Wisconsin, Madison, Wisc.)

detection characteristics of two common types of collimators are shown in Fig. 17.12. The curves are lines of equal counting efficiency (isocount lines).

The open or flat field collimator is useful for detecting gamma rays from a relatively large volume such as the thyroid or kidney. The focused collimator, which has a number of tapered holes oriented to point at a small volume 7.5 to 12.5 cm from it, is often used in nuclear medicine imaging since it permits the detector to efficiently detect gamma rays from a known small volume. Focused collimators do not actually focus the gamma rays in any way. Under laboratory conditions a good focused collimator can recognize two radioactive sources only 1 cm apart. Since high-energy gamma rays are very penetrating even in lead, focused collimators must be designed for particular gamma ray energies. Many are optimized for the 140 keV gamma rays from [99mTc].

The 24-hr uptake of radioactive iodine by the thyroid to evaluate

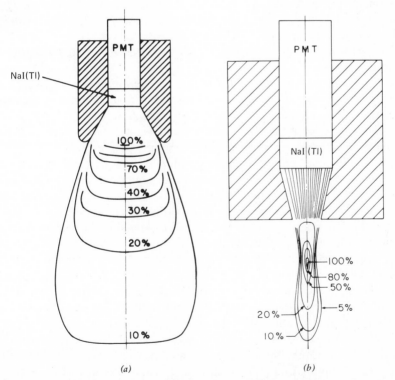

(a)   (b)

Figure 17.12.   Two common types of collimators used with scintillation detectors. (a) The open or flat field collimator. (b) The focused collimator.

thyroid function is one of the oldest nuclear medicine tests using a scintillation detector. The thyroid uses iodine in the production of hormones that control the metabolic rate of the body; a person with an underactive thyroid (hypothyroid) will take up less iodine than a person with normal thyroid function (euthyroid), and a person with an overactive thyroid (hyperthyroid) will take up more iodine. For the 24-hr uptake test, a small amount of ^{131}I, about 300 kBq ($\sim$ 8 μCi) in a liquid or capsule, is given by mouth, and 24 hr later the amount of ^{131}I in the thyroid is counted for 1 min (Fig. 17.13a). The same original amount of ^{131}I—the standard—is set aside at the beginning of the study, and 24 hr later it is placed in a neck phantom and also counted for 1 min (Fig. 17.13b). Since the ^{131}I in the patient and in the standard decay at the same rate, no correction needs to be made for the decay of the ^{131}I. After corrections are made for background counts, the ratio of the thyroid counts to the standard counts times 100 gives the percent 24-hr uptake. Values for euthyroids range from about 10 to 40%, with an average of around 20%. If the uptake is above 40%, the patient may be hyperthyroid. Patients with uptakes of less than 10% may be hypothyroid or may have recently taken in a lot of stable iodine and be temporarily oversupplied.

In recent years, new thyroid tests have been developed that do not require the administration of ^{131}I to the patient (an *in vivo* test); these new tests are done on blood samples in test tubes (*in vitro* tests). We do not discuss the principles of the *in vitro* tests, but you should note that they are safer since no radiation is given to the patient.

Kidney function is also often studied with scintillation detectors. About 7 MBq ($\sim$ 200 μCi) of ^{131}I-labeled hippuric acid is injected into the bloodstream, and as it is removed from the blood by the kidneys the radioactivity of each kidney is monitored. Open collimator scintillation detectors are used. The signals from each detector are fed to an electronic instrument called a *ratemeter* to record the change in radioactivity with time. The ratemeter averages the signals over a short period of time and indicates the count rate in counts per minute or counts per second. It is often connected to a strip chart recorder to make a permanent record of the count rate versus time, or a renogram. Figure 17.14 shows schematically the equipment involved in making a renogram. The renogram in Fig. 17.14 is a 30-min record of a typical normal kidney. The unevenness in the renogram is due to the statistical nature of the detected gamma rays.

The smoothness of the count rate curve depends on the length of time over which the ratemeter averages the counts, or the *time-constant* of the ratemeter. On most ratemeters the time-constant can be changed. Figure 17.15 shows the count rate curves that resulted when the same experiment was done using different time-constants; note that the curve for the longer

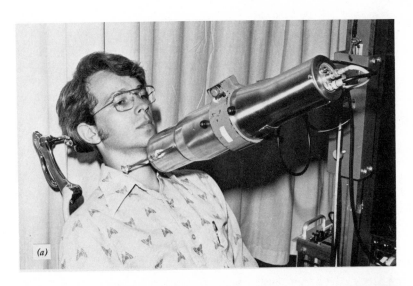

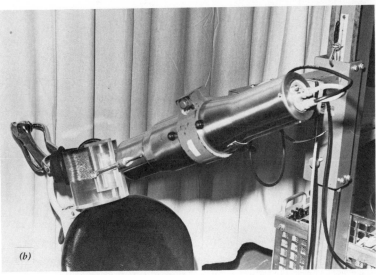

Figure 17.13. For the 24-hr thyroid uptake test, the patient takes about 3 kBq of [131]I. The uptake in the thyroid is measured 24 hr later (a) and compared to the counts from a standard in a neck phantom (b).

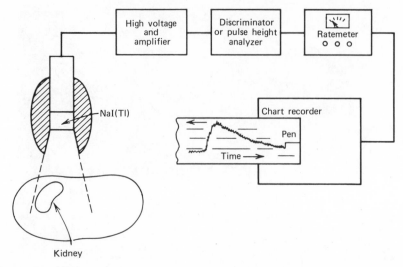

Figure 17.14. The components involved in making a renogram. The three components at the top are often mounted in a single cabinet. Usually a separate unit measures the function of the other kidney at the same time.

time-constant (Fig. 17.15*b*) is smoother and shows greater delay in responding to the increase in count rate that occurred at 10 sec. When it is important to see rapid changes, a short time-constant should be used.

The percent standard deviation %σ of a ratemeter is given by

$$\%\sigma = \frac{\sqrt{2\dot{N}\tau}}{2\dot{N}\tau} (100) \qquad (17.8)$$

where $\dot{N}$ is the count rate and τ is the time-constant. If τ is in seconds, $\dot{N}$ must be in counts per second. Note that for a given count rate increasing the time-constant by a factor of 4 decreases the percent standard deviation by $\sqrt{4}$ or by a factor of 2; increasing the count rate by a factor of 3 decreases the percent standard deviation by $\sqrt{3}$ (see Example 17.3).

Example 17.3
Calculate the standard deviations for the two count rates and time-constants shown in Fig. 17.15. Note that the counts per minute must be divided by 60 to get counts per second.

The lower count rate of 240 counts/min is 4 counts/sec. Using the shorter time-constant in Equation 17.8 we get

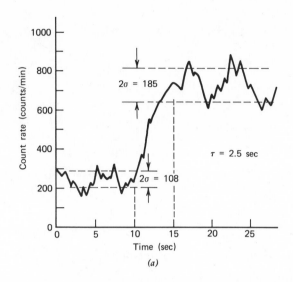

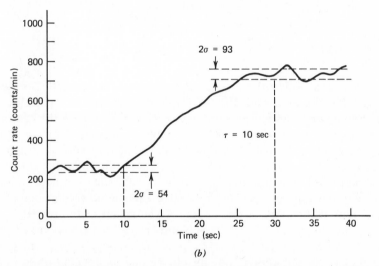

Figure 17.15. Graphs of count rate versus time for two different time-constants: (a) 2.5 sec and (b) 10 sec. In each case the count rate increased abruptly at 10 sec from an average of 240 counts/min to 720 counts/min. Note that it takes approximately 2τ sec for the ratemeter to indicate the new count rate. Increasing τ by a factor of four decreased the standard deviation by a factor of two but produced a much slower response.

$$\%\sigma = \frac{\sqrt{2(4)(2.5)}}{2(4)(2.5)} (100) = 22.4\%$$

$$22.4\% \text{ of } 240 = 53.8 \text{ counts/min}$$

For the same count rate and longer time-constant we get

$$\%\sigma = \frac{\sqrt{2(4)(10)}}{2(4)(10)} (100) = 11.2\%$$

$$11.2\% \text{ of } 240 = 26.8 \text{ counts/min}$$

The higher count rate of 720 counts/min is 12 counts/sec. For $\tau = 2.5$ sec we get

$$\%\sigma = \frac{\sqrt{2(12)(2.5)}}{2(12)(2.5)} (100) = 12.9\%$$

$$12.9\% \text{ of } 720 = 92.9 \text{ counts/min}$$

For the same count rate and $\tau = 10$ sec we get

$$\%\sigma = \frac{\sqrt{2(12)(10)}}{2(12)(10)} (100) = 6.5\%$$

$$6.5\% \text{ of } 720 = 46.5 \text{ counts/min}$$

For counting *in vitro*, it is customary to use a specially designed NaI(Tl) crystal with a well (Fig. 17.16). Well counters are very efficient for counting small samples since the activity is placed near the center of the sensitive crystal detector. Clinically, well counters are used to measure the blood volume of a patient following an accident or surgery by means of the dilution technique. (The straightforward physics approach of draining all the blood to determine its volume is obviously not acceptable.)

The dilution technique was used for measuring blood volume before the radioactivity technique was developed. A known amount of blue dye was injected into the blood, and after the blood and dye were well mixed (~15 min) a sample of blood was taken and the amount of dye in it was measured. From the known amount of dye and its concentration in the blood, the volume was calculated.

In the more reliable radioactivity technique, about 200 kBq (~5 μCi) of ^{131}I-labeled albumin is injected into an arm vein, and after about 15 min a blood sample is drawn from the other arm and counted. (If the patient's blood contains radioactivity from a previous study, a preinjection sample of the blood must also be drawn and counted.) The net count rate and volume of the sample is compared to the count rate and volume of the

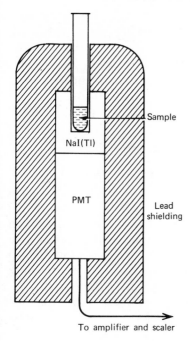
Sample

NaI(Tl)

PMT

Lead
shielding

To amplifier and scaler

Figure 17.16. A cross-section of a well counter showing a sample in a test tube recessed in the NaI(Tl) scintillation crystal. The crystal is well shielded with lead to reduce background counts.

injected material to determine the blood volume (see Example 17.4). Since the count rate of the injected material is usually too large to determine accurately, it is common to dilute an equal amount of radioactive material in a known volume of water and then count a sample after the water and material have mixed well. For example, if 5 ml of ^{131}I-labeled albumin, as used in Example 17.4, was diluted in 1 liter of water, it would be found that a 5 ml sample of the water would have a count rate of about 5×10^2 counts/sec.

Example 17.4

What is the blood volume of a patient if 5 ml of ^{131}I-labeled albumin with a net count rate of 10^5 counts/sec was injected into the blood and the net count rate of a 5 ml blood sample drawn 15 min later was 10^2 counts/sec?

Since the ratio of count rates is 10^3, the blood volume is 10^3 times greater than the injected volume, or 5000 ml.

In a liquid scintillation detector, inexpensive scintillation fluid replaces the NaI(Tl) crystal; it is mixed with the sample to be counted and then discarded after the measurement. The big advantage of the liquid scintillation detector is its ability to detect and count the weak beta rays from tritium (^{3}H) and ^{14}C, which are difficult to count with any other system. Because the signals are usually weak, the detector and samples are sometimes cooled to reduce background noise from thermal sources in the PMT. Liquid scintillation detectors are primarily used in research and for radiation protection purposes such as determining the amount of radioactivity in the urine of a worker who has been exposed to tritium.

17.5. NUCLEAR MEDICINE IMAGING DEVICES

While counting the radioactivity in an organ provides important information, it is often more useful to know how the radioactivity is distributed in that organ. Imaging, or producing pictures of the distribution of the radioactivity, is now the most important aspect of nuclear medicine; in 1976, approximately 80% of all *in vivo* clinical studies in nuclear medicine involved imaging. In this section we discuss in detail the two principal devices used to produce nuclear medicine images—the rectilinear scanner and the gamma camera. The less common positron camera is also described. The best images from these devices appear crude when compared to a good x-ray image, but they often provide information that cannot be obtained in any other way.

In the 1940s, nuclear medicine pioneers laboriously moved their insensitive GM counters over different parts of the body to get a crude idea of the distribution of radioactivity. In 1950, Cassens developed the first mechanical scanner, or *rectilinear scanner*. The NaI(T1) detector of a rectilinear scanner moves in a raster pattern over the area of interest, making a permanent record of the count rate, or a map of the radiation distribution in the body (Fig. 17.17). Scanners rapidly improved during the 1950s and 1960s, mainly due to the increasing availability of large NaI(T1) detectors. In the 1950s a crystal 5 cm in diameter by 5 cm thick was commonly used; in the 1970s a crystal 12.7 cm in diameter by 5 cm thick became more common. Most modern hospitals have one or more rectilinear scanners in their nuclear medicine laboratory.

During most scans, two images are made of the radiation distribution (Fig. 17.18). One image, on a storage oscilloscope or consisting of marks tapped on paper, is usually visible during the scan. The density, or sometimes the color, of the marks indicates the intensity of the detected activity in the corresponding areas of the patient. The other image is made by moving a small light source over a photographic film as shown in Fig. 17.17. The intensity of the light increases with an increase in activity, producing corresponding dark areas on the film, and can be modified

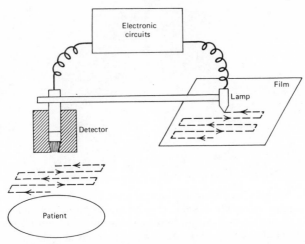

Figure 17.17. Principle of a rectilinear scanner. The scintillation detector is mechanically connected to a bulb that records the amount of radioactivity on the film. The electronic circuits consist of the usual components used with a scintillation detector plus controls for the mechanical scanner and for adjusting the intensity of the lamp. The lamp and film are in a light-tight container. The darkness of an area on the film indicates the amount of radioactivity over the corresponding area of the patient.

electronically to enhance count rate differences in areas of medical interest. It is also possible to record the data taken during the scan on a magnetic tape or disc; these data can later be analyzed by a computer to provide a quantitative image.

The spacing of the scan lines and the rate of movement of the scanning head can be adjusted to suit the organ size and the amount of activity in the organ. The dimensions of the scan area can also be adjusted. While most early scanners covered areas of up to 36 cm × 43 cm (14 in. × 17 in.), many modern scanners can be used for scanning the entire body; the resulting image is minified so that it can fit on a standard 36 cm × 43 cm film.

Rectilinear scanners use focused collimators, and thus they primarily measure the radiation distribution 7.5 to 12.5 cm from the end of the collimator. For this reason it is often necessary to scan both sides of the patient to determine the distribution of radioactivity. A few scanners have two detectors facing each other so that they can simultaneously scan both sides of the patient (Fig. 17.19); since some scans last over 30 min, this can be a decided advantage for the patient. While the patient is expected

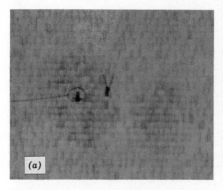 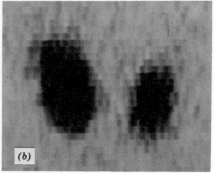

Figure 17.18. Scans or an essentially normal thyroid. (*a*) **The image recorded with a mechanical tapper, originally in color.** (*b*) **The image produced by a variable light source over a photographic film. This type of image can be electronically modified to reduce the effect of radiation from areas other than the thyroid.**

to remain motionless during a scan, he cannot hold his breath for more than 30 to 60 sec, and as a result scans of the liver include motion artifacts since the liver moves up and down about 2 cm during normal breathing.

While the time needed to obtain an image with a scanner can be reduced by using two or more detectors, scanning is nonetheless time consuming. A breakthrough in nuclear medicine imaging occurred in 1956 when Hal Anger of the University of California invented the *gamma camera*, which produces an image while stationary. Figure 17.20 shows a modern gamma camera. It has a large NaI(Tl) scintillation crystal about 1 cm thick and 30 to 45 cm in diameter (Fig. 17.21). The scintillations are viewed through a light pipe by an array of 19 or 37 PMTs. When a gamma ray interacts somewhere in the crystal the light from the scintillation produces a large signal in the closest PMT and weaker signals in PMTs further away. These signals are electronically processed to determine the (x,y) coordinates of the scintillation. Electronic circuits then deflect an electron beam in a cathode ray tube (CRT) to cause a momentary bright spot to appear at the corresponding (x,y) location of the CRT. The face of the CRT is usually viewed by a camera that records on film all of the spots from the CRT. After the gamma camera has recorded a preset number of counts (for example, 400,000), there will be enough spots on the film to satisfactorily show the distribution of the radioactivity (Fig. 17.22). A computer can be used in conjunction with a gamma camera to obtain quantitative information.

A parallel hole collimator is usually used with a gamma camera, but occasionally a diverging hole collimator is used to view a large area such

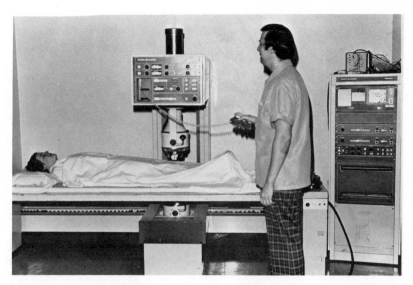

Figure 17.19. A whole body rectilinear scanner. This unit has two detectors—one above and one below the body. The images are produced in the component at the right. Bone scans such as those shown in Fig. 17.28 can be obtained with this unit.

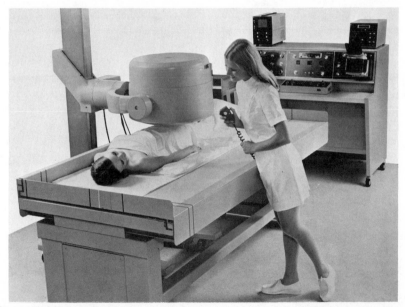

Figure 17.20. A gamma camera over the abdomen of a subject. The image is produced in the unit in the background. (Courtesy of Searle Radiographics, Inc., Des Plaines, Ill.)

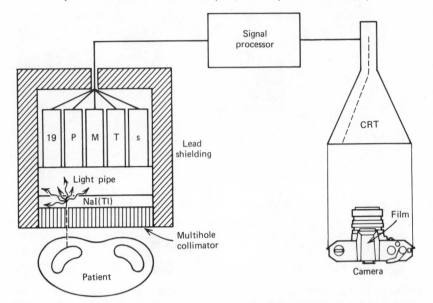

Figure 17.21. Components of a gamma camera. The detector and PMTs are in a shielded lead housing. A scintillation at a point in the crystal produces a large signal at the closest PMT and weaker signals in more distant PMTs. The signal processor determines the (x,y) location of the scintillation and causes a bright spot to appear at the corresponding (x,y) location of the CRT. This bright spot is then recorded on the film in the camera.

as the lungs. The original gamma camera had a pinhole aperture (Fig. 17.23), and this design is still used for some studies.

The resolution of the gamma camera is comparable to that of a rectilinear scanner—it can distinguish two sources about 5 mm apart when they are held close to the collimator—but the imaging time for a gamma camera is only 1 or 2 min. Because of the short imaging time involved, it is

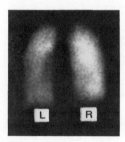

Figure 17.22. A gamma camera image of the lungs of a normal subject taken from the back. Approximately 600,000 counts were used to produce the image.

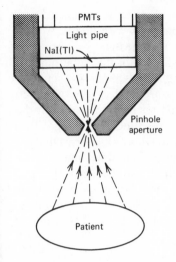

Figure 17.23. The original gamma camera used a pinhole aperture to produce an image. This design is still sometimes used to image the thyroid.

possible to get dynamic information with a gamma camera that cannot be obtained with a rectilinear scanner, and the equivalent of motion pictures can be produced. In addition, it is possible to use radionuclides with very short half-lives of 2 min or less.

While the distribution of a positron emitter in the body can be imaged with either a rectilinear scanner or a gamma camera, a special imaging

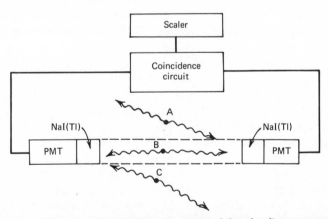

Figure 17.24. A positron camera takes advantage of the simultaneous emission of annihilation photons in opposite directions. In most cases only positrons located between the dashed lines can produce a coincidence of pulses in the 2 PMTs. Positrons at other locations, such as at A and C, normally produce pulses in only one of the PMTs; however, if annihilations occur at A and C simultaneously, an accidental coincidence could be produced.

device called a *positron camera* has been developed. As discussed previously, when a positron comes to rest it combines with an electron to emit two 0.511 MeV photons (annihilation radiation) that go in opposite directions, and the positron camera takes advantage of this simultaneous emission (Fig. 17.24). In a positron camera the signals from two scintillation detectors on opposite sides of the patient are analyzed by a coincidence circuit. The coincidence circuit records an event when both detectors produce pulses at essentially the same time from a positron annihilation in the volume between them, while it ignores all pulses that are not in coincidence. The positron camera image is sometimes made in the same way as a conventional scan, although it can be made by two gamma cameras located on opposite sides of the patient. While homemade positron cameras have been available for many years, they have not been used extensively in nuclear medicine because of the general shortage of suitable positron emitters.

17.6. PHYSICAL PRINCIPLES OF NUCLEAR MEDICINE IMAGING PROCEDURES

In this section we discuss the principles involved in obtaining images of the radioactive distribution in various organs. In general, the images can be obtained with either the gamma camera or the rectilinear scanner. The imaging characteristics of the two devices are quite similar, but since the gamma camera obtains the image in considerably less time it is usually preferred over the rectilinear scanner. When it is necessary to obtain dynamic information, the gamma camera is required.

As mentioned in Chapter 16 x-rays are absorbed similarly in tumor tissue and normal tissue, and thus x-ray images do not generally show tumors. However, tumors can often be seen on nuclear medicine images. The thyroid, one of the first organs to be imaged after the rectilinear scanner became available in 1950, often develops nodules or lumps that may be cancerous, and a lump that does not take up radioactivity, or is "cold," is more apt to be cancerous than a lump that functions like thyroid tissue and takes up radioactivity. Figure 17.25 shows scans of a normal thyroid and a thyroid with a cold nodule. When ^{131}I is used for a thyroid scan, about 4 MBq ($\sim$100 μCi) is given orally the day before the scan. ^{99m}Tc is also used for thyroid imaging since the pertechnetate ion ^{99m}TcO$_4^-$ is taken up by the same tissues that take up iodine. A typical dose of ^{99m}Tc is 150 MBq ($\sim$ 4 mCi). ^{99m}Tc is the preferred radionuclide since it gives a much smaller radiation dose to the patient. ^{123}I with its lack

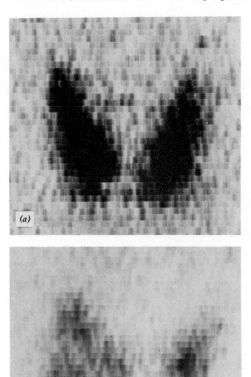

Figure 17.25. Scans of (a) an essentially normal thyroid and (b) an abnormal thyroid. Note the reduced radioactivity in b over areas corresponding to nodules.

of beta emission and a 13-hr half-life is also an excellent radionuclide for thyroid imaging; a dose of about 20 MBq is used.

Cancer often spreads, or *metastasizes*, to the liver, and this condition can often be detected on a liver scan. Normal liver tissue will filter radioactive submicroscopic particles from the blood, while a tumor in the liver will not and appears on the scan as an area of reduced radioactivity. For a typical liver scan, 200 MBq (~5 mCi) of ^{99m}Tc-labeled sulfur colloid with particles about 0.5 μm in diameter is injected into a vein, and the

476 Physics of Nuclear Medicine (Radioisotopes in Medicine)

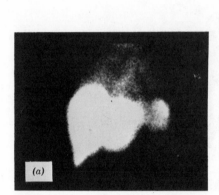
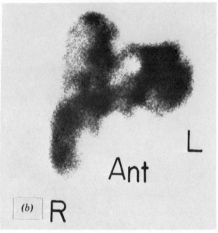

Figure 17.26. Gamma camera images of (a) a normal liver and (b) an abnormal liver. The abnormal liver shows areas of reduced radioactivity indicating cancer.

image is obtained after about 10 min. Figure 17.26 shows scans of a normal liver and a cancerous liver.

Since the symptoms of brain tumors and those of many less serious conditions are similar and fairly common, nuclear medicine techniques have been developed to identify brain tumors. When injected into the blood, different radioactive materials are taken up better by brain tumors than by the surrounding normal brain tissues. Typically about 500 MBq ($\sim$15 mCi) of ^{99m}Tc is given, and after about 2 hr four gamma camera images (the front, back, left, and right sides of the head) are taken. Figure 17.27 shows gamma camera images of a brain with a tumor.

Cancer often metastasizes to the bones, and bone scans are often more useful than x-rays for detecting it. A portion of the skeleton being destroyed by cancer will be trying to rebuild itself and will take up more of many elements than normal bone. Although there are no suitable radioisotopes of calcium for bone scans, several other radionuclides can be used. A common technique is to inject into the blood about 500 MBq ($\sim$15 mCi) of a phosphate compound labeled with ^{99m}Tc and make a scan about 3 hr later. The scan will show increased radioactivity in the areas of bone tumor or bone growth. The bone scans in Fig. 17.28 of a normal adult and a young person who is still growing were made with a whole body rectilinear scanner such as that shown in Fig. 17.19.

A number of medical centers use radioactive fluorine atoms (^{18}F), which fit into the bone crystal in place of OH$^-$ ions, for bone scans. ^{18}F has a half-life of only 110 min and must be produced close to its point of use.

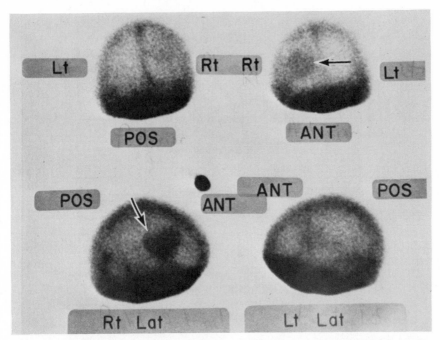

Figure 17.27. **Gamma camera images of a brain with a tumor in the right anterior portion (arrows). Note that the tumor is not clearly visible in the posterior view or in the left lateral view.**

Nuclear medicine imaging techniques are also used to evaluate kidney function, often after a kidney transplant. Usually a small amount of radioactive hippuric acid, which is cleared rapidly by the kidneys, is injected into the blood, the kidneys are viewed by a gamma camera, and sequential images are made every few minutes. By connecting a computer to the gamma camera, it is possible to obtain renograms while the images are being made. Figure 17.29*c* shows the renogram for the kidneys imaged in Fig. 17.29*b*. Note the lack of function in the left kidney.

A blood clot blocking a major artery in the lung, or a *pulmonary embolism*, is a fairly serious medical problem. In the nuclear medicine test for a pulmonary embolism, about 100 MBq ($\sim$ 3 mCi) of lumpy (macroaggregated) ^{99m}Tc-labeled albumin is injected into a vein. This material travels to the heart and then to the lungs; the lumps are too large to pass through the capillaries of the lungs and are temporarily trapped at the entrances to some of the functioning capillaries. The lumps block less than 1% of the capillaries and break up after an hour or two. A scan or gamma camera image taken immediately after the injection shows

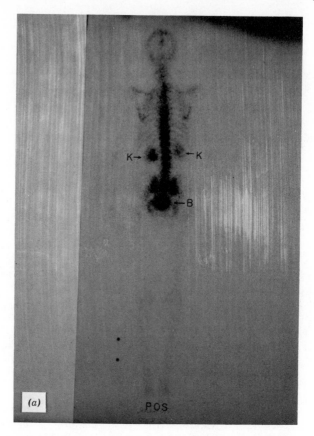

(a)

POS

K→   ←K

←B

radioactivity where the functioning capillaries are located and little radioactivity in the part of the lung that is blocked. Figure 17.30 shows gamma camera images of normal lungs and lungs with an embolism.

The air circulation system of the lungs can be studied with a radioactive gas such as radioactive xenon (^{133}Xe), which has a 5.3-day half-life. Both the distribution of the radioactivity and the length of time it remains in a given volume give diagnostic information.

The heart is one of the most difficult organs to study with nuclear medicine techniques because its continual beating limits the detail that can be seen in the image. However, the resolution of the image can be improved by using the ECG signal to activate the detector during the resting phase (diastole) of the heartbeat. Radiopharmaceuticals that will concentrate at the site of damage (infarct) following a heart attack are

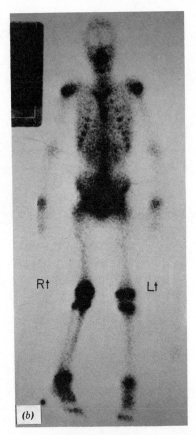

Rt Lt

(b)

Figure 17.28. Whole body scans. (a) The scan of an adult was taken from the back and shows the spine clearly. The kidneys (K) and the bladder (B) are visible. (b) The scan of a growing person was taken from the front and shows increased radioactivity in the areas of active growth at the ends of the long bones.

being developed, and it is hoped that the region and extent of heart attacks can be determined through studies with these drugs.

17.7. THERAPY WITH RADIOACTIVITY

Ionizing radiation is very useful in treating cancer, and this application of radioactivity is discussed in detail in Chapter 18. In this section we discuss other less common radiation therapy procedures and areas of interest.

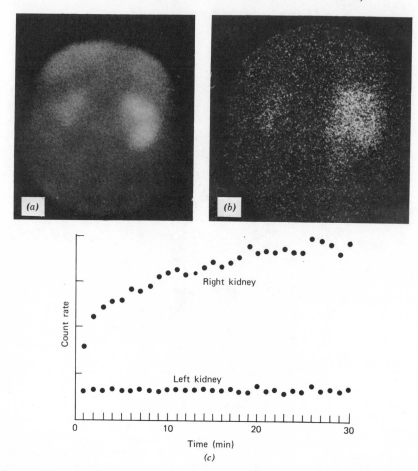

Figure 17.29. Initial (*a*) and 6-month follow-up (*b*) gamma camera studies of a patient with a defective left kidney. The function of the left kidney decreased noticeably in the time between the scans. (*c*) A 30 min graph of the radioactivity from each kidney versus time obtained when scan *b* was made.

Liquid radioactive substances can be taken by mouth for therapeutic purposes. ^{131}I has been used in quantities of 150 to 400 MBq (~4 to 10 mCi) for treating overactive thyroids, and it is also used in larger quantities of 1 to 3 GBq (~30 to 100 mCi) for treating some cancers of the thyroid. The therapeutic effect is largely due to the local absorption of beta particles.

^{32}P, a pure beta emitter, has been used in the treatment of polycythemia

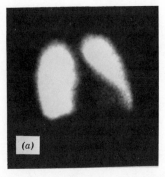 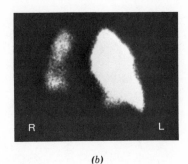

Figure 17.30. Lung scans of (*a*) a patient with normal lungs and (*b*) a patient with a severe pulmonary embolism in the right lung. The dark area on the lower left lung in scan *a* is due to the heart.

vera, a disease that causes an overproduction of red blood cells. While ^{32}P reduces red blood cell production, this form of therapy became less popular as the risks of ionizing radiation became better known.

Many researchers have tried to find a radioactive drug that would concentrate in a tumor in sufficient quantity to produce therapeutic effects. With the exception of radioactive iodine, which concentrates well in certain thyroid cancers, no such drug has been discovered. However, research is still underway in this area.

17.8. RADIATION DOSES IN NUCLEAR MEDICINE

In general, the radiation dose to the body from a nuclear medicine procedure is nonuniform since radioisotopes tend to concentrate in certain organs. While it is essentially impossible to measure the radiation received by a particular patient, it is possible to calculate the dose to various organs of a *standard man*. The characteristics of a standard man are given in Appendix C.

The organ receiving the largest dose during a procedure is sometimes referred to as the *critical organ* for that procedure. Table 17.6 gives some typical critical organ and gonad doses for some common nuclear medicine procedures. The gonad dose is used in estimating the possible genetic effects of the procedure. Keep in mind that these doses can vary considerably from one individual to another.

The dose to a particular organ of the body depends on the physical characteristics of the radionuclide—what particles it emits and their

482

Table 17.6. Critical Organ and Gonad Doses to a Standard Man From Some Common Nuclear Medicine Procedures

Procedure	Radiopharmaceutical	Amount Administered	Critical Organ Dose (Gy)[a]	Gonad Dose (mGy)[a]
Brain scan	^{99m}Tc pertechnetate	500 MBq (~ 15 mCi)	Intestine, 0.02	4
Liver scan	^{99m}Tc sulfur colloid	150 MBq (~4 mCi)	Liver, 0.02	0.85
Lung scan	^{99m}Tc macroaggregated albumin	100 MBq (~3 mCi)	Lungs, 0.009	0.3
Bone scan	^{99m}Tc pyrophosphate	500 MBq (~15 mCi)	Bladder, 0.06	4
Renogram	^{131}I hippuric acid	8 MBq (~200 μCi)	Bladder, 0.02	0.2
Thyroid uptake	^{131}I sodium iodide	300 kBq (~8 μCi)	Thyroid, 0.08	0.6
Thyroid scan	^{99m}Tc pertechnetate	150 MBq (~4 mCi)	Intestine, 0.01	0.8

[a]The gray (Gy) is defined in Section 18.1.

energies—and on the length of time the radionuclide is in the organ. Two factors determine the length of time the radionuclide is in the organ, or the effective half-life ($T_{1/2\ \text{eff}}$)—the physical half-life ($T_{1/2\ \text{phy}}$) and the biological half-life ($T_{1/2\ \text{bio}}$). The biological half-life of an element is the time needed for one-half of the original atoms present in an organ to be removed from the organ, and it is independent of whether the element is radioactive. Most elements are excreted at an exponential rate. The effective half-life can be calculated as follows.

$$T_{1/2\ \text{eff}} = \frac{(T_{1/2\ \text{bio}})\,(T_{1/2\ \text{phy}})}{T_{1/2\ \text{bio}} + T_{1/2\ \text{phy}}} \tag{17.9}$$

Note that if either the biological or physical half-life is much shorter than the other, the effective half-life is essentially equal to the shorter value (see Example 17.5). Whenever a new radiopharmaceutical is introduced, extensive studies are done on animals to determine the effective half-lives for the various organs.

Example 17.5

a. What is the effective half-life of ^{131}I in the thyroid if $T_{1/2\ \text{bio}} = 15$ days and $T_{1/2\ \text{phy}} = 8$ days?
    From Equation 17.9

$$T_{1/2\ \text{eff}} = \frac{(15\ \text{days})(8\ \text{days})}{15\ \text{days} + 8\ \text{days}} = 5.2\ \text{days}$$

b. What is the effective half-life of ^{131}I hippuric acid (used for renograms) if half of it is excreted in the urine in 1 hr (i.e., $T_{1/2\ \text{bio}} = 1$ hr)?

$$T_{1/2\ \text{eff}} = \frac{(1\ \text{hr})(192\ \text{hr})}{1\ \text{hr} + 192\ \text{hr}} = 0.99\ \text{hr}$$

c. What is the effective half-life of ^{18}F in bone if $T_{1/2\ \text{bio}} = 7$ days and $T_{1/2\ \text{phy}} = 110$ min? (7 days $\simeq 10^4$ min)

$$T_{1/2\ \text{eff}} = \frac{(110\ \text{min})(10^4\ \text{min})}{110\ \text{min} + 10^4\ \text{min}} \simeq 109\ \text{min}$$

BIBLIOGRAPHY

Baum, S., and R. Bramlet, *Basic Nuclear Medicine*, Appleton-Century-Crofts, New York, 1975.

Behrens, C. F., E. R. King, and J. W. J. Carpender (Eds.), *Atomic Medicine*, 5th ed., Williams and Wilkins, Baltimore, 1969.

Blahd, W. H. (Ed.), *Nuclear Medicine*, 2nd ed., McGraw-Hill, New York, 1971.

Early, P. J., M. A. Razzak, and D. B. Sodee, *Textbook of Nuclear Medicine Technology,* Mosby, St. Louis, 1969.

Freeman, L. M., and P. M. Johnson (Eds.), *Clinical Scintillation Scanning,* Hoeber, New York, 1969.

Hine, G. J. (Ed.), *Instrumentation in Nuclear Medicine,* Vol. 1, Academic, New York, 1967.

Hine, G. J., and J. A. Sorenson (Eds.), *Instrumentation in Nuclear Medicine,* Vol. 2, Academic, New York, 1974.

Keyes, J. W., W. H. Beierwaltes, D. C. Moses, and J. E. Carey, *Manual of Nuclear Medicine Procedures,* 2nd ed., CRC Press, Cleveland, 1973.

Lange, R. C., *Nuclear Medicine for Technicians,* Year Book Medical, Chicago, 1973.

Maynard, C. D., *Clinical Nuclear Medicine,* Lea and Febiger, Philadelphia, 1969.

MIRD Pamphlets, Medical Internal Radiation Dose Committee, *J. Nucl. Med.,* Suppl. 1 to 11, 1968 to 1975.

Powsner, E. R., and D. E. Raeside, *Diagnostic Nuclear Medicine,* Grune and Stratton, New York, 1971.

Quimby, E. H., S. Feitelberg, and W. Gross, *Radioactive Nuclides in Medicine and Biology,* 3rd ed., Lea and Febiger, Philadelphia, 1970.

Sodee, D. B., and P. J. Early, *Technical Interpretation of Nuclear Medicine Procedures,* 2nd ed., Mosby, St. Louis, 1975.

Spiers, F. W., *Radioisotopes in the Human Body: Physical and Biological Aspects,* Academic, New York, 1968.

Wagner, H. N. (Ed.), *Nuclear Medicine,* Hospital Practice, New York, 1975.

Wagner, H. N. (Ed.), *Principles of Nuclear Medicine,* Saunders, Philadelphia, 1968.

REVIEW QUESTIONS

1. Give the essential characteristics of alpha, beta, and gamma rays.
2. Beta particles are characterized by a spread of energies up to a maximum. What happens to the difference in energy between the actual beta energy and the maximum beta energy?
3. What is the distinction between *radioisotope* and *radionuclide*?
4. How many neutrons does $^{131}_{53}$I have in its nucleus?
5. What does a metastable radionuclide emit?
6. What emission is characteristic of a radionuclide that decays by electron capture?
7. If a radionuclide has a decay constant λ of 0.001 days^{-1},
   (a) what is its half-life?
   (b) what is its mean life?
   (c) how much do you have left after 24 hr if you have 10 MBq at $t = 0$?
8. How is ^{99m}Tc obtained for most nuclear medicine procedures?
9. Why do most radionuclides produced in reactors emit beta rays while many radionuclides produced in cyclotrons emit positrons?
10. If a sample is counted for 10 min and gives 3700 counts and a 10 min

background count gives 800 counts, what is the net count rate and its standard deviation?

11. Give one advantage and one disadvantage of a solid-state semiconductor detector over a NaI(Tl) detector.
12. What is an MCA?
13. Describe the steps in the 24-hr thyroid uptake test.
14. What is a renogram?
15. What are the general characteristics of the two most common instruments used for nuclear medicine imaging?
16. Describe the principles of the nuclear medicine technique used to detect a pulmonary embolism.
17. What is the effective half-life? How is it related to the physical and biological half-lives?
18. What is meant by *critical organ?*

CHAPTER 18

Physics of Radiation Therapy

Within a year of Roentgen's discovery of x-rays in 1895 it became obvious that x-rays could produce biological damage in the form of reddened skin, ulcers, and so forth. In 1896 Lister in England and Grubbe and Ludlam in the United States suggested that x-rays might be useful in treating cancer. Early attempts were not a great success; however, today radiation therapy is recognized as an important tool in the treatment of many types of cancer.

The probability of your having cancer during your lifetime is about 1 out of 4. Each year about 1 million people in the United States get cancer. Currently three major methods are used alone or in combination to treat cancer: surgery, radiation therapy, and drugs (chemotherapy). About half of all cancer patients receive radiation as part or all of their treatment. The success of radiation therapy depends on the type and extent of the cancer; the skill of the radiotherapist, the physician who specializes in the treatment of cancer with radiation; the kind of radiation used in the treatment; and the accuracy with which the radiation is administered to the tumor. The last factor is the responsibility of the radiological physicist. Before 1950 radiation therapy was used for many diseases besides cancer, and it is still used occasionally for a nonmalignant condition that does not respond to other treatment. However, because of the possibility of radiation-induced cancer (see Chapter 19) it is generally safer to use other types of treatment when possible.

Before 1940 the radiation energies used for therapy were not much higher than those used in diagnostic radiology, and in the United States most radiation therapy was given to patients by general radiologists. In the late 1940s much higher radiation energy became available and started the trend toward megavoltage therapy. The betatron, which was developed by Kerst in the early 1940s, can accelerate electrons to an energy of 45 MeV and produce high-intensity very penetrating x-rays. It was first used for treating cancer in 1948. Since then there has been a steady growth in the use of high-energy beams for radiation therapy. The advantages of these beams are discussed in this chapter. As a result of the improved characteristics of these megavoltage machines and the improved training of radiotherapists, cures of many types of cancer with radiation have doubled since the premegavoltage era.

There is evidence that an error of 5% to 10% in the radiation dose to the tumor can have a significant effect on the results of the therapy. Too little radiation does not kill all the tumor; too much can produce serious complications in normal tissue. Figure 18.1 shows schematically how local cure and severe complications depend on the dose in the treatment of cancer with radiation therapy. The solid line shows the percent of cure of a local tumor as a function of dose for a large number of patients with the same type of tumor. The dashed line shows the percent of severe complications as a function of dose. Note that giving a dose just 5% lower than 50 Gy reduces the chance of local cure by a factor of nearly 2 while giving a dose 5% higher does not greatly increase the cure rate but does greatly increase the number of major complications. (The gray is defined in Section 18.1.) These curves, which differ from tumor to tumor, demonstrate why accurate physical measurements by radiological physicists are important in radiation therapy. In 1976, over half of the about 1000 medical physicists employed in hospitals in the United States were working in the field of radiation therapy physics.

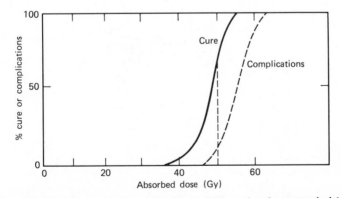

Figure 18.1. For many tumors treated with radiation, the dose needed to give a reasonable chance of cure (control of the local tumor) is only slightly lower than the dose that will cause severe complications. This graph does not represent data for a particular tumor but shows the general trend.

The *radiotherapist* is a physician who has had three or four years of training in oncology (the study of cancer) and the treatment of cancer with ionizing radiation. In a modern medical center one or more *radiotherapy technologists* work with him and give the treatment; they have had two or three years of training after high school. In addition, a *radiological physicist* and a *dosimetry technician,* often called a dosimetrist, look after the calibration of the therapy equipment and aid in planning the treatment.

In this chapter we discuss the physical aspects of radiation therapy.

18.1. THE DOSE UNITS USED IN RADIOTHERAPY—THE RAD AND THE GRAY

In the very early days of radiation therapy the unit used to measure the amount of radiation to the patient was the *erythema dose*—the quantity of x-rays that caused a reddening of the skin. Starting in about 1930 radiation to the patient was measured in Roentgens (r),* a unit based on ionization in air. The term *exposure* had not yet been introduced. Since the Roentgen was defined in terms of ionization in air it was an inappropriate unit to use for radiation absorbed by a part of the body. In about 1950 the quantity *absorbed dose* was introduced and the unit *rad* was defined to measure it. From 1950 to 1975 the rad was the official unit of absorbed dose.

The rad is defined as 100 ergs/g. That is, a radiation beam that gives 100 ergs of energy to 1 g of tissue gives the tissue an absorbed dose of 1 rad. (The terms *dose* and *absorbed dose* are used interchangeably in radiotherapy.) The rad can be used for any type of radiation in any material; the roentgen (R) is defined only for x-rays and gamma rays in air. The rad can be related to the exposure in roentgens. The rad was defined so that for x-rays and gamma rays an exposure of 1 R would result in nearly 1 rad of absorbed dose in soft tissue (water). In bone the ratio of rads to roentgens depends on the energy of the x-ray photons. At the energies used in diagnostic radiology the ratio of rads to roentgens in bone is about 4; that is, 1 R exposure results in about 4 rads of absorbed dose. At the high energies used in modern radiotherapy the ratio of rads to roentgens is nearly 1 for both bone and soft tissue (Fig. 18.2).

In 1975 the International Commission on Radiological Units (ICRU) adopted the gray (Gy) as the international (SI) unit of dose. 1 Gy = 1 J/kg. Since a joule is 10^7 ergs, a gray equals 100 rads. This unit was named after

*In about 1960 the accepted style became roentgen (R).

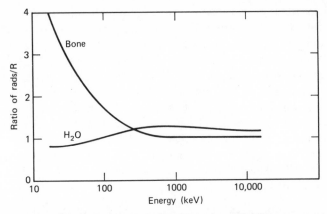

Figure 18.2. **The energy absorbed in tissue depends on the tissue composition and the photon energy. High Z materials such as calcium absorb more energy per roentgen than water for x-ray and gamma ray energies below 200 keV.**

Harold Gray, the British medical physicist who discovered the oxygen effect (see Section 18.2).

18.2. PRINCIPLES OF RADIATION THERAPY

The basic principle of radiation therapy is to maximize damage to the tumor while minimizing damage to normal tissue. This is generally accomplished by directing a beam of radiation at the tumor from several directions so that the maximum dose occurs at the tumor. (The details of treatment planning are discussed in Section 18.3.) Some normal tissues are more sensitive to radiation than others, and this is taken into account when the therapist and physicist plan the treatment.

Ionizing radiation, such as x-rays and gamma rays, tears electrons off atoms to produce positive and negative ions. It also breaks up molecules; the new chemicals formed are of no use to the body and can be considered a form of poison. The biochemical process of how x-rays kill cells is still being studied. We can for the present accept the simple theory that toxic chemicals formed in the cell by the breakup of the normal molecules kill the cell.

How much radiation does it take to kill a cell? This is analogous to the question: How many bullets does it take to kill a person? One bullet may do the job, but a person may survive an onslaught of 10 if the bullets do not strike a critical organ. Factors that determine how much radiation is required are the type of radiation, the type of cell, and the environment of the cell, for example, its blood and oxygen supplies. Also, the nucleus of the cell is more sensitive to radiation than the surrounding cytoplasm.

Some types of radiation are more effective in killing cells, or have a higher relative biological effect (RBE). The RBE of a given radiation is defined as the ratio of the number of grays of 250 kVp x-rays needed to produce a given biological effect to the number of grays of the test radiation needed to produce the same effect. Radiation that produces dense ionization generally is more lethal and has an RBE greater than 1. The RBE depends on the biological experiment used to measure the effect and is not the same for all tissues. Table 18.1 lists some approximate RBE values for several different types of radiation.

Table 18.1. Some Approximate RBE Values for Different Types of Radiation

Particle	RBE
Electrons or beta rays	1
X-rays or gamma rays	1
Fast neutrons	5
Alpha particles	>10

Figure 18.3 shows the percent of surviving Hela cells* for various doses of radiation. Note that at low doses the curve for x-rays and gamma rays has a shoulder that is not present when alpha particles are used. About 1.4 Gy from x-rays will kill half of a population of Hela cells, but it only takes about 1.0 Gy to kill half of those remaining. The quantity of radiation that will kill half of the organisms in a population (cells, mice, people, etc.) is called the *lethal dose for 50%* or LD_{50}. It is assumed that the radiation is given uniformly over the entire organism in a short period of time. This quantity is sometimes modified to include the time factor. For example, the amount of radiation that will kill 50% of the organisms in 30 days is called the LD_{50}^{30} or $LD_{50(30)}$. The LD_{50}^{30} for humans is about 4.5 Gy; fortunately, experiments have not been performed to get an accurate value.

The oxygen effect was discovered by Gray in England in about 1955. Gray noticed that cells irradiated in the presence of oxygen were much easier to kill than cells of the same type irradiated without oxygen. The oxygen effect may play a role in the recurrence of cancer. The cells near the center of a large tumor have a poor blood supply and thus a poor oxygen supply. When the tumor is irradiated the "healthy" cancer cells

---

*Hela cells originated from a human tumor. They can be grown *in vitro* (in glass) and are convenient to use in research on the biological effects of radiation (radiobiology).

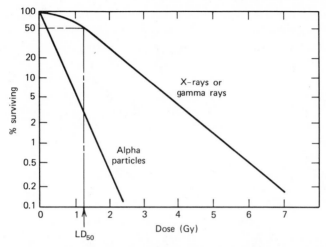

Figure 18.3. **The percent of surviving Hela cells as a function of the dose of x-rays or gamma rays and alpha particles. Note that the first grays of x-rays are not very lethal. The alpha particles have a greater biological effect per gray.**

with a good blood supply are killed and many of the more poorly oxygenated, radioresistant tumor cells remain alive. These cells may later divide and permit the tumor to regrow. This explanation could account for the ability of some tumors to regrow even after they have received large doses of radiation. All radiation beams with RBEs greater than 1 (such as fast neutrons) eliminate some or all of the oxygen effect.

It was suggested that the radioresistance of the poorly oxygenated cells in the center of large tumors could possibly be overcome by increasing the oxygen supply to the entire body. The hypothesis was that improving the supply of oxygen to the oxygen-starved cells would make them easier to kill. Hyperbaric oxygen tanks were therefore developed for radiotherapy, and patients were treated with radiation while in sealed tanks with 3 atm of pure oxygen. This development is discussed in Chapter 6. The results of the hyperbaric oxygen technique were inconclusive, and most of the studies have been discontinued.

18.3. A SHORT COURSE IN RADIOTHERAPY TREATMENT PLANNING

The physicist working in radiotherapy has three important functions:

1. To determine how much radiation is being produced by a given therapy

machine under standard conditions, that is, to calibrate the machine. Calibration includes determining not only the output at the treatment distance but also the grays per minute throughout the volume being irradiated under different operating conditions.

2. To calculate the dose to be administered to the tumor and any normal tissues in the patient. This is not easy, and many radiotherapy departments use computers to aid with this computation. The calculation should take into account irregularities in the shape of the patient and nonuniformities within the patient such as bone and air spaces (e.g., the lungs).

3. To confirm that the correct amount of radiation was really administered to the patient at the proper locations.

Calibrating the output of the therapy unit requires the careful use of an ionization chamber to measure the roentgens per minute at the usual treatment distances. Such a chamber should be annually compared to a standard instrument at a standards laboratory (e.g., the National Bureau of Standards). Corrections must be made for such things as air temperature and pressure. The accuracy of this measurement is typically about 2%, while the precision is usually better than 1%.

To determine the radiation distribution in the beam a small radiation detector is moved around by remote control inside a water phantom (Fig. 18.4). (A phantom is a tank of water or any other substance that is used to represent a patient.) The beam shape is usually rectangular and two distributions in the middle of the beam are measured at 90° to each other. The points of equal radiation dose (isodose) are plotted as continuous curves. The point of maximum radiation intensity is usually assigned to be 100%, and isodose curves of 90%, 80%, . . ., 10% are made. Figure 18.5 shows the isodose curves of a radiation beam in a water phantom.

The process of determining the best combination of radiation beams and their orientation is called *treatment planning*. It is usually done cooperatively by the radiotherapist and the radiological physicist. The dosimetrist aids with this process (i.e., does most of the work).

Before radiotherapy treatments are begun the patient is carefully examined to determine the location and extent of the tumor. In some cases the therapist will palpate (feel with his fingers) the tumor and surrounding tissues. Since many tumors are deep inside the body it is often necessary to use diagnostic x-ray, ultrasound, and nuclear medicine techniques to aid in determining the location and size of the tumor as well as to help determine the optimum position and size of the radiation beams to use in the treatment. Several special-purpose diagnostic x-ray devices have been developed to assist in these determinations.

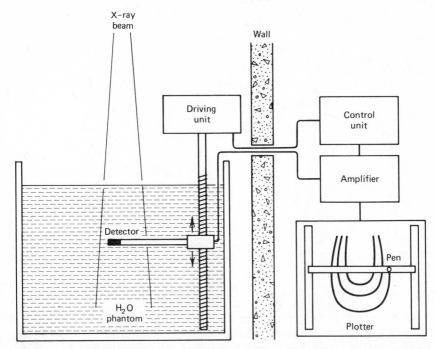

Figure 18.4. A small detector in a water phantom measures the radiation intensity in the x-ray beam. It can be controlled electronically to plot the isodose curves as it moves throughout the phantom.

A special diagnostic x-ray device—an axial transverse tomographic unit—is often used to make pictures of "slices" of the body at the location of the tumor (Fig. 18.6). The films obtained help determine the location of the tumor relative to normal structures. With whole body computerized tomography, which was developed in 1975, tumors can be located even more accurately (see Section 16.6).

In a modern radiation therapy center, a simulator is used to allow the radiation therapist planning the treatment to see what normal structures will be in the treatment beam. A simulator is a special fluoroscopic diagnostic x-ray unit with an image intensifier and TV screen. The x-ray unit is positioned in the same physical arrangement as the therapy unit that will give the treatment; that is, it simulates the therapy unit. The therapist can change the location and size of the beam by remote control while he watches the image on the TV screen. After he has determined the best size and location of the beam, he marks the beam location with colored ink on the patient's skin (Fig. 18.7).

The cross-section outline of the patient in the region to be treated is

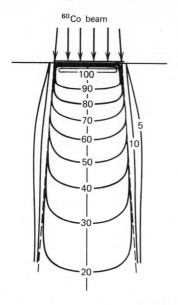

^{60}Co beam

Figure 18.5. The radiation distribution of a cobalt 60 beam in a water phantom with isodose lines for each 10% change of beam intensity (usually drawn full scale).

determined by one of several methods. A simple technique is to wrap a strip of lead (or other soft metal) around the patient at the level of the tumor. This must be done with the patient in the position he will be in for the treatment, since the body changes shape when it changes position.

Isodose curves are then superimposed on this outline of the patient to determine how the radiation will be distributed inside the body. Usually several beams are used; isodose curves for each beam are superimposed in their appropriate places and the sum of all the beams is determined. This addition can be done by hand, but it is much faster to use a computer (Fig. 18.8). The isodose curves for beams of various sizes are stored on magnetic tape and can easily be transferred into the computer memory for use. The radiotherapist can see the beams on a TV screen and can try different combinations of beams and manipulate their orientations to see how well they give an adequate treatment of the tumor and minimize radiation to normal tissues. The computer adds up the beams and shows the isodose curves of the combined beams. When the radiotherapist has decided on the best arrangement, the computer draws out a full-size copy of the treatment plan to be used.

Figure 18.6. Axial transverse tomography. (a) A schematic of the orientation of the patient, x-ray source, and film. The motions of the x-ray source and film produce sharp shadows from structures in the slice and blurred shadows from other structures. (b) A transverse axial tomogram through the chest. The chest wall is thinner on the left than on the right because due to cancer, part of it has been surgically removed. (Courtesy of Toshiba International Corporation, Tokyo, Japan.)

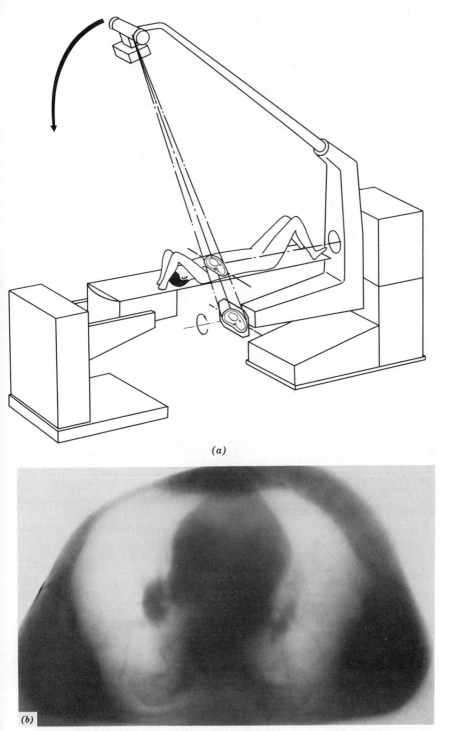

(a)

(b)

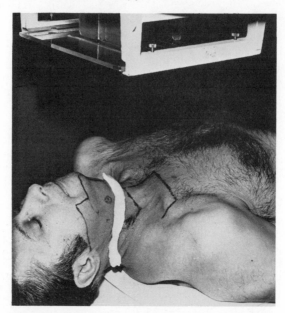

Figure 18.7.   A patient being treated for head and neck cancer with a linear accelerator. The marks on his skin show the location of the radiation field. Lead blocks, which are just visible on the plastic tray, limit the radiation to the area indicated. The treatment distance is measured to the circled dot on the patient's skin.

The reason it is customary to irradiate the tumor from several directions is that this reduces the damage to normal tissues. Figure 18.9*a* shows a computer-calculated distribution for a three-beam method of treating a tumor in the esophagus. The radiation beams are aimed to reduce the radiation to the spinal cord. Notice that the radiation is greatest at the tumor and that it is quite uniform over the tumor. Figure 18.9*b* shows a more realistic picture of how the radiation is distributed after computer corrections have been made for the lower density of the lungs. The calculated radiation to the tumor has increased about 20% (from 53 to 62 units).

Since patients come in all shapes and sizes and tumors vary in extent, it would be surprising if simple rectangular beams suited all situations. The beam must often be modified to protect normal organs. Protecting the lungs in the treatment of Hodgkin's cancer, which involves the lymphatic system, is a common problem. Large fields must be used to include all areas to be treated. Figure 18.10 shows the radiation fields used in the treatment of a Hodgkin's tumor. Lead blocks, custom-made for each patient, reduce the radiation to the lungs. The proper treatment of

Figure 18.8. A small dedicated computer used to calculate radiation distributions in radiotherapy patients. Computing programs and data are stored on the magnetic tape in the center. (Courtesy of Artronix, Inc., St. Louis, Mo.)

Hodgkin's cancer has resulted in a dramatic increase in its cure rate since 1960.

Precise calculation of the treatment on a computer does not ensure that the tumor really receives the intended radiation. Special care must be taken to position the patient properly according to the treatment plan. A typical treatment schedule consists of daily sessions of several minutes each for about one month. Various alignment devices such as light beams and mechanical pointers are used in conjunction with marks on the skin to help reposition the patient each day (Fig. 18.7). Plaster casts are used for precise repositioning of the head (Fig. 18.11). Some therapy centers have developed systems that shut off the beam if a patient moves more than a few millimeters during a treatment.

The shape of the body depends on its position. The internal organs shift when a patient who is lying down turns over. In order to allow accurate treatment of a tumor, modern radiation therapy units are usually designed to rotate about the treatment couch. Thus a patient can remain in one position while the beam is aimed at his body from different angles. In some cases it is possible to treat a tumor while the therapy unit is rotating

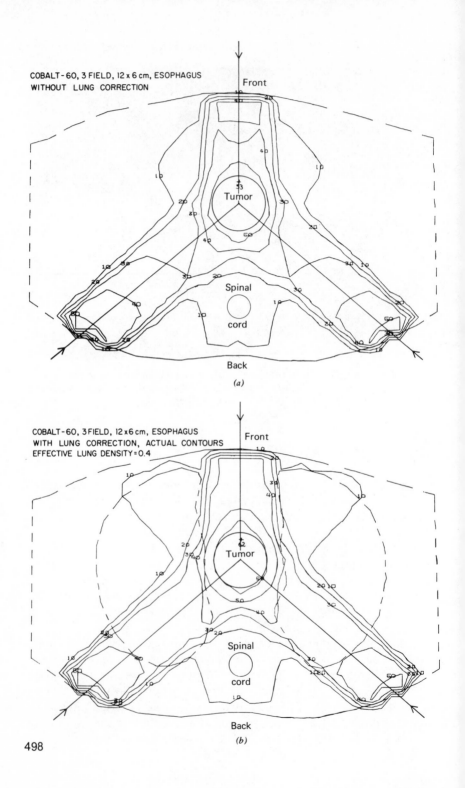

COBALT-60, 3 FIELD, 12 x 6 cm, ESOPHAGUS
WITHOUT LUNG CORRECTION

Front

Tumor

Spinal

cord

Back

(a)

COBALT-60, 3 FIELD, 12 x 6 cm, ESOPHAGUS
WITH LUNG CORRECTION, ACTUAL CONTOURS
EFFECTIVE LUNG DENSITY=0.4

Front

Tumor

Spinal

cord

Back

(b)

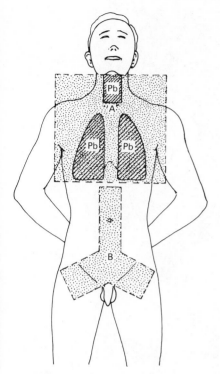

Figure 18.10. The treatment of Hodg-kin's cancer with two large radiation fields (A and B) that cover the major lymphatic nodes. In field A a lead shield blocks radiation to the larynx (voice box) and custom-made lead shields protect the lungs.

with the beam on. Rotational treatment is suitable for certain centrally located tumors.

Even the best computers only do what they are told, and they are not always told the truth since it is not always known. Thus it is desirable to actually measure the radiation being received in different parts of the patient. Small radiation detectors such as thermoluminescent dosimetry crystals (see Chapter 19) can be used at locations of interest, for example, in the tumor if it is accessible and at normal organs the radiotherapist is trying to protect (e.g., the eyes).

To check on the position of the beam during treatment, it is common to place an x-ray film where the beam exits from the body and take a "port film" exposure. The image obtained does not have the detail of a diagnostic x-ray film, but sufficient internal anatomy is visible to show if the beam is correctly located.

Figure 18.9. A three-beam method for treating a tumor in the esophagus. (a) The radiation distribution calculated assuming the lungs are water density. (b) The more realistic radiation distribution calculated assuming the lungs have a density of 0.4. (Courtesy of John S. Laughlin, Memorial Sloan-Kettering Cancer Center, New York.)

Figure 18.11.   Plaster cast used to precisely position the head for radiation therapy. (From W.T. Murphy, *Radiation Therapy*, 2nd ed., © W.B. Saunders Company, Philadelphia, 1967, p. 304.)

A modern radiotherapy center may have equipment worth over $1 million. This limits the use of good radiation therapy to large, well-equipped medical centers. Smaller medical centers sometimes use older medium-voltage (orthovoltage) therapy units to relieve suffering in those cases where they believe the cancer is so widespread that a cure is not possible. This palliative use of radiation is sometimes a disservice to the patient since cancers that were considered incurable a few years ago may now be cured with the best methods.

A small medical center may have a modern therapy unit but lack the trained personnel and related equipment needed to use it properly. The best chance for a cure increases rapidly above 40 Gy to the tumor (Fig. 18.1), but the probabilities of severe complications and of malpractice suits also rise rapidly above this level. The physician in a smaller medical center sometimes undertreats the tumor in order to avoid serious complications. The public is generally not aware of this type of malpractice.

18.4. MEGAVOLTAGE THERAPY

Before 1940 the radiotherapist had little choice in the radiation source he used for treating cancer. Most external therapy was given with orthovoltage x-ray units that had maximum potentials of 250 kVp or less (Fig. 18.12); a few centers had 400 kVp units or the new 1000 kVp or 1 million volt machine. After World War II several significant developments were made in therapy machines. Kerst developed the betatron, which accelerates electrons to high energy (Fig. 18.13). The electrons can be used directly or they can be used to produce a high-energy x-ray beam. The isodose curves of an x-ray beam from a 25 MeV betatron are shown in Fig. 18.14a. The betatron helped open a new radiation therapy era—the supervoltage or megavoltage era.

The application of man-made radioactive sources to radiation therapy was another important postwar development. Although cancer had been treated for decades with radioactive radium sources placed directly in or on the tumor (see Section 18.5), there was insufficient radium to produce a useful external beam of gamma rays. However, with the development of nuclear reactors during and after World War II it became possible to produce many artificial radioisotopes in undreamed of quantities. (Some people might consider this a nightmare!)

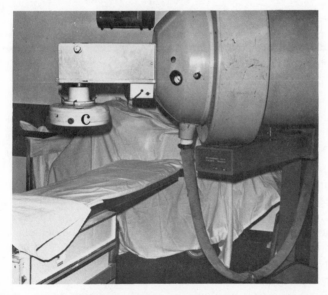

Figure 18.12.   A typical 250 kVp radiotherapy unit with an adjustable collimator C attached.

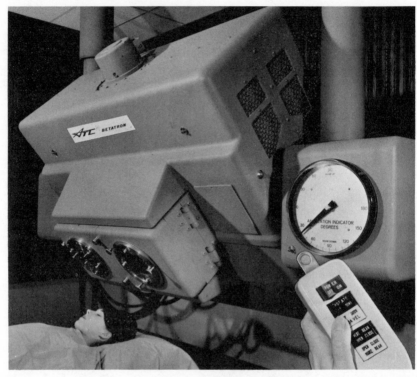

Figure 18.13.   The 25 MeV betatron effectively treats deep tumors using high-energy x-rays and tumors near the surface using an electron beam. (Photographed at Milwaukee County General Hospital, courtesy of ATC Betatron Corporation, West Allis, Wisc., successor to the Allis-Chalmers Betatron Department.)

One of the radioactive sources easy to produce in a reactor is ^{60}Co. Cobalt 60 emits penetrating gamma rays of about 1.25 MeV energy. These rays are about as penetrating as the x-rays from a 3 million volt x-ray machine, but the ^{60}Co unit is much more compact (Fig. 18.15). The first ^{60}Co therapy unit was made in Canada by Harold Johns in 1951. About 1000 ^{60}Co units (also called *cobalt teletherapy* and *cobalt bomb units*) were sold in the United States during the 1950s and 1960s. Many of the ^{60}Co units are designed to rotate around the patient. A large metal beamstop absorbs the radiation that passes through the patient and reduces the amount of shielding needed in the walls.

The high-energy gamma rays emitted by ^{60}Co are absorbed by the tissue and produce high-energy electrons, most of which move in the same general direction as the original beam. As the gamma ray beam penetrates the first few millimeters beneath the skin, the number of electrons in-

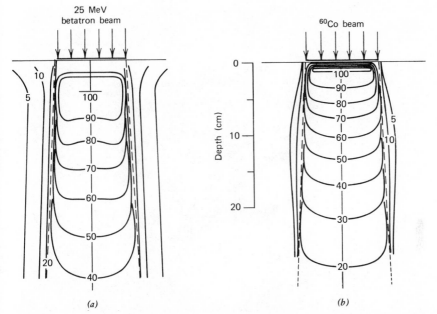

Figure 18.14. Isodose curves for (a) a 25 MeV betatron x-ray beam and (b) a ⁶⁰Co beam. Note that the maximum dose occurs below the surface.

creases and the energy deposited by them increases (electron build-up); the dose maximum occurs about 5 mm below the skin. The relatively low dose at the skin from high-energy x-rays or gamma rays is referred to as the *skin-sparing effect*. Figure 18.14*b* shows typical isodose curves for ^{60}Co.

Very intense sources of ^{60}Co—up to 370 TBq ($\sim 10^4$ Ci)—are usually used, and it is necessary to absorb the unwanted radiation since the disintegrations cannot be turned off. For this reason the source is contained in a large lead shield weighing several tons. This container is mechanically designed so that the source can be positioned at the beam port to turn on the beam (Fig. 18.16). Since ^{60}Co is continually decaying, the intensity of the source decreases at a rate of about 1% a month (or 50% in 5.3 yr—its half-life). The source is usually replaced after 5 to 10 yr.

The radiation output of a 370 TBq ^{60}Co source is about 200 R/min at 1 m from the source. A typical treatment in radiotherapy consists of 3 Gy (less than 2 min) each day for about 20 days (excluding weekends).

In about 1970 a breakthrough in the design of electron linear accelerators, or *linacs*, led to the development of a compact 4 million volt linac that is about the same size as a ^{60}Co unit (Fig. 18.17). It can produce an intense beam all the time (well, almost all the time), and the radiation

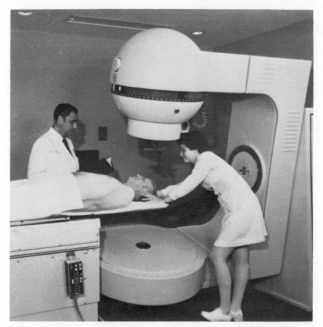

Figure 18.15. Picker ⁶⁰Co teletherapy unit. The source is above the patient and the beamstop, which reduces the amount of shielding needed, is below the patient. (Photographed at The Cleveland Clinic, courtesy of Picker Corporation, Cleveland, Ohio.)

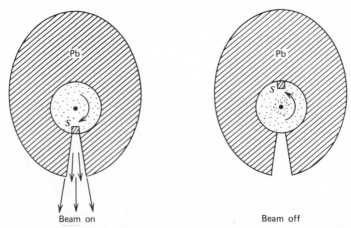

Beam on   Beam off

Figure 18.16. The radiation from a ⁶⁰Co teletherapy unit is emitted continuously. The unit is turned on by rotating the source wheel until the source is over the beam port in the large lead shield. It is turned off by rotating the source wheel until the source is shielded by the wheel.

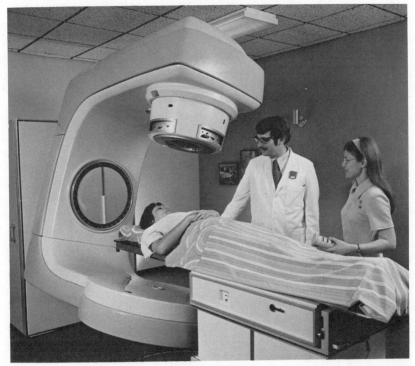

Figure 18.17.   A compact 4 million volt linear accelerator. Note its similarity to the ⁶⁰Co unit shown in Fig. 18.15. (Courtesy of Varian Radiation Division, Palo Alto, Calif.)

comes from an area smaller than a ⁶⁰Co source, resulting in sharper isodose distributions at the edges of the beam.

Megavoltage therapy has three major advantages: (1) the maximum dose occurs below the skin, and this skin-sparing effect greatly reduces the pain from the treatment; (2) the high energy is almost completely in the Compton effect region and, unlike 250 kVp radiation, does not give a large dose to the bone; and (3) the greater penetrating ability permits better treatment of tumors deep inside the body. (The Compton effect is discussed in Chapter 16.)

18.5. SHORT-DISTANCE RADIOTHERAPY OR BRACHYTHERAPY

In her 1904 doctoral thesis Madame Curie described a biological experiment in which she placed a capsule containing radium on her husband's arm and left it for several hours. She said it produced a sore that took over a month to heal. This sore was not a surface "burn"; the damage went much deeper. The possibility of using radium to destroy cancer was recognized almost immediately, and a method was developed in which sources of radium were put into or on the surface of tumors. This short-distance therapy, or *brachytherapy*, is still a standard treatment method for certain types of cancer, especially cancer of the female reproductive organs.

The advantage of brachytherapy is that it gives a very large dose to the tumor with minimum radiation to the surrounding normal tissue. Its main disadvantage is the nonuniformity of the dose since the radiation is much more intense near the source, although using a number of sources helps make the dose more uniform. Another disadvantage concerns radiation safety. The therapist must be close to the sources while they are being put in place. The patient is a "radiation source" during the days the sources are in place, and nurses and others are thus exposed to the radiation. The radiation to the therapist has been greatly reduced by the "after loading" technique. The therapist carefully places hollow tubes in the patient and later rapidly places the radioactive sources in the tubes. The radiation to nurses can be kept at reasonable levels if they use proper methods: stay near the patient only when necessary, keep as great a distance as possible from the sources, and use shielding where possible. Sometimes nurses from other wards care for brachytherapy patients; this spreads the radia-tion risk over a larger work force. A pregnant nurse should not care for brachytherapy patients because of possible radiation damage to the sensi-tive fetus (see Chapter 19).

Figure 18.18*a* shows a pelvic x-ray of a patient with radioactive sources

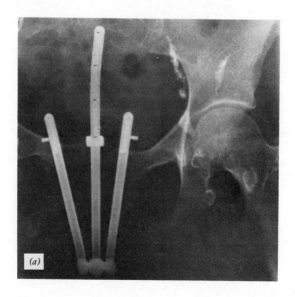

(a)

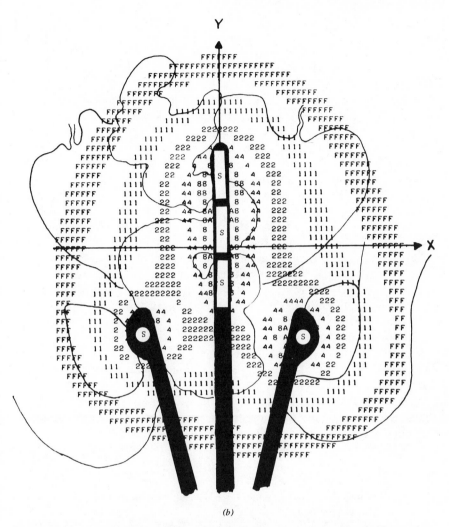

(b)

Figure 18.18.   (a) X-ray of a patient with radioactive sources in place in the vagina and uterus. (b) Isodose curves around a radium cervix applicator. The sources are indicated by S. The various letters and numbers indicate the dose in the plane chosen. For example, 2 indicates a calculated dose between 20 and 28 Gy and F indicates a dose between 5.0 and 7.0 Gy. (Courtesy of John S. Laughlin, Memorial Sloan-Kettering Cancer Center, New York.)

in place in the vagina and uterus. Each source used is typically the equivalent of 10 or 20 mg of radium. The radiation distribution around a typical treatment arrangement is shown in Fig. 18.18*b*. This type of radiation therapy is often supplemented with external radiation beams. The length of the treatment is usually limited by the radiation to the surrounding normal tissues—in this case, the rectum and the bladder.

Although radium is still often used, it is slowly being replaced by man-made radioactive sources such as cesium 137 (^{137}Cs). These sources are easier to shield than radium, and a break in the container does not present as serious a hazard. Table 18.2 lists a number of radionuclides that are sometimes used in brachytherapy. Note that radium has a half-life of 1620 years; this means that the strength of this source is essentially constant from year to year.

In one common type of brachytherapy needles containing radionuclides are forced into the tumor area. Figure 18.19 shows an x-ray of radium needles in place to treat a tumor of the cheek. This therapy looks primitive, but it is often the treatment of choice when it can be used. The radioactive strengths of the needles and their arrangement are chosen to give a relatively uniform dose over the extent of the tumor. The needles are usually left in place 3 to 7 days.

In an alternative type of brachytherapy "seeds" of radioactive materials containing relatively short-lived nuclides are used. The original technique used the natural radioactive gas radon (^{222}Rn), which is the daughter product of radium (^{226}Ra) and has a half-life of 3.8 days. It decays into the same series of daughter products as radium, and thus the radiation it gives off is essentially identical to that of radium. Radon is obtained by placing radium in solution. When used for seeds the radon gas was sealed in small pieces of pure gold tubing (1 mm in diameter × 3 mm long). These

Table 18.2. Radionuclide Sources That Are Sometimes Used for Brachytherapy

Source	Half-Life	Principal Emission
^{226}Ra	1620 yr	Gamma rays
^{137}Cs	30 yr	Gamma rays
^{60}Co	5.3 yr	Gamma rays
^{192}Ir	74 days	Gamma rays
^{222}Rn	3.8 days	Gamma rays
^{198}Au	2.7 days	Gamma rays
^{125}I	60 days	X-rays
^{90}Sr-^{90}Y	28 yr	Beta rays
^{252}Cf	2.6 yr	Neutrons

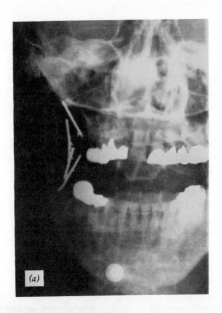

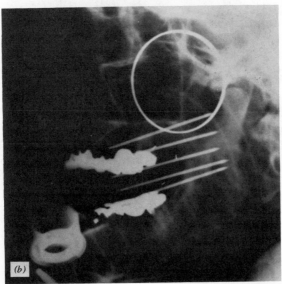

Figure 18.19.   Radium needles in place to treat a tumor of the cheek. (a) Front view.
(b) Lateral view. A metal ring placed on the patient's skin allows the magnification of
the x-ray image to be determined.

radon seeds were then forced into the tumor and left permanently (Fig. 18.20). After 38 days (10 half-lives) the radioactivity was about 0.1% of its original intensity.

Because handling radium in solution and making radon seeds is hazardous, man-made sources were sought for permanent implants. Radon seeds have been largely replaced by radioactive gold (^{198}Au) grains. The ^{198}Au is encased in a nonradioactive cover of platinum or stable gold. These grains are about the same size as radon seeds but they can be made more uniform. The half-life of ^{198}Au (2.7 days) is similar to that of ^{222}Rn (3.8

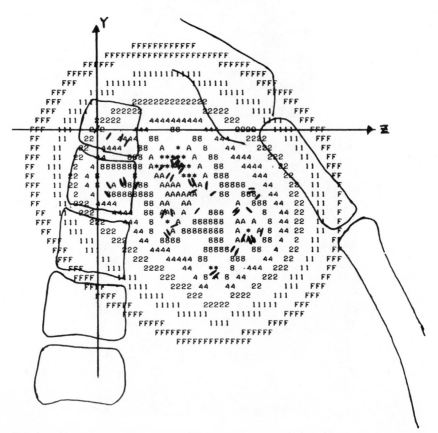

Figure 18.20. An outline drawn from a lateral x-ray of a patient with many ^{125}I seeds implanted in a cancer of the upper lobe of the lung. (The breastbone is on the right.) The seeds are indicated by black marks. A computer has calculated the isodose curves. The symbols *, A, 8, 4, 2, 1, and F indicate decreasing levels of total dose delivered during a year. (Courtesy of John S. Laughlin, Memorial Sloan-Kettering Cancer Center, New York.)

days). Longer-lived radioactive materials such as iodine 125 (60 day half-life) are sometimes used for permanent implants. An obvious disadvantage of all permanent implants is that the radiation distribution cannot be modified once the sources are in place.

One special type of brachytherapy uses beta particles, fast electrons emitted from certain radionuclides. The beta source is held in contact with the area to be treated (Fig. 18.21). The most common source of these beta particles is yttrium 90 (^{90}Y). This nuclide has a half-life of only 64 hr. It is continuously being formed by the decay of strontium 90 (^{90}Sr), which has a half-life of 28 yr. Strontium 90 emits beta rays with a maximum energy of 0.54 MeV and ^{90}Y emits beta rays with energies of up to 2.27 MeV. These latter rays can penetrate about 4 mm and are almost ideal for treating lesions on the eye (Fig. 18.21). Most rays stop before reaching the radiation-sensitive lens. The dose at the lens is usually less than 5% of the surface dose. The dose rate is quite high, so treatments of only seconds are needed. Neither ^{90}Sr nor ^{90}Y emits gamma rays, but if the source is placed near a high Z material such as lead ($Z = 82$) the beta rays can produce high-energy x-rays. The production of x-rays is very inefficient

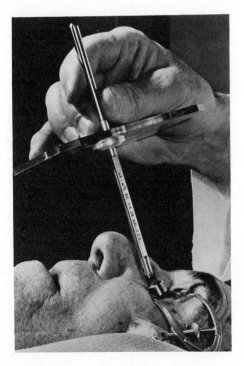

Figure 18.21. A ^{90}Sr-^{90}Y eye applicator. The beta rays penetrate about 4 mm and do not give radiation to the lens. The treatments usually last less than a minute. (From W.T. Murphy, *Radiation Therapy*, 2nd ed., © W.B. Saunders Company, Philadelphia, 1967, p. 333.)

for low Z materials, so the ^{90}Sr-^{90}Y applicator is stored surrounded by low Z plastic; the plastic absorbs the beta rays, and few x-rays are produced.

18.6. OTHER RADIATION SOURCES

Therapeutic uses of fast electrons from ^{90}Y were just discussed. Since electrons, unlike gamma rays and x-rays, have a fixed range, it was natural to consider machine-made electron beams for radiotherapy. Most modern radiotherapy centers now have electron beam therapy capability. Figure 18.22 shows the dose as a function of depth in tissue for electron beams of various energies. Notice that the radiation at low electron energies is relatively uniform over most of its travel. This advantage is gradually lost in going to higher electron energies.

Several high-energy accelerators such as betatrons and linear accelerators operate with either electron or x-ray beams. The electrons pass through a thin window and strike a scatterer that makes the electron beam larger and more uniform. A convenient way to measure the extent and distribution of the electron beam is to place a piece of film on edge

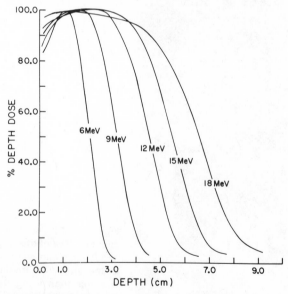

Figure 18.22. Depth dose curves for electron beams of 6, 9, 12, 15, and 18 MeV. Notice the sharp drop at the end of the range. (Courtesy of Bhudatt Paliwal, University of Wisconsin, Madison, Wisc.)

between sheets of pressed wood and take a brief exposure in the electron beam. Figure 18.23 shows the isodose curves from 6, 12, and 18 MeV electron beams that are 10 cm × 10 cm at the surface. There is some skin sparing since the maximum dose occurs below the surface, but the amount of skin sparing is small compared to that from high-energy x-ray beams where the skin dose may be less than 50% of the maximum dose.

Other particles besides electrons are used for radiotherapy. The size, cost, and complexity of the machines used to produce these more exotic particles prohibit their general use at present. An advantage of very high energy (greater than 100 MeV) protons and alpha particles (helium 4 nuclei) is the way they lose energy. Near the end of their travel they deposit most of their energy in a region called the *Bragg peak* (Fig. 18.24). They have a rather low dose until they get to the tumor and thus extend the skin-sparing effect. Only a few medical centers in the world use these beams on a routine basis.

The use of π^- (pi minus) mesons for radiotherapy is being investigated.

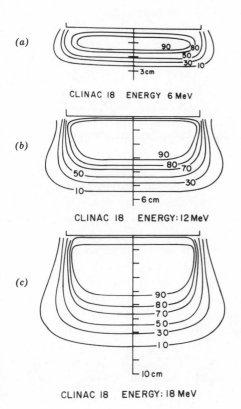

Figure 18.23. Isodose curves for electron beams of (a) 6 MeV, (b) 12 MeV, and (c) 18 MeV. All beams were 10 × 10 cm and 100 cm from the source. (Courtesy of Bhudatt Paliwal, University of Wisconsin, Madison, Wisc.)

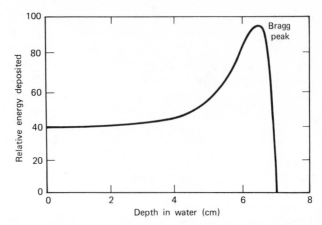

Figure 18.24. Pattern of energy loss of very high energy alpha particles.

These particles play an important role in the nucleus of the atom, but their lifetime is only 10^{-8} sec. They can be produced in the high-energy accelerators used for physics research. Until recently, not enough π^- mesons have been available for radiotherapy. The recent construction of the Los Alamos Meson Physics Facility (LAMPF) at a cost of 5.6×10^7 makes possible the production of π^- meson beams of sufficient intensity for therapy. The LAMPF will be used for both physics research and π^- meson radiotherapy. A proton beam of about 1 mA will be accelerated to an energy of about 800 MeV. The power in the beam will be $P = IV = (10^{-3}) (8 \times 10^8) = 8 \times 10^5$ W, or almost 1 megawatt!

The protection problems of handling a beam of this energy are impressive. The π^- meson beam will have to be shaped to the desired size with a huge collimator that will become very radioactive in the process. About 6 m of concrete shielding will be necessary to protect the patient in the treatment room from the main beam area.

You might wonder if it is worth it. We will not know for some time; we may never know. The unique feature of the π^- meson beam is the way it deposits most of its energy at the end of its path (Fig. 18.25). The π^- meson ionizes as it moves through tissue. At the end of its path it enters the nucleus of an atom and literally causes the nucleus to explode, or form a "star." The fragments deposit about 30 MeV of very dense ionization energy very close to the "accident." This intense source of energy explains the very high energy peak seen in Fig. 18.25. The biological effect of these fragments is about 3.3 times greater than that of the same amount of energy deposited by ionization from electrons.

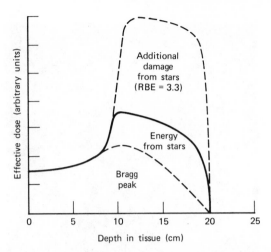

Figure 18.25. The energy contribution from the Bragg peak and stars for a 68 MeV π^- meson beam. The additional biological damage is due to the dense ionization of the star fragments.

Neutrons were first used for radiotherapy by Stone in the 1940s for treating cancer. Stone used neutrons from a cyclotron developed by E. O. Lawrence. The results of this early work were discouraging. The patients developed serious complications similar to those that are caused by excessive radiation. In fact, we now know that the cause of the complications *was* excessive radiation. Stone did not realize that neutron treatments given in small doses have a much larger effect than a single large dose of the same total amount. Radiobiological studies of fast neutrons at the London Postgraduate Medical College in the 1960s clarified many of the problems. Fast neutrons ($\sim$ 7 MeV average) are again being used in the treatment of cancer—this time with promising results.

Accelerators other than cyclotrons can be used to produce fast neutrons. The accelerator that appears most promising for routine neutron therapy is the deuteron-tritium (D-T) generator. Cyclotrons capable of producing useful beams of neutrons for radiotherapy are much larger than D-T generators. The D-T generator takes advantage of the energetic ($\sim$ 14 MeV) neutrons produced when a relatively low-energy ($\sim$ 200 keV) deuteron collides with a triton. The neutrons spread in all directions from the target. In order to treat a patient in a reasonable time (e.g., $\sim$ 10 min), a generator must produce 10^{13} neutrons/sec. The conventional D-T generator has an output of about 10^{11} neutrons/sec. Considerable effort is being spent on developing a machine with 100 times this output.

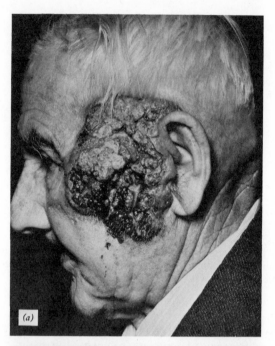

(a)

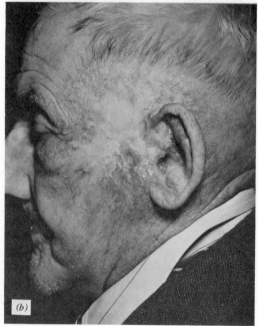

(b)

It is also possible to use neutrons in implant radiotherapy. A man-made radionuclide, californium 252 (^{252}Cf), emits fast neutrons as well as gamma rays. Clinical tests with this source were started in 1973; it will be several years before it is known if ^{252}Cf has advantages over other implant sources using only gamma rays.

Fast neutrons primarily deposit their energy in tissue by colliding with protons (hydrogen nuclei) which then produce dense ionization as they come to a stop. The RBE from fast neutrons appears to be about 5.

The ability of densely ionizing radiation such as π^- mesons and fast neutrons to eliminate some or all of the oxygen effect is expected to make these types of radiation more effective in treating cancer (see Section 18.2).

18.7. CLOSING THOUGHT ON RADIOTHERAPY

Until an effective cure for cancer is found we can expect scientists and politicians to be willing to spend time and money to carry on cancer research. Progress has been made, but it has been slow. The combination of the talents of basic scientists and medical scientists has accounted for much of the progress to date. Modern radiation therapy is a good example of this cooperation.

Some people think that radiation therapy is a temporary method of treating cancer that will eventually be replaced by a "cancer cure." We all hope that such a cure will be found, but in the meantime the proper use of radiation therapy offers many cancer victims their best chance of survival. In some types of cancer, radiation therapy is the treatment of choice. When used in the early stages of the disease it has a cure rate (5-yr survival) of over 90%. It is not as disfiguring as surgery, and it is often chosen for cancers of the head and neck for this reason. Like any medical specialty it can be practiced best by a specialist who has been properly trained in its use. The radiotherapist should have access to modern therapy equipment and should work in close cooperation with a radiological physicist to assure the accuracy of the administered dose. The result of a successful radiotherapy treatment can be dramatic (Fig. 18.26).

Figure 18.26. Successful radiotherapy treatment of squamous cell carcinoma in a 77-year-old man. (a) Before treatment; (b) following treatment. (From W.T. Murphy, *Radiation Therapy*, 2nd ed., © W. B. Saunders Company, Philadelphia, 1967, p. 132.)

BIBLIOGRAPHY

Barnes, P., and D. Rees, *A Concise Textbook of Radiotherapy*, Faber and Faber, London, 1972.

Fletcher, G. H., *Textbook of Radiotherapy*, Lea and Febiger, Philadelphia, 1966.

Hendee, W. R., *Medical Radiation Physics*, Year Book Medical, Chicago, 1970.

Johns, H. E., and J. R. Cunningham, *The Physics of Radiology*, 3rd ed., Thomas, Chicago, 1969.

Stanton, L., *Basic Medical Radiation Physics*, Appleton-Century-Crofts, New York, 1969.

REVIEW QUESTIONS

1. To what general radiation energy does orthovoltage refer?
2. List three advantages of supervoltage or megavoltage therapy.
3. What does LD_{50} mean?
4. Define RBE.
5. Define *rad*. How is the rad related to the roentgen?
6. In radiotherapy, what is a phantom?
7. What is the purpose of a simulator?
8. Why are multiple radiation beams often used for radiotherapy?
9. What is brachytherapy? Give one major advantage and one major disadvantage of brachytherapy.
10. Why is radium being replaced by other sources for brachytherapy?
11. How is a ^{60}Co unit turned off?
12. What has largely replaced radon seeds?
13. What does a ^{90}Sr-^{90}Y source emit, and what is it used for?
14. What does a ^{252}Cf source emit?
15. Why is there an interest in fast-neutron radiotherapy?
16. Give one advantage of electron-beam therapy.
17. Why are π^- meson beams of interest in radiotherapy? Give one advantage and one disadvantage of π^- meson therapy.

CHAPTER 19

Radiation Protection in Medicine

Modern living is filled with a variety of hazards, and one of the less significant is the hazard from medical uses of radiation. Nonetheless it is an occupational hazard for medical workers and a slight hazard for the about one-half of the United States population who have medical or dental x-rays each year. The science of protecting workers and the public from unnecessary radiation is known as *radiation protection*. It involves the accurate measurement of radiation to radiation workers and the public and the design and use of methods to reduce this radiation.

The world had no need for radiation protection before 1895, when x-rays were discovered. Concentrated quantities of natural radioactivity did not become available until the early 1900s. That is not to say that people were not being irradiated. They were. Ionizing radiation has always been present on the earth. All of our ancestors lived in a continuous sea of ionizing radiation, and so do we. This natural radiation, or *background radiation,* comes from several sources (Table 19.1). (The unit rem is defined in Section 19.2.) About 20% of this background radiation comes from natural radioactivity in our bodies—primarily potassium 40 (^{40}K). Since a sizable amount of this radiation comes from the soil and building materials, it should not be surprising that background radiation varies with geographical location (Table 19.2). There are a few places, including one area in Brazil and one area in India, where the background radiation is a factor of 3 to 40 times greater than average. Typically, 30 to 40% of the background radiation comes from cosmic rays from outer space. The atmosphere acts as a shield to absorb some of this radiation as well as hazardous components of the ultraviolet spectrum. At high elevations some of this protection is lost, and cities such as Denver ($\sim$ 1600 m) and Mexico City ($\sim$ 2300 m) have a higher than average background due to

Table 19.1. Typical Sources of Background Radiation

Source	Background Radiation (mrem/yr)
External irradiation:	
Cosmic rays (including neutrons)	50
Terrestrial radiation (including air)	50[a]
Internal irradiation:	
^{40}K	20
Other radionuclides	5
Total	125

[a]Depends on geographical location.

this effect. The yearly background at 3000 m elevation is about 20% higher than that at sea level.

One source of natural radiation is the air we breathe. Radon gas, one of the radioactive daughter products of the radium family, is present in the air. It decays with a half-life of about 4 days into a solid that is deposited on any surface that is handy, such as dust particles. As we breathe, some of these radioactive dust particles stick to the lining of the lungs, and thus the lungs receive more radiation than the rest of the body. On the average, the annual dose equivalent to the lungs is about nine times the amount to the rest of the body. (Dose equivalent is defined in Section 19.2.) The amount of radon in the air inside a building depends on the composition of the building. The radiation to the lungs due to radon in a wooden building is half that in a brick building and one-third that in a concrete building. Tobacco leaves collect radioactivity while drying. If you are a cigarette smoker (and we hope you are not!), the smoke that enters your lungs may give them up to five times the radiation a nonsmoker receives. This may be a contributing factor to lung cancer in cigarette smokers.

Table 19.2. Typical Values of Background Radiation for Various Locations[a]

Location (and Type of Soil)	Background Radiation (mrem/yr)
Normal populated regions	125
France (granite)	190
Kerala, India (monazite)	830
Brazil (monazite)	315

[a]Adapted from data in the World Health Organization Technical Report No. 166, 1959.

Medical radiation exposures come from therapeutic uses of x-rays and radioactivity, diagnostic uses of radioactivity (nuclear medicine), and diagnostic uses of x-rays. In 1976, about 0.25% of the population of the United States received radiation therapy, about 1% received radiation from nuclear medicine sources, and over 50% received radiation from diagnostic x-ray studies.

The contributions of natural and man-made sources of radiation to the average United States resident in 1970 are shown in Table 19.3. If we wish to reduce radiation we obviously must look closely at the largest man-made source—medical x-rays. In some other countries with good medical care the contribution from x-rays is considerably smaller than in the United States. A report by a committee of specialists appointed by the National Academy of Sciences—the Advisory Committee on the Biological Effects of Ionizing Radiation (BEIR)—states that "unnecessary radiation from medical x-rays can and must be reduced."

The responsibility for medical x-ray safety has been left to the states. Most states have few, if any, laws controlling the use of x-rays on humans except that x-ray exposures must be made by or under the supervision of a medical practitioner. Unfortunately, most medical practitioners have had little or no training in radiation safety. In 1976 only three states (New York, New Jersey, and California) had effective laws requiring x-ray operators (radiologic technologists) to be licensed.

In about 1960, many states started inspection programs in which medical and dental x-ray units are checked about every three years. This was a big step forward and greatly reduced the radiation exposure to the public. However, there is still much room for improvement. One aspect of the

Table 19.3. Estimates of Radiation Received by the Average United States Resident in 1970

Source	Average Dose Equivalent (mrem/yr)
Environmental	
Natural	125
Fallout from bomb tests	4
Nuclear power	0.003
Medical	
Diagnostic x-rays	75
Nuclear medicine	1
Occupational	1
Miscellaneous	2
Total	~208

problem is that in most states the responsibility for establishing rules and regulations for the control of medical x-rays belongs to the board of health, which is made up of members of the medical profession, resulting in a conflict of interest.

The United States government through the Nuclear Regulatory Commission (NRC)* establishes limits for radiation to the public from nuclear power plants and medical uses of radioactive materials. Decisions on who can use radioactivity for nuclear medicine purposes in the United States are made by the NRC or the states that have entered into agreements with the NRC. There were 25 agreement states in 1976. Physicians must have received training in the use of radioactivity in medicine before they are permitted to use radionuclides on humans. However, only a few states require nuclear medicine technicians to receive formal training. In general, the radiation protection aspects of nuclear power and nuclear medicine are much stricter than those of medical x-rays. If the 1976 limits for radiation to the public from nuclear power plants were imposed on medical x-ray installations, many would have to be closed down. If x-rays had been discovered recently it is likely that much stricter rules would apply to their production and use.

Specialists in radiation protection are called *health physicists*—a title given to them during the atomic bomb project of World War II. Health physicists usually have a B.S. degree in a physical science and then specialize in radiation protection in graduate school. Most receive an M.S. degree, although a few do research and earn a Ph.D. The Health Physics Society in the United States had about 3500 members in 1976. The society publishes the *Health Physics Journal,* which contains specialized articles dealing with radiation protection. Most health physicists work for industry and government laboratories. Major medical centers sometimes employ health physicists, but usually the responsibilities of radiation protection are taken care of by radiological physicists who have had training in radiation protection.

Various organizations make recommendations in the field of radiation protection, but the preeminent one is the National Council on Radiation Protection and Measurement—NCRP. This private, semiofficial group of scientists publishes its recommendations as reports. Several are listed in the bibliography at the end of this chapter.

The basic solution to the radiation protection problem in the medical field can be stated in one word—education. The medical users are often unconcerned by the hazards. X-rays are so common to them that they have a tendency not to worry about them, just as we do not worry about

*The NRC is the successor to the Atomic Energy Commission (AEC) as of 1975.

having an automobile accident every time we get into a car. On the other hand, some patients are overly concerned with the risks of x-rays. It is easy to overemphasize the hazards of x-rays. It is possible that hazards that are not detected by any of our senses, such as x-rays, are more frightening than those that are detectable by several of our senses, such as cigarette smoke.

There has been much controversy over the amount of radiation that is "safe" for the public. It is generally believed that even small amounts of radiation may cause harm, just as it is thought that one cigarette or even one puff on a cigarette may be harmful. There is probably no absolutely safe amount of radiation or smoke. Of course, we expect the hazard of a milliroentgen to be much less than that of a roentgen, just as we expect the hazard from one puff on a cigarette to be much smaller than that from smoking a carton of cigarettes. The problem is to balance the benefits against the risks. When we go for a ride in a car or a plane we realize there are risks. We accept these risks in exchange for the benefits of the trip. Similarly, we should balance the ability of an x-ray to detect a medical problem against the slight risk from the radiation.

To better understand the relative risks of radiation and smoking, let us imagine that Congress in its wisdom decided to limit cigarette smoking in order to reduce lung cancer. As a guide to establishing the MPS (maximum permissible smokes), Congress could decide to use the MPD (maximum permissible dose) guidelines established for radiation (see Table 19.6). It would probably assume that the main hazard from both smoking and radiation is their cancer-inducing property. If Congress assumed that cancer data from large amounts of smoking and radiation could be extended linearly to small amounts of these hazards, it would legislate that each member of the population could smoke one "unnecessary" cigarette a month. People who must work with cigarettes would be allowed the greater risk of 10 cigarettes a month! If Congress used as a guideline the strict radiation limits near nuclear power plants, it would allow one cigarette every 10 years near cigarette factories! (We are in favor of MPDs. We use this example to emphasize that many other common hazards are not as well controlled.)

19.1. BIOLOGICAL EFFECTS OF IONIZING RADIATION

Biological effects of x-rays were first noted in 1896 in some of the early x-ray workers, and the damaging effects of concentrated radioactivity were first observed in around 1900. The early x-ray workers noticed that the hair fell off areas that had been exposed to x-rays. The skin became red, and if it received large amounts of radiation it became ulcerated. Workers found that skin cancer often developed after a number of years on the areas that had been exposed. In addition, Madame Curie reported in her Ph.D. thesis in 1904 that a radium source she had placed on her husband's arm for a few hours had produced a painful sore that extended well below the skin and was slow to heal.

The biological effects of ionizing radiation have been studied since about 1900. The scientists concerned with this research area are called radiobiologists. Although many areas of radiobiology are well understood, research efforts are still continuing, especially in those areas dealing with low levels of ionizing radiation.

The biological effects of ionizing radiation are of two general types—*somatic* and *genetic*. Somatic effects affect an individual directly (loss of hair, reddening of the skin, etc.), while genetic effects consist of mutations in the reproductive cells that affect later generations. Since genetic effects occur only when reproductive cells are irradiated, the gonads should be shielded during x-ray studies when possible (see Section 19.4).

In order to evaluate the genetic effects of x-rays on the population, the concept of *genetically significant dose* (GSD) is useful. The GSD due to an exposure depends on the dose to the individual's ovaries or testes and the individual's age, which determines the probability of that person becoming a parent in subsequent years. Thus x-raying women over 50 yr old, who normally have little chance of having babies, contributes very little to the GSD of the population. Exposure of the reproductive organs of children results in the maximum contribution to the GSD of the population, since their potential for producing offspring is at maximum. In the United States, x-raying young men results in a major contribution to the GSD. An x-ray that includes the testes, which have little shielding tissue, results in a larger genetic dose than a similar x-ray that includes the ovaries. On the other hand, it is easy to shield the testes without losing any diagnostic information, while this is not true for the ovaries. The 1972 BEIR report estimated that in the United States the GSD due to medical x-rays could be reduced by 50% by proper limitation of the beam and use of gonadal shielding.

Somatic effects depend on the amount of radiation, the part of the body irradiated, and the age of the patient. In general, the younger the person, the more hazardous the radiation. In fact, the most dangerous period to receive radiation is before birth. At certain periods in the development of the fetus, radiation can produce deformities. Figure 19.1 shows the effects of a 100 R x-ray exposure on baby mice that were irradiated *in utero* 8 days after conception. This is the most sensitive time during the 20-day gestation period, and it corresponds to the 23rd day of human development. Anomalies in mice have been produced with as little as 25 R.

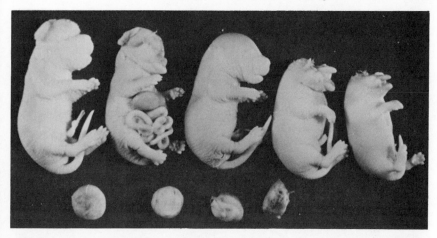

Figure 19.1. An entire litter of nine mice which were exposed to x-rays at 8 days gestation. Of the five baby mice in the upper part only the middle one is apparently normal, but it is likely to have internal damage. The two on the left both have brain hernia; the second from the left has the gut outside the body (evisceration). The two on the right have only partially developed heads. The four objects below are the remains of four mice that were killed as embryos by the x-rays. (Courtesy of Roberts Rugh, Bureau of Radiological Health, Rockville, Md.)

Much information on the effects of radiation on various organs has been obtained as a by-product of the treatment of cancer with radiation. It is usually the somatic effects of radiation therapy to normal tissues near the tumor that limit the amount of radiation that can be given to the tumor. Some of the common somatic effects of radiation are reddening of the skin (erythema), loss of hair, ulceration, stiffening (fibrosis) of the lung, formation of holes (fistulas) in tissues, reduction of white blood cells (leukopenia), and induction of cataracts in the eyes.

Perhaps the most feared somatic effect of radiation is *carcinogenesis,* the induction of cancer. It has been found that radiation can induce many types of cancer besides skin cancer. Radiation to the thyroid has caused thyroid cancer, and radiation to the blood-forming organ (bone marrow) has caused blood cancer, or leukemia. In the 1940s enlarged thymus glands in infants were often treated with therapeutic amounts of x-rays. In the 1950s a number of young people who had received these treatments as infants developed cancer of the thyroid, which lies very close to the thymus at the base of the neck. The survivors of the atomic bomb explosions in Hiroshima and Nagasaki, Japan, in 1945 have been studied extensively by the United States Atomic Bomb Casualty Commission (ABCC) to determine the effects of the radiation. There were 117 cases of leukemia in the 117,000 survivors during the period 1950 to 1966, about

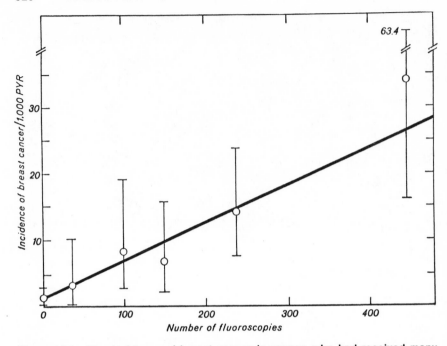

Figure 19.2. The incidence of breast cancer in women who had received many fluoroscopic examinations of the chest. The circles show the average values and the bars indicate the standard deviations. PYR signifies person year at risk. (From *The Effects on Populations of Exposure to Low Levels of Ionizing Radiation*, The Advisory Committee on the Biological Effects of Ionizing Radiation, National Academy of Sciences, Washington, D.C., 1972, p. 138.)

twice the normal rate. The incidence in survivors who had received the greatest radiation dose (~ 4 Gy) was about 40 times the normal rate.

Breast cancer can also be caused by radiation. Figure 19.2 shows the incidence of breast cancer in women who had received many fluoroscopic examinations of the chest. The amount of radiation involved is greater than that currently being used for x-raying the breast (mammography).

19.2. RADIATION PROTECTION UNITS AND LIMITS

In Chapter 16 we introduce you to the roentgen (R), a unit for measuring exposure of x-rays or gamma rays. You also may remember from Chapter 16 the rap, a unit for exposure times area that is useful as a measure of patient radiation in diagnostic radiology. In Chapter 17 we define units of radioactivity, the becquerel (Bq) and the curie (Ci). In Chapter 18 we discuss the units for absorbed dose, the gray (Gy) and the rad. You are probably fed up with radiation units, but there is one more important unit—the *rem* (Rad Equivalent Man), which is used in radiation protection. Fortunately, it is simply related to the rad, and for most purposes rads and rems are numerically equal.

The rem is a unit for the quantity *dose equivalent* (DE). The DE is defined as the rads times the quality factor (QF) of the radiation, or DE = (rads) (QF). The QF takes into account the increased damage done by certain types of radiation. A rad of densely ionizing radiation does much more damage to a cell than a rad of x-rays, gamma rays, or beta rays and is assigned a larger QF.

The QF is related to the relative biological effect (RBE) discussed in Section 18.2. Both the RBE and the QF are due to the increased biological effects of densely ionizing radiation. However, the RBE for a particular radiation is often different for different types of cells, while the QF is arbitrarily defined to be a constant for a particular radiation.

Table 19.4 lists the QFs for various types of radiation. The QF for x-rays, gamma rays, and beta rays (electrons) is unity, so for these common types of radiation rads equal rems. Also, for x-rays and gamma rays the exposure in roentgens is almost equal to the dose in rads in soft tissue. Thus for the most important man-made radiations roentgens, rads, and rems can be considered to have the same values. Table 19.5 summarizes the various radiation units.

Federal laws restrict the amount of man-made radiation to the public (excluding medical x-rays) to an average of 0.17 rem/year. Individual members of the public may receive up to 0.5 rem/year, and radiation workers are allowed 5 rems/year. These values apply to the reproductive

Table 19.4. The Quality Factors for Various Types of Radiation[a]

Radiation Type	Rounded QF
X-rays, gamma rays, electrons and positrons, energy > 0.03 MeV	1
Electrons and positrons, energy < 0.03 MeV	1
Neutrons, energy < 10 keV	3
Neutrons, energy > 10 keV	10
Protons	10
Alpha particles	20
Fission fragments, recoil nuclei	20

[a]Adapted from *Basic Radiation Protection Criteria,* National Council on Radiation Protection and Measurements, Report No. 39, Washington, D.C., 1971, p. 83.

Table 19.5. Summary of Various Radiation Units

Unit	Recommended Use	Definition
becquerel (Bq)	The SI unit for radioactivity	1 Bq = 1 nuclear transformation/sec
curie (Ci)	This unit for radioactivity is being replaced by the SI unit Bq.	1 Ci = 3.700×10^{10} nuclear transformations/sec
gray (Gy)	The SI unit for absorbed dose	1 Gy = 1 J/kg
rad	This unit for absorbed dose is being replaced by the SI unit Gy.	1 rad = 0.01 J/kg = 100 ergs/g
roentgen (R)	The unit of exposure for x-rays and gamma rays	1 R = 2.58×10^{-4} coulomb/kg of air
rem	The unit of dose equivalent	The dose equivalent in rems is numerically equal to the absorbed dose in rads multiplied by the quality factor and any other future modifying factors (such as the radiation rate).
rap	A unit of exposure-area product that measures the radiation insult to the patient from diagnostic x-rays. This unit is not yet generally accepted.	1 rap = 100 R cm^2

Table 19.6. Recommended Maximum Permissible Doses (MPDs) for Radiation Workers and the General Population, Excluding Intentional Medical Exposures[a]

Type of Exposure	MPD (rems/yr)
Radiation workers:	
Whole body	5
Gonads or lenses of eyes	5
Skin of whole body	30
Hands and feet	75
Population:	
Whole body	0.5
Gonads	0.167

[a]The Nuclear Regulatory Commission (NRC) has established the MPDs from radioactive sources produced in nuclear reactors. The detailed rules are published in *The Federal Register*, 10 CFR 20.

organs and the eyes. Less radiation-sensitive organs such as the hands and feet are allowed much more, up to 75 rems/year for workers. These values, referred to as the maximum permissible doses (MPDs), are given in Table 19.6. They have been the subject of much controversy since there is no assurance that these amounts are safe. The MPDs have been judged by radiation experts to be acceptable risks. There has been a tendency to reduce the MPDs with time. For example, the MPD to the population from a nuclear power plant is now only 0.005 rem/year. Since background radiation is typically 0.125 rem/year, variations in background in an area with a nuclear power plant may exceed the MPD.

19.3. RADIATION PROTECTION INSTRUMENTATION

In order to reduce a hazard we need a way to measure it. Your senses alert you to many hazards—for example, you can see and smell air pollution, it hurts your eyes, and it makes your breathing painful. Ionizing radiation in normal quantities does not alert any of the senses, so instrumentation is especially necessary. In this section we describe some of the common instruments used in radiation protection to monitor radiation. These instruments are often called *radiation monitors*. Those worn by radiation workers are called *personnel monitors;* larger monitors are called *portable monitors* if they can be easily carried and *laboratory monitors* or *area monitors* if they cannot.

Ionizing radiation is fairly energetic and produces a number of physical and chemical effects. Some of these effects are ionization, darkening of film, production of light (scintillation), breakdown of molecules and production of new combinations, storage of energy in a crystal (which can later be released as light—thermoluminescence), change in the conductivity of certain solids, a slight temperature increase in any material, and change in the colors of certain dyes. All of these effects are currently being used to measure radiation. Methods that measure the ionization or the energy deposited (calorimetry) are considered fundamental, or absolute. All methods based on the other effects must be compared directly or indirectly to a fundamental method. The national standardizing laboratories such as the National Bureau of Standards (NBS) in the United States maintain standards for measuring radiation. While great accuracy is needed for radiation therapy, an accuracy of 20% is usually satisfactory for radiation protection purposes. Thus radiation protection instruments are seldom sent to NBS for calibration but are usually calibrated on radiation sources in secondary laboratories.

One of the first physical effects of x-rays noted by Roentgen was their

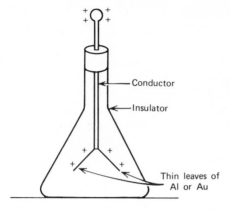

Figure 19.3. The simple electro-scope. The same type of charge on the two leaves causes them to repel each other. Ionization can reduce the amount of charge and allow the leaves to move closer to each other.

ability to discharge an electroscope (Fig. 19.3). A common instrument used in radiation protection is a modification of this extremely simple instrument. Figure 19.4*a* shows the principle of operation of this instrument, called a *portable ionization chamber* or "Cutie Pie." The ionization charges produced in the air-filled chamber are collected, with the electrons going to the positive center electrode. This weak current is amplified and displayed on a meter. The ionization chamber has a good response for x-rays, gamma rays, and beta rays, but it is relatively insensitive and cannot detect background radiation. It is used to look for relatively high radiation levels ranging from 0.01 to 5 R/hr.

Another popular version of the ionization chamber is a pocket ionization chamber that can be charged up to a few hundred volts (Fig. 19.5). Ionization produced by the radiation reduces the charge on the chamber, and the voltage remaining after irradiation indicates the amount of radiation to which the chamber has been exposed. Two versions of this device are available: one must be placed in a "reader," which measures the remaining voltage; the other has a small electrometer and magnifier built into the case and can be read directly by looking through it while holding it to the light. This latter type is useful for a worker in an area of high or variable radiation fields since it allows him to easily keep track of his radiation. However, the pocket ionization chamber is generally not used for an official record of a worker's exposure. Unless the chamber is of a special "low-energy" design, it does not accurately measure x-rays in the diagnostic region. The maximum reading of a pocket ionization chamber is generally 0.2 R.

In Chapter 17 we discuss the principle of operation of Geiger–Mueller (GM) counters and their use in early nuclear medicine measurements. Although GM detectors are now seldom used in scientific research, they

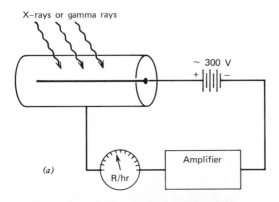

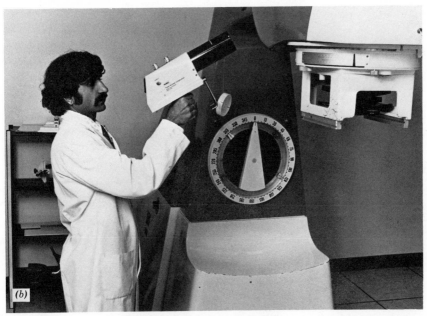

Figure 19.4. The portable ionization chamber (Cutie Pie). (a) The instrument consists of an ionization chamber, a voltage supply, and a device for measuring the ionization current produced. (b) A commercial Cutie Pie being used to monitor the radiation leakage from a ^{60}Co teletherapy source. (Courtesy of Texas Nuclear Corporation, Subsidiary of G.D. Searle & Co., Austin, Tex.)

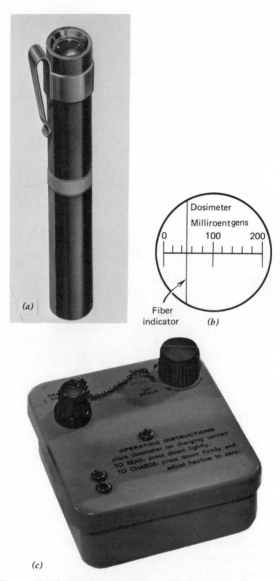

Figure 19.5. The pocket ionization chamber (a) is about the size of a fountain pen. The scale of this model (b) can be read by looking through the chamber while holding it to the light. The chamber must be charged on a charger (c).

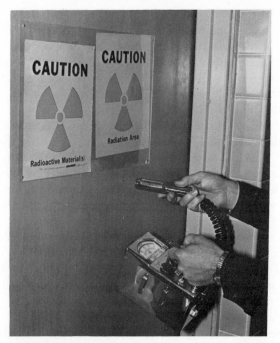

Figure 19.6.   A portable GM detector being used to monitor the radiation outside a door to a radium storage room. Note the radiation warning signs on the door. The symbols are red on yellow backgrounds.

are often used as portable radiation protection monitors (Fig. 19.6). The GM detector gives a count for each ionizing event it detects, whether the event is of low or high energy. It is much more sensitive than the ionization chamber and can easily detect background radiation, typically 200 to 300 counts per minute. Although GM detectors do not measure the ionization produced in air as ionization chambers do, GM detectors often have scales that read in milliroentgens per hour. Such a scale is useful and may be fairly reliable for the type of radiation for which the unit was calibrated, but it can be seriously in error for other types of radiation since counts per minute cannot be equated to milliroentgens per hour for different types of radiation. All GM detectors have a limited life and usually behave erratically near the end (they are almost human!).

The GM detector discharges with each ionizing event that occurs in its sensitive volume and must recover before the next event (Fig. 17.6). It is possible to have such a high radiation field that the detector remains in a continuous state of discharge, registering no events and giving a reading of zero. The level at which this occurs is about 0.2 R/hr (5000 counts/sec) but depends on the condition of the detector. Some modern GM detectors have fail-safe circuits so that they cannot remain in a state of discharge.

GM detectors can be made to fit into the pocket and provide an audible chirp for every few hundred counts (Fig. 19.7). Such detectors are very useful as personnel monitors since they quickly alert the wearer when there is an increase in the radiation field.

It was natural that the scintillation detectors developed for nuclear physics would be adopted for radiation safety work since they are highly sensitive and can determine the energies of gamma rays. The operation of the scintillation detector is described in Section 17.4 and will not be repeated here. The radiation protection expert uses it when he wishes to identify an unknown radioactive contaminant or count two different radionuclides in one sample.

The most common way of monitoring the radiation received by workers is to use film badges (Fig. 19.8). Film badges are compact, simple, inexpensive, and provide a permanent record. They can measure beta-ray dose as well as x-ray and gamma-ray exposure. An additional advantage is that wearing a film badge reminds the worker that he is working in a radiation area.

Film badges have several disadvantages. The film responds better to low-energy x-rays than to high-energy gamma rays; this disadvantage can be overcome by using special filters to help determine the type of radia-

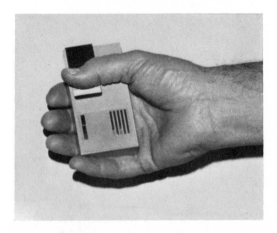

Figure 19.7. A pocket GM detector that emits a chirp for every few hundred detected events. An increased frequency of chirping alerts the wearer that the radiation rate has increased. Under normal conditions, the detector chirps about once a minute.

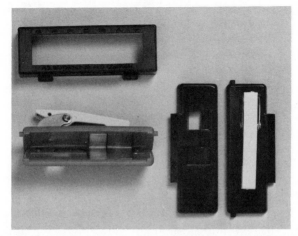

Figure 19.8. This film badge has three plastic parts shown assembled at the lower left without the film in its light-tight covering in place. The metal shields cover different areas of the film to permit an estimate of the energy of the radiation to which it was exposed. The film is usually replaced once a month. (Courtesy of R.S. Landauer, Jr. & Co., A Tech/Ops Company.)

tion producing the film blackening. If the film badge is worn in a warm humid environment, the signal fades rapidly (~ 50% in 1 week). Although the accuracy of the film badge under laboratory conditions is satisfactory, sizable errors (up to 600%) have been detected in a study in which commercial film badge suppliers read film badges that had been given a known exposure.

Employers often have workers wear film badges so that they can later prove that the workers had not been exposed to excessive amounts of radiation. Employers are legally required to "badge" workers that are expected to receive more than about 25 mrem per week, but only about 1% of the present hospital film-badge wearers fall into this category. It is generally required that radiation workers *not* wear their film badges when receiving medical and dental x-rays.

Many crystalline materials give off light when heated after being exposed to radiation. This light is called *thermoluminescence* (TL). Some natural minerals like fluorite (calcium fluoride) give off large amounts of TL when heated due to the large exposure they have accumulated from natural radioactivity over centuries. Since all materials have some natural radioactivity, the phenomenon is very common and was probably observed over 2000 years ago when fluorite was used in making lead. Madame Curie was aware of TL; in her Ph.D. thesis she described how the TL of calcium sulfate doped with manganese ($CaSO_4$:Mn) could be restimulated by exposure to radium sources. However, she did not use the technique for measuring the radiation.

In 1950 Farrington Daniels proposed using TL for measuring radiation and demonstrated how thermoluminescent dosimetry (TLD) could be used for measuring radiation from atomic bomb tests. The principles involved in measuring TL are shown in Fig. 19.9. A small crystal of lithium fluoride (LiF) is exposed to ionizing radiation. Less than 1% of the energy is stored in the crystal by lifting electrons to higher energy levels where they stay in metastable states until the crystal is heated. When the LiF is heated the electrons are released and "fall" back to their original energy levels. Some of the energy released in the process is emitted as light, and the amount of light emitted is proportional to the original exposure. The process can be repeated. Table 19.7 lists some of the properties of the currently most popular TLD phosphor, lithium fluoride. Note the large range over which it can measure radiation.

A commercial TLD meter is shown in Fig. 19.10 and a personnel monitor using TLDs is shown in Fig. 19.11. TLD has a wide variety of applications. In addition to personnel dosimetry, TLD is useful for environmental and area monitoring. The dosimeters are left in an area for a period of time and then read out to determine the ambient radiation where they were located. Also, TLD dosimeters are used extensively for

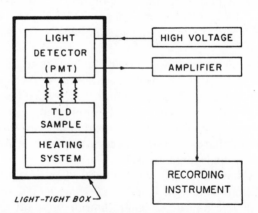

Figure 19.9. The basic components for measuring the TL from a dosimeter. An exposed LiF crystal is placed in the light-tight box and heated to about 300°C in less than 1 min. The emitted light is converted into an electric current by the photomultiplier tube (PMT). The current is integrated to give a measure of the TL emitted. (From J.R. Cameron, N. Suntharalingam, and G.N. Kenney, *Thermoluminescent Dosimetry*, The University of Wisconsin Press, Madison, Wisc., 1968, p. 76.)

Table 19.7. Some of the Characteristics of the TLD Phosphor Lithium Fluoride[a]

Characteristic	LiF
Density	2.64 g/cc
Effective atomic no.	8.2
TL emission spectra range	350–600 nm
Temperature of main TL glow peak	195°C
Useful range	10 mR–10^4 R
Fading	< 5%/12 wk
Light sensitivity	essentially none
Physical form	powder, extruded, Teflon-embedded, glass capillaries

[a]Adapted from J.R. Cameron, N. Suntharalingam, and G.N. Kenney, *Thermoluminescent Dosimetry,* The University of Wisconsin Press, Madison, 1968, p. 33.

Figure 19.10. A commercial TLD system. The digital display on the right can be adjusted to read directly in R or mR. (Courtesy of Harshaw Chemical Company, Solon, Ohio.)

Figure 19.11. A TLD personnel monitor containing 2 LiF dosimeters. The badge is about the size of a dental film. (Courtesy of Harshaw Chemical Company, Solon, Ohio.)

measuring radiation to patients undergoing radiation therapy treatments; the dosimeters are placed inside body cavities such as the bladder, rectum, and vagina.

The instruments described in this section are probably used for more than 99% of all radiation protection measurements. However, many other physical and chemical techniques can be used to measure radiation. *Radiation Dosimetry,* edited by Attix et al., describes these techniques (see the bibliography at the end of this chapter).

19.4. RADIATION PROTECTION IN DIAGNOSTIC RADIOLOGY

Medical x-rays are the largest source of man-made radiation in the United States. However, it has been estimated that by using presently known techniques, radiation from medical x-rays could be cut by more than half without losing any medical information. In this section we discuss recommended techniques for minimizing unnecessary radiation in diagnostic radiology.

Figure 19.12 indicates that there is a wide variation in the amount of radiation a "standard" patient receives when he has his lower (lumbar sacral) spine x-rayed in different kinds of medical facilities in the United States. These data, which are for 1972 through 1974, are from a cooperative study by the Bureau of Radiological Health (BRH) and the radiation control directors of various states. The study is called the Nationwide Evaluation of X-ray Trends (NEXT).

The first column of Fig. 19.12 shows the range of exposures from lumbar sacral spine radiographs. Note that the exposures range from about 50 to 5500 mR. Several factors contribute to the larger exposures. If an x-ray unit does not have sufficient filtration in the beam, the low-energy x-rays that would normally be removed by the filter enter the body and contribute to the patient exposure. Occasionally a low-energy beam is useful, such as in mammography (x-raying of the breast), but even when such a beam is used there should be some filtration in the beam. Most x-ray units should have at least 3 mm of aluminum filtration. The amount of filtration in the beam can be estimated by measuring the half-value

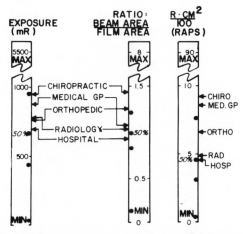

Figure 19.12. The exposures, ratio of beam area to film area, and the exposure-area products in raps for lumbar sacral spine radiographs measured in the NEXT program. The dots indicate the minimum and first, second, and third quartiles for the 634 facilities measured. The lowest dot in each column indicates the minimum value. The dot above it indicates the first quartile—25% were below this value. The fourth dot indicates the third quartile. The arrows indicate the mean values for the various types of facilities. (Courtesy of J. Wochos, University of Wisconsin, Madison, Wisc.)

layer (see Chapter 16, p. 397). Unnecessary exposure may also be due to improper film development. If the film is being underdeveloped, overexposures are necessary to achieve the proper density. Also, when the film and screen used for the image are insensitive, a larger than normal exposure is needed.

The second column in Fig. 19.12 shows the ratio of x-ray beam area to film area found in the NEXT study. Ideally this ratio should be less than 1. Note that the beam size used by chiropractors is on the average 1.5 times larger than the film and that the largest beam size recorded is 8 times larger than the film. In a nationwide study by BRH in 1964, it was found that on the average the x-ray beams used were more than twice as large as the film. Every properly functioning x-ray unit should have a device (collimator) to limit the size of the beam to be no larger than the film. A federal law that took effect in August 1974 requires manufacturers to incorporate devices on x-ray equipment that prevent the x-rays from being turned on if the beam is larger than the film being used. However, the law does not apply to the many x-ray units made previously that will continue to be used for years.

The third column in Fig. 19.12 shows the distribution of exposure-area products, which are a good measure of the total radiation insult to the patients. The exposure-area product is obtained by multiplying the exposure in roentgens by the size of the beam in square centimeters. For convenience, this value is divided by 100 to give raps. Note that the average patient having an x-ray of the lower spine receives fewer raps in a hospital than in a private practitioner's office.

The retaking of x-rays is a common cause of unnecessary radiation and may be done for several reasons. If the x-ray operator is not well trained, he may make a mistake in setting the various controls on the x-ray unit or in positioning the patient. A repeat x-ray is sometimes necessary because the x-ray machine or film processor is not working correctly. A surprisingly large number of x-ray units, even in large medical centers, do not function as they should. Through a quality assurance program, the functioning of each portion of the x-ray system should be routinely checked. Other unfortunate reasons for retaking x-rays are the loss of the original films and the refusal of another medical center to loan recently taken films of a patient.

Because of the importance of reducing genetically significant radiation, it is recommended that the gonads be shielded whenever possible. This is easier to do on males than on females. Figure 19.13 shows an x-ray of a young female with a special gonadal shield in place. In taking x-rays of infant chests, it is desirable to shield the entire lower abdomen.

A special problem in radiation protection in diagnostic radiology in-

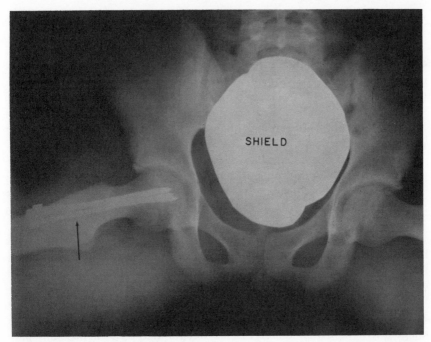

Figure 19.13. An x-ray of a young woman showing a lead gonadal shield. Note the metal pins in her right hip (arrow).

volves pregnant and possibly pregnant women. As mentioned in Section 19.1, the biological effects of radiation on young organisms are more severe than the effects on older organisms. Many hospitals and radiologists have a policy of not x-raying female patients who are or may be pregnant except in emergency situations. In an emergency, they make every effort to shield the fetus from radiation. Since a woman may not know she is pregnant until she misses a menstrual period, it is a common policy to take x-rays only during the 10 days following the last period. This is known as the *10-day rule*.

A considerable amount of radiation to patients comes from fluoroscopic examinations. In fluoroscopy as in conventional radiology, there should be sufficient filtration in the beam to remove the low-energy x-rays and the radiation beam should be collimated to the area of interest. Inexperienced "fluoro" operators often use too large a radiation field because they do not want to miss anything. An x-ray beam larger than necessary produces more scatter than a beam of the proper size, and since scatter seriously degrades the image, the patient gets more radiation and the operator gets less information.

The image intensifiers on modern fluoroscopy units (see Chapter 16, p. 423) often have automatic brightness stabilizers that adjust the brightness of the image for the radiologist. While such a device is useful, if there is a defective component in the apparatus the automatic brightness stabilizer may increase the radiation to the patient in an effort to keep the image brightness constant, resulting in an unnecessary radiation exposure to the patient.

Every fluoroscopy unit has a timer that automatically signals the operator or interrupts the beam when 5 min has elapsed to remind the operator of the length of the examination. The purpose of the timer is sometimes defeated by an alert and cooperative technologist who resets the timer before it runs out.

New electronic image storage devices should help reduce radiation from fluoroscopic examinations and may also play a role in conventional radiology (see Chapter 16). The amount of radiation needed to produce one electronic image is about 1% of that needed to produce an image with the conventional screen-film combination. The quality of the image is not as good as with film, but it is satisfactory for many situations and the image is available immediately. Electronic radiography will probably play

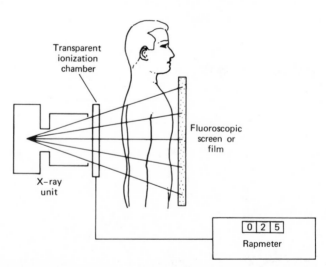

Figure 19.14.   The principle of a rapmeter. The exposure-area product to a patient is measured by a flat transparent ionization chamber on the collimator. The current from the chamber is integrated to give the exposure-area product.

a large role in the future, not because it reduces radiation, but because it is more convenient than using film.

The last time you had an x-ray you may have asked the x-ray operator how much radiation you were receiving. Unfortunately, many radiologic technologists simply do not know how much radiation they are giving. As you may remember from Chapter 16, the radiation received depends on many factors—kVp, mA, time, filter, distance, and area exposed—and the calculation is not easy. However, a simple device called a *rapmeter* can be attached to an x-ray unit to measure the exposure-area product to the patient in raps (R cm²/100). A rapmeter is a flat ionization chamber (Fig. 19.14) that is attached to the end of the collimator, and it does not touch the patient or interfere with the operation of the x-ray unit. After an exposure the reading is readily visible. If all x-ray units and fluoroscopic units had rapmeters attached, x-ray operators would become more aware of the amount of radiation they are giving to patients.

19.5. RADIATION PROTECTION IN RADIATION THERAPY

Since the radiation therapy area of a hospital contains intense radiation sources, it is typically surrounded by concrete walls about 0.5 m thick. Also, the scattered radiation during a radiotherapy treatment is large. To protect individuals who might inadvertently enter the room during a treatment, the door has a switch (interlock) that turns off the machine when the door is opened.

The radiation sources themselves must be adequately shielded. We can calculate the amount of shielding that is needed to reduce the gamma radiation from a ^{60}Co source to a safe level. We first explain how to calculate the exposure rate from the gamma rays from a particular radionuclide source. The basic equation for I_γ, the radiation intensity in roentgens per hour, is

$$I_\gamma = \frac{\Gamma N}{D^2} \tag{19.1}$$

where D is the distance in meters from a source of N megabecquerels or curies of radioactivity and Γ (gamma) is a constant for the particular radionuclide that depends on the number and energies of the gamma rays emitted per disintegration (e.g., 3.5×10^{-5} R m²/MBq hr for ^{60}Co). The Γ factors for some common radionuclides are given in Table 19.8. Equation 19.1 tells us that the radiation at a given distance from a radioactive source is proportional to the intensity of the source and inversely proportional to the distance squared. Using this equation we can calculate,

Table 19.8.   Gamma Factors, Linear Attenuation Coefficients, and Half-Value Layers for Some Common Radionuclides

Radionuclide	Γ (R m²/MBq hr)	Γ (R m²/Ci hr)	μ_{Pb} (cm^{-1})	HVL (cm Pb)
^{137}Cs	8.4×10^{-6}	0.31	1.2	0.58
^{60}Co	3.5×10^{-5}	1.3	0.58	1.2
^{198}Au	6.2×10^{-6}	0.23	2.3	0.3
^{131}I	5.9×10^{-6}	0.22	2.3	0.3
^{203}Hg	3.2×10^{-6}	0.12	2.3	0.3
^{99m}Tc	1.9×10^{-6}	0.07	23	0.03
^{18}F	1.5×10^{-5}	0.57	0.76	0.91

for example, that the radiation intensity at 1 m from a 185 TBq ($\sim$ 5000 Ci) ^{60}Co teletherapy source is about 6000 R/hr (see Example 19.1a).

The amount of shielding that is necessary to reduce the radiation to a safe level—2 mR/hr at 1 m—can be calculated in several ways. Perhaps the easiest involves using the thickness of lead that is a half-value layer (HVL) for the radiation involved (Table 19.8). Each half-value layer reduces the intensity by a factor of 2. Thus using 5 HVLs reduces the intensity by a factor of 2^5, or 32, and using 10 HVLs reduces the intensity by a factor of 2^{10}, or 1024. Once you know by what factor you wish to reduce the radiation intensity (in this case, 3×10^6), you can simply calculate the number of half-value layers that are necessary (see Example 19.1b). A graph (Fig. 19.15) can be used to determine the number of half-value layers needed when the attenuation factor is known. If the source size is doubled, the amount of shielding is not doubled but an additional half-value layer must be added.

An alternate way to calculate the thickness of the shielding is to use the equation for exponential attenuation (see Appendix A)

$$I = I_0 e^{-\mu x} \qquad (19.2)$$

where I is the intensity after passing through a thickness x; I_0 is the initial intensity at the same point without any absorber; e, which equals 2.718, is the base of the natural logarithm; and μ is the linear attenuation coefficient, which is related to the half-value layer by the equation

$$\mu = \frac{0.693}{\text{HVL}} \qquad (19.3)$$

Equation 19.2 can be rewritten in a more convenient form to calculate shielding (see Example 19.1c).

$$x = \frac{2.3 \log(I_0/I)}{\mu} \qquad (19.4)$$

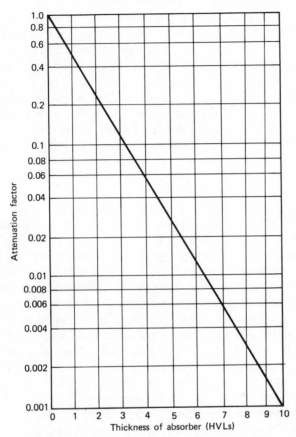

Figure 19.15. The number of half-value layers needed to achieve a given attenuation.

Example 19.1

a. Calculate the radiation intensity at 1 m from a 185 TBq ($\sim$ 5000 Ci) ^{60}Co teletherapy source (see Table 17.4).

Using Equation 19.1, we obtain

$$I_\gamma = \frac{(3.5 \times 10^{-5} \text{ R m}^2/\text{MBq hr}) \,(185 \times 10^6 \text{ MBq})}{1 \text{ m}^2} \simeq 6500 \text{ R/hr}$$

assuming no self-attenuation in the source. Attenuation in the source would reduce the intensity to about 6000 R/hr.

b. Calculate the thickness of lead necessary to shield the source in Example 19.1*a* to a radiation level of less than 2 mR/hr at 1 m.

The intensity must be reduced from 6000 R/hr to 2×10^{-3} R/hr, or by a factor of 3×10^6. Since each half-value layer reduces the intensity by a factor of 2, reducing the intensity by 3×10^6 would require 22 HVLs ($2^{22} \simeq 4.2 \times 10^6$). The half-value layer for ^{60}Co is about 1.2 cm of lead; therefore, the shielding thickness would have to be about 26 cm.

c. Calculate the shielding needed in Example 19.1*b* using Equation 19.4.

The thickness *x* for the shielding necessary is

$$x = \frac{2.3 \log (3 \times 10^6)}{0.58} = 25.7 \text{ cm}$$

We have assumed that the shielding is for a narrow beam of radiation going through a flat sheet of absorbing material. When the shielding must be put in the form of a sphere of lead surrounding the source it must be somewhat greater. A typical ^{60}Co source shield is about 60 cm in diameter and may weigh several tons.

The most serious radiation hazards in radiotherapy involve internal radiation sources. Patients containing such radiation sources stay in the hospital for several days to a week during the radiation treatment. Their rooms should be well supervised from a radiation safety standpoint, and nurses taking care of them should be adequately informed on simple radiation safety measures. There are three simple and obvious ways for nurses to reduce their radiation: minimize the time near the radiation source, maximize the distance to the source, and use absorbing material (shielding) where possible. The radiation intensity decreases inversely with the square of the distance from the source, and thus if the nurses increase their distance from the source by a factor of 3, the radiation intensity will be reduced by a factor of 9. The nurses should wear personnel monitors such as film badges.

If the patient contains removable sources such as radium needles, care should be taken that none is lost; radium needles occasionally come out accidentally or with a patient's help. Careful inventory of radium sources should be maintained so that the staff can look for a missing source soon after it disappears.

Patients with permanently implanted radioactive "seeds" of radon or gold 198 and patients who have received therapeutic doses of ^{131}I are kept in the hospital until the radiation level surrounding them has fallen to a safe level. Patients who have received therapeutic quantities of ^{131}I need special handling. Since much of the ^{131}I is excreted in the urine, it is customary to collect the "hot" urine in special jugs for monitoring and proper disposal by the health physicist.

19.6. RADIATION PROTECTION IN NUCLEAR MEDICINE

During the 1960s the use of nuclear medicine procedures in the United States grew at a rate of about 25% per year. In 1976 it was estimated that the amount of radiation to the public from nuclear medicine was about 1% of that received from medical x-rays. While the amount of radiation received by a patient from a nuclear medicine procedure is comparable to that received during an x-ray study, the techniques used to reduce radiation have to be somewhat different.

The nuclear medicine physician should:

1. Determine whether the study is necessary or desirable.
2. Use the right radiopharmaceutical; sometimes different radiopharmaceuticals are in similar containers.
3. Use the right amount of the radiopharmaceutical; for ^{99m}Tc it is necessary to have a calibration device.
4. Give the radiopharmaceutical to the right patient; in a busy department it is easy to make mistakes.
5. Make sure the detection equipment is working correctly; there are standard test procedures for gamma cameras and scanners.

In the 1950s the most common radionuclide used in nuclear medicine was ^{131}I. It emits gamma rays of several energies, but most of the gamma rays have an energy of 362 keV so they easily escape from the body. In addition, ^{131}I emits beta rays, which are absorbed in a few millimeters of tissue and contribute a large fraction of the radiation received by the patient. For example, in a thyroid uptake study the thyroid receives about 0.08 Gy of dose from the 300 kBq of ^{131}I given to the patient, and about 90% of this radiation is from the beta rays.

There are radionuclides that emit only gamma rays; they are called *isomers*. The radioactive isomer ^{99m}Tc, introduced in the 1960s, has almost ideal physical and chemical properties for many nuclear medicine studies, and in 1976 it was the most common radionuclide used. The *m* identifies it as an isomer. Technetium 99m decays to ^{99}Tc with a half-life of 6 hr. It emits 140 keV gamma rays. Because it emits no beta rays, the radiation to the patient is much lower than that from ^{131}I, and much larger quantities may be used to produce a better image. However, the larger doses—up to 500 MBq ($\sim$ 15 mCi) have resulted in larger radiation exposures to personnel working in nuclear medicine laboratories. This radiation can often be reduced by using proper shielding equipment such as lead syringe holders during injections (Fig. 19.16).

Practically all radionuclides used in nuclear medicine emit penetrating gamma rays that can be detected outside the body. Most of the shielding

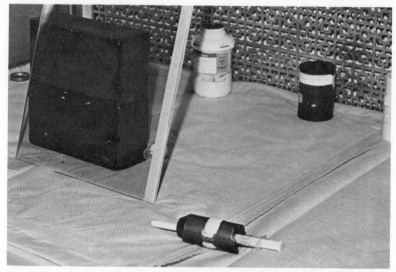

Figure 19.16. A syringe containing ^{99m}Tc is placed inside a lead cylinder to shield the physician giving the injection. The needle is covered with sterile plastic until it is used. In the background is the loading area; note the 5-cm thick lead bricks and the sheet of lead glass to protect the loader.

used in a nuclear medicine laboratory is for absorbing the gamma rays. The radiation intensity from a gamma ray emitter can be calculated using Equation 19.1 (see Example 19.2a). The shielding needed can be calculated in several ways (see Examples 19.2b and c). Figure 19.15 can be used to estimate the shielding needed. If an attenuation of 1/46 or 0.02 is desired, you simply read off the value of 5.5 HVLs.

Example 19.2
a. What is the radiation intensity 0.35 m from an unshielded 7.5 GBq ($\sim$ 200 mCi) source of ^{198}Au?

From Table 19.8, Γ (^{198}Au) is (6.2×10^{-6}) R m^2/MBq hr.

$$I_\gamma = \frac{(6.2 \times 10^{-6})(7.5 \times 10^3)}{0.35^2} \simeq 0.4 \text{ R/hr}$$

b. How much shielding is necessary to reduce the radiation from the source in Example 19.2a to a safe level? Assume that an employee must work at 1 m from the source for 10 hr each week and we want to limit his radiation from the source to less than 10 mR/week (i.e., < 1 mR/hr at 1 m). Also assume that new shipments of sources keep the average strength at 7.5 GBq.

The radiation intensity at 1 m is

$$I_\gamma = \frac{(6.2 \times 10^{-6})(7.5 \times 10^3)}{1^2} = 0.046 \text{ R/hr or } 46 \text{ mR/hr}$$

From Table 19.8, the half-value layer for this radiation is about 0.3 cm Pb. Since each half-value layer reduces the intensity by a factor of 2 and we wish to reduce the intensity by a factor of 46, it is necessary to use between 5 and 6 HVLs. Thus 1.8 cm Pb will provide adequate shielding.
c. Calculate the thickness of lead needed in Example 19.2b using Equation 19.4.
The linear attenuation coefficient in Pb for ^{198}Au given in Table 19.8 is 2.3 cm^{-1}.

$$x = \frac{2.3 \log (46)}{2.3 \text{ cm}^{-1}} = \frac{(2.3)(1.7)}{2.3} \simeq 1.7 \text{ cm}$$

In a nuclear medicine laboratory the many containers of radioactivity are generally placed behind a wall of 5-cm-thick lead bricks. There is usually a laboratory monitor in the work area to indicate the radiation level on a meter and also to give an audible signal of the radiation level. It is also customary to place film badges at a few places in the storage area ("hot lab") to monitor radiation levels.

19.7. RADIATION ACCIDENTS

In this section we discuss some general rules for handling radiation accidents. "Typical" radiation accidents occur when a ^{60}Co therapy source refuses to turn off, a sealed radiation source (such as a radium needle) is broken, or a patient who has just swallowed 3.5 GBq ($\sim$ 100 mCi) of ^{131}I vomits. The usual handlers of radioactivity should be aware of elementary procedures to follow in case of a radiation accident. In addition, each medical center should have a radiation safety specialist who is prepared for most radiation accidents. He should store a supply of plastic gloves, shoe covers, and clean-up materials in one or more convenient locations such as the nuclear medicine receiving and dispensing room. He should have monitoring equipment readily available and check it regularly. When there is a spill of a liquid radioactive material, the activity should be limited to a single area if possible. The expert should be called to supervise the clean-up operation.

When a ^{60}Co unit fails to shut off, there is not time to call the expert. The patient should be removed from the room immediately, and the collimator should be closed as much as possible to restrict the beam. Each therapy unit should have emergency procedures posted at the control panel.

Perhaps the most serious radiation accident is the breakage of a radium needle, which can cause the contaminated area to be shut down for years. The recommended maximum permissible amount of radium in the body is 3.7 kBq (0.1 μCi), and a typical 110 MBq (3 mg) needle has enough radioactivity to give 30,000 people this amount. Radium needles and other radium sources should be inspected regularly and checked for leakage every 6 months. When checked, the needle is placed in a test tube with a wad of cotton. Any escaping radon gas will deposit some of its daughter products on the cotton, and after about a day the cotton is placed in a well counter to determine whether it is contaminated. Leaking radium needles should be returned to the manufacturer for repair or better yet replaced with needles containing newer, safer radionuclides such as cesium 137 (^{137}Cs).

BIBLIOGRAPHY

Attix, F. H. (Ed.), *Topics in Radiation Dosimetry, Radiation Dosimetry Supplement 1,* Academic, New York, 1972.

Attix, F. H., and W. C. Roesch (Eds.), *Radiation Dosimetry, Vol. I: Fundamentals,* 2nd ed., Academic, New York, 1968.

Attix, F. H., and E. Tochilin (Eds.), *Radiation Dosimetry, Vol. II: Instrumentation,* 2nd ed., Academic, New York, 1966.

Attix, F. H., and E. Tochilin (Eds.), *Radiation Dosimetry, Vol. III: Sources, Fields, Measurements, and Applications,* 2nd ed., Academic, New York, 1969.

Blatz, H., *Introduction to Radiological Health,* McGraw-Hill, New York, 1964.

Cameron, J. R., N. Suntharalingam, and G. N. Kenney, *Thermoluminescent Dosimetry,* University of Wisconsin Press, Madison, 1968.

Casarett, A. P., *Radiation Biology,* Prentice-Hall, Englewood Cliffs, N. J., 1968.

Eisenbud, M., *Environmental Radioactivity,* McGraw-Hill, New York, 1963.

Henry, H. F., *Fundamentals of Radiation Protection,* Wiley-Interscience, New York, 1969.

Morgan, K. Z., and J. E. Turner, *Principles of Radiation Protection: A Textbook of Health Physics,* Wiley, New York, 1967.

NAS-NRC, *The Effects on Populations of Exposure to Low Levels of Ionizing Radiation,* Report of the Advisory Committee on the Biological Effects of Ionizing Radiations, National Academy of Sciences—National Research Council, Washington, D.C., 1972. (BEIR Report.)

NCRP Report No. 32, *Radiation Protection in Educational Institutions,* National Council on Radiation Protection and Measurements, Washington, D.C., 1966.

NCRP Report No. 39, *Basic Radiation Protection Criteria,* National Council on Radiation Protection and Measurements, Washington, D.C., 1971.

NCRP Report No. 48, *Radiation Protection for Medical and Allied Health Personnel,* National Council on Radiation Protection and Measurements, Washington, D.C., 1976.

NCRP Report No. 49, *Structural Shielding Design and Evaluation for Medical Use of*

X-Rays and Gamma Rays of Energies Up to 10 MeV, National Council on Radiation Protection and Measurements, Washington, D.C., 1976.

Radiological Health Handbook, compiled and edited by the Bureau of Radiological Health and the Training Institute Environmental Control Administration, revised ed., Public Health Service, Rockville, Md., 1970.

REVIEW QUESTIONS

1. What are the sources of background radiation?
2. What is the typical amount of background radiation each year to a resident of the United States?
3. About what percent of residents in the United States have a medical or dental x-ray each year?
4. What governmental agency has responsibility for x-ray safety?
5. What is the NCRP?
6. What is meant by GSD?
7. What is the MPD to the general population from a nuclear power plant in the United States?
8. List five physical effects that can be used to measure radiation.
9. What is thermoluminescence? How is it used to measure radiation?
10. What is the 10-day rule?
11. If you have a 10 MBq ^{137}Cs source, at what distance from it will the exposure rate be 0.01 R/hr?
12. A 185 TBq ^{60}Co source has adequate shielding with 26 cm of lead. If the activity of the source is doubled, what thickness of lead shield would be necessary for the same degree of safety?
13. Calculate the exposure rate 0.3 m from a syringe containing 500 MBq of ^{99m}Tc for a brain scan.
14. What is the recommended maximum permissible amount of radium in the body?
15. How often should radium sources be checked for leakage?

CHAPTER 20

Computers in Medicine

Richard Friedman
Department of Medicine
University of Wisconsin

Computers are playing a growing role in our lives, and it should not be surprising that they are also playing an increasingly important role in medicine. There is not yet a medical specialty that deals solely with computers in medicine, but each year there are several scientific meetings where their medical use is the main topic. Hospitals first used computers extensively in their business offices, but computers now have important applications in many medical specialties. A modern medical center often has over 10 computers in operation. In this chapter we discuss some of their uses.

Let us first define a few of the common terms of computer jargon. A computer is usually considered to have two major parts—the *hardware* and the *software*. The hardware is the machine itself and all accessories attached to it. The software consists of the instructions, or *programs,* that tell the computer how to carry out its functions. The programs are constructed or written by a programmer. Just as it is possible to give different directions on how to travel from one city to another, it is possible to write different programs for performing the same computation. A programmer skilled in writing compact and efficient programs is of great value. Of course, the programmer sometimes makes mistakes; when he does, the program has "bugs" in it. "Debugging," or editing, the program consists of finding and correcting the mistakes.

The information fed to the computer is called the *input,* and the results are the *output.* The input may be fed to the computer through a keyboard like that on a typewriter, or it may be the output of an instrument, such as an ECG unit. The output may be printed on paper by a teletypewriter, or it may be displayed on a cathode ray tube (CRT). Other inputs and outputs are also used. The devices used for input and output are called peripherals.

A computer that is used for a single purpose is called a *dedicated computer,* while a *time-shared computer* is a large computer that serves several uses simultaneously. Which type is more economical has yet to be resolved. In the 1960s the trend was toward time-sharing, but in the 1970s dedicated computers appear to be more popular. Sometimes a dedicated computer is connected to a large time-shared computer that "subcontracts" the really big jobs. This arrangement is called *distributed intelligence. Microcomputers* such as the computing elements of hand-held electronic calculators will undoubtedly play a significant role in medicine in the future.

Nearly all modern computers are digital computers, that is, they work with numbers. *Analog* computers use electrical signals such as voltages and currents as inputs. When a digital computer is used to analyze data that consist of electrical voltages, such as ECGs, it must first convert the ECG voltages into numbers, or determine the voltage as a function of time. This conversion is the opposite of plotting a graph—in this case the curve is given and the (x,y) values that fall on the curve are "read" from it. This is called *digitizing* the data, and it is done by an analog to digital converter (ADC). The hardware of many computers includes ADCs.

For all practical purposes a computer is simply a piece of equipment capable of performing a specified series of operations of logic or arithmetic upon data presented to it. A programmer must enter a complete set of instructions on how each task should be accomplished. The advantage of the computer is, therefore, not that it can make new or novel decisions—it cannot—but rather that it can gather, record, and retrieve data according to predetermined formulas at a speed millions of times faster than was previously possible. Because of this great speed, it can do calculations that would otherwise not be possible, such as those made in computerized tomography (see Chapter 16, p. 427).

In the practice of medicine, a physician must gather, store, and retrieve data on his patients. On the basis of these data the physician makes decisions using predetermined medical "facts" he learned in medical school, during internship and residency, and throughout long experience. It was once the desire of computer scientists to program a computer with all the information-gathering, data-storing, and decision-making rules known by the physician so that it could duplicate many of these tasks. This goal has proven to be impractical for a number of reasons. First, the physician's knowledge is far too vast to program in a single computer; second, the physician's decisions are often based on experience rather than on documented laws; and third, a computer does not have the unique intuitive capability of humans.

Nevertheless, the mushrooming size and complexity of medical information have made it increasingly clear that a physician cannot hope to learn it all, and computers have been enlisted as aids to assist the physician in gaining some measure of control over a broad area of this information. In this chapter we discuss several of their medical applications.

20.1. HISTORY TAKING

Perhaps no part of a clinical examination is more important than the gathering of a good medical history. No physician can adequately treat a patient before he knows what hurts, when the pain started, what relieves the pain, and so forth. Medical scientists estimate that 85% of a diagnosis is based on the patient's medical history. A good medical history includes the identification of the patient's chief complaint (e.g., chest pain on the left side); a detailed chronological history of this complaint (when it started, etc.); a review of all past and present complaints about all parts of his body; a past medical history including previous hospitalizations, medications, and travel; a full family medical history going back to his grandparents; and finally, a discussion of his habits and environment. Obtaining all this information requires from 30 min to 1 hr. Most physicians, while acknowledging the need for this information, would admit that they cannot spend this much time with each patient.

The computer has thus been enlisted to assist the physician in data gathering, and a number of different approaches have been used. At the Kaiser-Permanente Clinics in California, each patient is given a deck of prepunched computer cards. Each card has one question printed on it (e.g., Do you have hay fever?). The patient sorts the cards by putting each card into the *yes* box or the *no* box. The cards are then collected and run through the computer, and a printed summary is generated. In a variation of this approach, the patient completes a printed questionnaire by marking the *yes* box or the *no* box for each question. The data are then keypunched on cards and fed to a computer, which generates the medical history. The questionnaire can be printed on special mark-sense compatible forms. The patient fills out the questionnaire by darkening the appropriate boxes; the form is then optically scanned by the computer, the results are recorded, and a summary is generated.

The card-sort history and medical questionnaire are easy to use and inexpensive but require that the patient answer all questions, including those that do not apply to him. In addition, with this type of history it is impossible to present detailed questions on a particular subject. To overcome these handicaps, in 1963 Dr. Warner Slack at the University of Wisconsin Hospitals conceived of placing the patient directly in contact with the computer via a computer terminal (Fig. 20.1). With this system, the computer records each answer and on the basis of the patient's response decides what to ask next. For example, if a patient responds *no* to the question, Do you have chest pain? the computer will ask him a question on a different subject. If the patient answers *yes*, the computer will ask him more detailed questions about the chest pain such as, Does it radiate to the back? The patient also can respond with *I don't know* to indicate that he does not know the answer to the question and *HELP* to indicate that he needs an explanation of the question, which the computer can provide. The computer can thus obtain a detailed history in some areas while not bothering the patient with numerous questions that are not relevant to his problems. At the end of the "interview" the computer prints out a medical history summary (Fig. 20.2).

Computers are now used in many hospitals and clinics to obtain medical

Figure 20.1.   A medical history can be taken by placing a patient directly in contact with a computer via a computer terminal.

UNIVERSITY HOSPITALS PATIENT QUESTIONNAIRE

ADMISSION DATE: 02-03-75   DATE: 10-29-74
HX NUMBER: 111111   NAME: MARY SMITH

CENTRAL NERVOUS SYSTEM
    RARELY CRY EASILY; OCCASIONALLY INDECISIVE
        NO ATAXIA; NO TREMOR; NO PARALYSIS; NO BLACKOUTS; NO CONVULSIONS
        NEVER HAD NERVOUS BREAKDOWN; NO INSOMNIA; NEVER AWAKE AT NIGHT
        NO NIGHTSWEATS; NO NIGHTMARES; NEVER ARISE EARLY,CANNOT RETURN
        TO SLEEP; NEVER CONSULTED COUNSELOR; NEVER DEPRESSED; RARELY
        TAKEN ADVANTAGE OF; RARELY IRRITABLE; NEVER HAVE AUDITORY HALLUCINATIONS
        NOT FORGETFUL; NOT EMOTIONALLY UNSTABLE; NEVER SUICIDAL; NEVER
        DESPONDENT

PAST MEDICAL HISTORY
    SURGICAL OPERATIONS APPENDECTOMY OR TONSILLECTOMY
        NO SEVERE INJURY; NO PROBLEMS FROM INJURY; NO BLOOD TRANSFUSIONS
        NO HOSPITAL ADMISSION NOT FOR SURGERY; NO REJECTED BY MILITARY
        NO REFUSED LIFE INSURANCE

HISTORY OF DISEASE
    TONSILLECTOMY; GERMAN MEASLES; MEASLES; CHICKEN POX;MUMPS; NORMAL
    CHEST X-RAYS; NORMAL BONE X-RAYS
    TUMOR OR CANCER: YES:DIAGNOSED AGE 24
        NEVER STOMACH ULCER; NEVER ULCER OTHER THAN STOMACH; NEVER HAD
        LIVER OR GALLBLADDER DISEASE; NEVER GALLSTONES; NO KIDNEY
        OR BLADDER DISEASE; NEVER KIDNEY STONES; NEVER HERNIA; NEVER
        HEMMORHOIDS; NEVER APPENDECTOMY; NEVER RHEUMATIC FEVER; NEVER
        POLIO; NEVER WHOOPING COUGH; NEVER DIPTHERIA; NEVER TYPHOID
        NEVER MALARIA; NEVER ARTHRITIS; NEVER GLAUCOMA; NEVER CATARACTS
        NO TROUBLE DISTINGUISHING COLOR; NO HAY FEVER; NO ASTHMA; NO
        SYPHILIS; NO HAD VENEREAL DISEASE; NO MOUTH SORES; NO TUBERCULOSIS
        NO THYROID DISEASE OR GOITER; NEVER DIABETES; NO STROKE; NO
        STOMACH X-RAYS; NO GALL BLADDER X-RAYS; NO KIDNEY X-RAYS; NO
        FLUOROSCOPY; NO ULTRAVIOLET LIGHT; NO RADIOTHERAPY; NO CHEMOTHERAPY
    OTHER MEDICATIONS:

MEDICATIONS
        NO ASPIRIN; NO ANTACIDS; NO LAXATIVES; PRESCRIPTION PAINKILLERS
        PREVIOUSLY TAKEN; NO PRESCRIPTION SLEEPING PILLS; NO NONPRESCRIPTION
        SLEEPING PILLS; NO PRESCRIPTION DIET PILLS; NO NONPRESCRIPTION
        DIET PILLS; NO PRESCRIPTION TRANQUILIZERS; NO NONPRESCRIPTION
        TRANQUILIZERS; NO PRESCRIPTION ANTIHISTAMINES; NO NONPRESCRIPTION
        ANTIHISTAMINES; NO PRESCRIPTION VITAMINS; NONPRESCRIPTION VITAMINS
        PREVIOUSLY TAKEN; NO CORTISONES; NO THYROID MEDICATION; NO FEMALE
        REPLACEMENT HORMONES; NO MEDICINE TO PREVENT SEIZURES; NO HYPERTENSION
        MEDICATION; NO CARDIAC MEDICATION; NO ANTICOAGULANTS; NO DIURETICS
        NO ARTHRITIS MEDICATIONS; NO B-12 INJECTIONS

Figure 20.2.   A portion of a computer-generated medical history summary.

histories. These histories save the physician time, are more complete than those taken by the physician, and are neatly typed and easy to read. Also, the patient and physician are not embarrassed by delicate questions (e.g., Have you had a venereal disease?), the patient is not rushed and is more apt to ask for an explanation of a medical term that he or she does not understand, and the computer does not accidentally skip questions. Studies have shown that medical histories obtained by computer are accurate and that patients enjoy the experience. Given a choice, many patients would rather have their medical history taken by a computer than by a physician.

20.2. LABORATORY AUTOMATION

Over the past ten years there has been a tremendous increase in the number of laboratory tests available to physicians. Recently developed automated devices such as the Technicon SMA/12 that makes 12 separate tests on one blood sample and the Coulter Counter that counts the number of cells per unit volume of blood (see Chapter 8, p. 157) have made it easy to perform numerous chemistry and hematology tests on a single blood sample. While these tests are done rapidly and automatically with a minimum of human intervention, the resulting mass of data can literally overwhelm the laboratory staff. The laboratory for a 1000-bed general hospital handles 10,000 to 20,000 individual test results each day.

The computer has been introduced into the clinical laboratory to collect these data, collate them, store them, and report them to the physician accurately and rapidly. The laboratory results are often fed directly to the computer from the analyzing machine. A clinical laboratory computer is shown in Fig. 20.3. Figure 20.4 shows a typical computer-generated lab report. Often a small dedicated laboratory computer such as that made by

Figure 20.3. A clinical laboratory data processing system—the Technicon SMAC High-Speed Computer-Controlled Biochemical Analyzer. The automated analyzers are on the left, and the computer is on the right. (Reproduced by courtesy of Technicon Instruments Corporation.)

UNIV OF WIS HOSPITAL
NAME: ▮
H# : 714933
R# : 266163

CHART COPY   CUMULATIVE REPORT   PG 3
N: 5B   ROOM: 536
DR: PULM   AGE: 53YR   4MO   1104PM 5/26/1975

				B L O O D G A S E S				
TEST:	PO2	PH	PCO2	BASE EX	BUF.BAS	STDHCO3	AC.HCO3	TOT.CO2	O2 SAT
UNITS:	MM HG		MM HG	MEQ/L	MEQ/L	MEQ/L	MEQ/L	MEQ/L	%
LO-HI:	80-90	7.36-7.44	34-46	-2.5-2.5	43-49	22-26	22-26	23-27	
5/02 2:36P	46.0* PLEA	7.50*	28.0*	PLUS .1	48.0	24.0	21.1*	21.0*	84.0
5/03 11:46A	52.0*	7.48*	30.0*	MNUS .0	47.7	23.9	21.6*	22.5*	87.5
5/05 12:20P	59.0*	7.52*	29.0*	PLUS 1.9	49.6*	25.5	22.9	23.8	91.5
5/07 10:52A	61.0*	7.50*	34.0	PLUS 3.7*	51.2*	27.0*	25.6	26.6	92.5
5/08 11:14P	78.0*	7.48*	28.0*	MNUS 1.4	46.0	22.9	20.2*	21.6*	96.5

Figure 20.4. A portion of a cumulative laboratory report on blood gases generated by a clinical laboratory data processing system. The date and time of each measurement are given in the left column.

the Digital Equipment Corporation, LABCOM, SPEAR-LABSYS, or DNA-AUTOLAB is used. In other cases the laboratory is connected to a time-shared computer system such as MEDITEC or IBM-SYSTEM 7.

Studies have shown that the direct reading of laboratory equipment by computer can cut laboratory errors by as much as 18%. Automated sorting and generating of laboratory data by computer have significantly reduced the time needed to send a test result to the physician. In addition, the computer can check for equipment malfunction and can test for calibration errors with quality control programs.

20.3. ELECTROCARDIOGRAM INTERPRETATION

The electrocardiogram (ECG) is a graphic recording of the electrical potentials produced in association with the automatic rhythmic contractions of the heart (see Chapter 9, p. 196). The ECG is recorded by attaching electrodes connected to an ECG machine to various locations on the body.

The ECG gives information on structural and electrical changes in the heart. By interpreting the ECG a cardiologist can detect heart damage (e.g., as caused by a heart attack), electrical conduction defects, and abnormal impulse generation (such as an irregular heartbeat, or *arrhythmia*). The interpretation of the ECG is based on definite rules, and in most cases trained cardiologists given the same ECG would make the same interpretation. Unfortunately, not all physicians learn or remember these numerous and complicated rules, and often ECGs must be read in emergency situations when no cardiologists are available. Also, it has been recommended that an ECG should be a routine part of the yearly physical examination of all patients over 30 years of age, and there are simply not enough cardiologists available to interpret the millions of ECGs that would result.

In an attempt to solve these problems, researchers have developed numerous computer programs to automatically interpret ECGs. The best known are those developed by the Mayo Clinic, Mount Sinai Hospital in New York, the United States Public Health Service, and the Veterans Administration. These programs take the analog signals directly from the ECG machine, digitize the signals, and then using programmed rules interpret the signals to detect abnormalities. These programs measure many parameters on the ECG waveform—peak-to-peak time interval, wavelength, wave slope, and so forth (Fig. 20.5). Electrocardiogram interpreting programs are now in routine use in many hospitals across the country. In most cases the ECGs are interpreted by the computer as soon

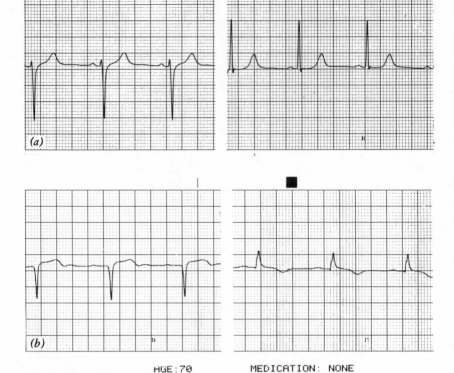

```
                        HGE:70           MEDICATION: NONE
VENTRICULAR RATE: 064/MINUTE    PR INTERVAL=.22    QRS INTERVAL=.09

CARDIAC POSITION:
 -INDETERMINATE

RHYTHM:
 -NORMAL SINUS RHYTHM
  WITH A REGULAR RR INTERVAL

IMPRESSION:
 -SEPTAL WALL INFARCTION
  (EXACT RECENCY CANNOT BE DETERMINED)
 -TRANSMURAL POSTERIOR WALL INFARCTION
  (EXACT RECENCY CANNOT BE DETERMINED)
 -ANTERIOR WALL TRANSMURAL ISCHEMIA

REVIEWER'S COMMENTS:



UW MEDICAL CENTER ECG COMPUTER PROGRAM      03/20/75
```

Figure 20.5. Portions of ECGs from (*a*) a patient with a healthy heart and (*b*) a patient with heart disease. The computer analysis of *b* identifies the abnormalities. (Courtesy of Dr. Wasserberger and Mr. W. Mueller, University of Wisconsin, Madison, Wisc.)

as they are completed and the interpretations are available within a few minutes. A cardiologist reviews these interpretations within 24 hr and then submits a final report.

It has been found that the computer programs are accurate in identifying normal ECGs and ECGs that show structural abnormalities such as those due to heart attack. However, the programs make occasional errors in interpreting abnormal rhythm patterns. As experience grows, these deficiencies are being improved.

20.4. PATIENT MONITORING

It is often necessary to monitor the arterial and venous blood pressures, pulse, temperature, breathing rate, ECG signals, cardiac output, and other vital functions of critically ill patients on a second-to-second basis. In an intensive care unit nurses and physicians must know immediately of any change in a patient's condition. Unfortunately, there are not sufficient trained personnel available to watch the monitor screens constantly, and even if there were, the task is sufficiently boring that fatigue would quickly decrease the ability of observers to note changes.

Researchers have, therefore, utilized computers for patient monitoring. At many medical centers around the country, monitoring signals from critically ill patients are sent directly to small dedicated computers that are programmed to detect changes in any of the patients' vital functions and to notify the personnel of these changes. In some centers the computers have been programmed to give life-saving medications by means of servo syringes in the important first seconds of a crisis while the medical personnel are still being mobilized.

At present, the main problem with computer monitoring systems involves the data fed to the computer. Electrical noise picked up by the monitoring leads can indicate to the computer that the patient's condition has changed drastically. This noise can be generated by nearby electrical equipment or by motion of the patient. While the computers can be programmed to ignore noise pulses of short duration, erroneous, or false positive, signals are still a serious nuisance. If they occur often, the nurse may start to ignore them and not respond to a true positive signal. The little boy who cried wolf has a modern counterpart in patient-monitoring equipment.

20.5. DRUG-TEST INTERACTIONS

A by-product of the recent revolution in drugs is the rapid growth in the number of side effects and complications due to the drugs. When two different drugs are given to the same patient, they may combine to produce undesirable effects (drug-drug interactions). Also, certain drugs may cause serious errors in laboratory test results (drug-test interactions). Each year hundreds of new interactions are discovered; by 1976, well over 30,000 such interactions were reported. It is clear that physicians cannot be expected to remember this volume of data.

The computer, with its ability to store and recall rapidly huge quantities of data, has been called upon to keep track of these interactions for the physician. A group at the National Institutes of Health headed by Dr. Donald Young and one at Stanford University headed by Dr. Stanley N. Cohen have developed computer files listing all such known interactions and have utilized them in actual clinical case situations. At the University of Wisconsin Hospitals Dr. Young's listing of over 20,000 drug-test interactions has been placed in a computer file (Fig. 20.6). Each day a list of all medicines each patient is taking is entered into the computer along with a list of any abnormal laboratory test results. The computer then checks the master file to determine whether any of the patient's medications could have caused the abnormal test results. If such an interaction is discovered, a note for the physician is automatically printed describing the interaction, listing its probable mechanism, and giving the pertinent literature reference.

20.6. PRESCRIBING DRUG DOSAGE

More and more is being learned about how drugs interact with other drugs, how drugs break down in the body, and how some drugs may have multiple effects, some beneficial and some deleterious. This information explosion has made it difficult for the physician to maintain even a cursory knowledge of the ever-expanding number of drugs at his disposal.

To assist physicians in intelligently prescribing some of the most commonly used medications, researchers have developed computer programs for drug dosage. Two of the most widely used programs are the program developed by Dr. Lewis B. Sheiner at the National Institutes of Health for prescribing coumarin, a blood thinner, and the program developed by Dr. Roger W. Jelliffe of the University of California at Los Angeles for prescribing digitalis, a heart stimulant. A physician using Dr. Jelliffe's program enters the blood level of digitalis he wishes to achieve along with the dosages of any previously given drugs and the results of tests measuring the patient's kidney function. The program then computes how much digitalis the physician should have the patient take each day for the next month so that the desired level will be achieved.

Drug	Lab Test		Interaction	Reference	
MEPERIDINE	S	SGPT INC	P	MAY CAUSE RISE IN INTRABILIARY PRESSURE	R13	107
MEPERIDINE	S	SGOT INC	P	MAY CAUSE RISE IN INTRABILIARY PRESSURE	R1610	108 *
SODIUM SALTS	S	CALCIUM INC	P		R1096	102 *
CHLOROTHIAZIDE	S	URIC ACID DEC	P	IF GIVEN I.V. IN LARGE DOSES HAS URICOSURIC EFFECT	R347	100
CHLOROTHIAZIDE	S	URIC ACID INC	P	DECREASED URATE CLEARANCE	R13	100
CHLOROTHIAZIDE	S	CALCIUM INC	P	IMPAIRED EXCRETION	R13	102
CHLOROTHIAZIDE	S	ALKALINE PHOSPHATASE INC	P	MAY CAUSE CHOLESTATIC JAUNDICE	R1696	104
CHLOROTHIAZIDE	S	SGPT INC	P	MAY CAUSE CHOLESTATIC JAUNDICE	R1696	107
CHLOROTHIAZIDE	S	SGOT INC	P	MAY CAUSE CHOLESTATIC JAUNDICE	R1096	108
CHLOROTHIAZIDE	S	GLUCOSE TOLERANCE DEC	P	DIABETOGENIC-LIKE ACTION OF DRUG	R1140	122
CHLOROTHIAZIDE	S	GLUCOSE INC	P	DIABETOGENIC PROPERTIES OF DRUG AFFECT GTT	R13	122
CHLOROTHIAZIDE	S	POTASSIUM DEC	P	DIURETIC ACTION	R13	124
MAGNESIUM SALTS	S	CALCIUM DEC	P	`	R235	102
ALLOPURINOL	S	URIC ACID DEC	P	INHIBITION OF XANTHINE OXIDASE	R3	100
ALLOPURINOL	S	ALKALINE PHOSPHATASE INC	P	REVERSIBLE CLINICAL HEPATOTOXICITY NOTED	R33,R3	104
ALLOPURINOL	S	SGPT INC	P	REVERSIBLE CLINICAL HEPATOTOXICITY REPORTED	R441	107
ALLOPURINOL	S	SGOT INC	P	REVERSIBLE CLINICAL HEPATOTOXICITY REPORTED	R441	108
ALLOPURINOL	S	UREA NITROGEN INC	P	AZOTEMIA AS SENSITIVITY REACTION IN 2 CASES	R1625	121
ALLOPURINOL	S	GLUCOSE DEC	P	HEPATOTOXIC EFFECT	R441	122
MORPHINE	S	ALKALINE PHOSPHATASE INC	P	ASSOCIATED WITH ABNORMAL LIVER FUNCTION	R1696	104 *
MORPHINE	S	SGPT INC	P	MAY CAUSE RISE IN INTRABILIARY PRESSURE	R13	107
MORPHINE	S	SGOT INC	P	MAY CAUSE RISE IN INTRABILIARY PRESSURE	R13	108
MORPHINE	S	GLUCOSE INC	P	MINOR, CLINICALLY INSIGNIFICANT INCREASE	R46	122
BISACODYL	S	POTASSIUM DEC	P	ASSOC WITH STEATORRHEA IF USED IN EXCESS	R1446	124 *
MAGNESIUM SALTS	S	CALCIUM DEC	P		R235	102

Figure 20.6. A small portion of a list of drug-test interactions.

20.7. PULMONARY FUNCTION TESTING

In order to determine the condition of a patient's lungs, a physician performs pulmonary function tests. These tests include determining the patient's abilities to take a deep breath, to breathe in and out rapidly over a timed period, and to breathe out against resistance. The measurements are made with a device called a spirometer that measures the air inspired and expired through a large drum and plots this airflow against time on a sheet of paper (see Chapter 7, p. 131). Using a series of formulas, the physician determines values for the pulmonary function tests from these plots and compares them with results for healthy persons of the same age and body size.

During the past few years, a number of researchers have developed systems that utilize computers for pulmonary function testing. In such a system, the spirometer is attached directly to a small computer. The analog results are converted to numbers by an ADC and these numbers are then compared to computer-stored normal data. The computer then generates a complete printed pulmonary status report.

The use of the computer in pulmonary function testing frees the physician from calculating the test results. It also decreases observer error.

20.8. MEDICAL RECORD SYSTEMS

Traditional handwritten hospital charts are often unmanageable, poorly organized, illegible, and difficult to obtain in an emergency. Also, when the increasingly large amounts of computer-generated data are added to the already large amounts of data on the traditional hand-written chart, the physician is faced with an overwhelming amount of information on each patient.

In order to allow the physician to efficiently use patient records, researchers have been working on computerized patient-record systems in which most information is entered directly into computers where it is stored in an easy-to-read form and available for rapid retrieval. Pioneering efforts have been undertaken by Dr. Lawrence Weed at the University of Vermont (the PROMIS system), by the Technicon Corporation at El Camino Hospital in California (the MIS system), and by the Harvard University Prepayment Medical Clinic (MUMPS). In each case, the patient's record is kept in the computer and notes by physicians, nurses, and laboratory personnel are entered directly into the computer via terminals on the hospital floor (Fig. 20.7). When doctors order tests at the terminals, requisitions are automatically printed at the respective laboratories, the tests are scheduled, and the necessary personnel are notified. When the

Figure 20.7. **A computerized patient record system permits a physician (in this case, Dr. Richard Friedman) to enter or retrieve data on a particular patient from the nursing station. The computer terminal is attached to the computer via a conventional telephone line.**

tests are completed, the results are automatically sent back to the terminal on the floor and a permanent entry is made in the patient's computer file.

Preliminary studies show that a computerized medical-record system can more than pay for itself in reduced personnel and clerical costs. In addition, it provides complete, legible, and accessible records.

20.9. HOSPITAL BOOKKEEPING

Over the past 10 years the number and cost of hospital services have risen dramatically, resulting in increasingly complex hospital bookkeeping operations. Hospitals are now multimillion dollar operations with large payrolls, huge operating budgets, and extensive patient billing. Many patients are covered by third party insurance carriers (Blue Cross, Blue Shield, Medicare, Medicaid, etc.), and hospitals are required to send detailed itemized accounts to these agents before bills can be paid. Often both direct costs (e.g., lab tests, consultation fees, and medication) and indirect costs (e.g., administrative costs, building maintenance, and security) must be carefully detailed. It is apparent that manual bookkeeping techniques are no longer sufficient. In many cases bookkeeping operations are totally dependent upon computers.

"Standard" bookkeeping operations are only one part of the job hospitals must perform. Recent federal and state legislation has put new constraints on hospital management. These regulations now require that hospitals supply detailed justifications for new equipment purchases, plant expansion, length of patient hospitalization, and test procedures or therapy given for specific disorders. The hospital administrators and health care providers must have access to a substantial amount of "management information" to justify their actions. The computer-based administrative bookkeeping systems make available on a daily basis data on hospital bed occupancy, patient diagnosis, patient medication, patient diet, and so forth. There has been a rich spinoff of computer-generated census sheets, diet sheets, pharmacy inventory forms, patient diagnostic profiles, and so forth from the "bookkeeping" operation.

Computer-based bookkeeping systems are used by individual physicians as well as large hospitals. Physicians are similarly confronted with the need to complete detailed itemized reimbursement forms for third party insurance payment of patients' bills. Many have turned to service bureaus that supply computerized bookkeeping services for individual physicians. In addition, many larger clinics and group practices have developed their own data-processing systems.

The most extensive use of computers in medicine has been in bookkeeping. As the cost and complexity of medical care continue to increase, the need for computer-based bookkeeping systems will grow.

20.10.   OTHER USES OF COMPUTERS IN MEDICINE

This chapter does not cover all the uses of computers in medicine. Some of the other uses are mentioned elsewhere in this book. In Chapter 16 we discuss how computers are used in diagnostic radiology to help diagnose brain tumors. The uses of computers in nuclear medicine are discussed in Chapter 17. Chapter 18 illustrates how computers are used to calculate the radiation distribution in the body when radiation is used to treat cancer. In addition, new uses for computers in medicine are being explored.

Many currently used systems could be improved by computerizing them. For example, in Toronto, Canada, patients with cardiac pacemakers are able to phone the Toronto General Hospital and have their ECG recorded. The ECG is transmitted as sound by a small box held near the telephone mouthpiece. An electrode is held under each armpit, and the sweat from the armpits gives good electrical contact. The accumulated ECG signals are reviewed every hour or two by a cardiologist, who calls the patient if further studies are necessary. A computer could be programmed to give the patient reassurance if the ECG is normal or to put a physician on the line if it is not.

The computer could help you decide whether you should see a physician—perhaps the most important decision a typical patient must make about his health. There may be a time in the near future when you have a bothersome symptom and call your local medical center computer to get advice. The computer could be programmed to ask you your medical history. (Computers are already able to make intelligible speech.) You would respond by pushing the buttons on your phone—1 for *yes*, 2 for *no*, 3 for *I don't know*, 4 for *I don't understand*, 5 for *I refuse to answer*, and so forth. With simple attachments the computer could "listen" to your heart and breathing rates.

BIBLIOGRAPHY

Barnett, G. O., "Computers in Patient Care," *New Engl. J. Med.*, **279**, 1321–1327 (1968).
Davis, L., M. Collen, L. Rubin, and E. Van Brunt, "Computer Stored Medical Record," *Computers Biomed. Res.*, **1**, 452–469 (1968).
Enlander, D., *Computers in Laboratory Medicine*, Academic, New York, 1975.
Rose, J., and J. H. Mitchell (Eds.), *Advances in Medical Computing*, Churchill Livingstone, Edinburgh, Scotland, 1975.
Slack, W. V., G. P. Hicks, C. E. Reed, and L. Van Cura, "A Computer-Based Medical History System," *New Engl. J. Med.*, **274**, 194–198 (1966).
Stacy, R. W., and B. D. Waxman (Eds.), *Computers in Biomedical Research*, Academic, New York, 1974.
Young, D. S., R. B. Friedman, and L. C. Pestaner, "Automatic Monitoring of Drug-Laboratory Test Interactions," in P. L. Morselli, S. Garattini, and S. N. Cohen (Eds.), *Drug Interactions*, Raven, New York, 1974.

REVIEW QUESTIONS

1. Define or describe the following computing terms:

a. software	e. output
b. hardware	f. dedicated computer
c. programs	g. distributed intelligence
d. debugging	h. ADC

2. What are some advantages of using a computer for obtaining a patient's medical history?
3. How many individual laboratory tests are done daily in a 1000-bed general hospital?
4. What is the main problem with monitoring patients in intensive care units with computers?
5. How many known drug-test interactions are there?
6. How can a computer assist in determining the correct dosage of digitalis for a heart patient?
7. List six common uses of computers in a modern hospital.

APPENDIX A

Exponentials and Logarithms

Mathematics is an essential tool of science, and each scientific field uses mathematical tools that are adapted to its special needs. In medical physics we seldom use differential or integral calculus, although the concepts are useful in many situations. In this appendix we review exponentials and logarithms, which do occur frequently in medical physics.

A.1. EXPONENTIAL BEHAVIOR

Exponential behavior occurs when the rate of change of a quantity depends on the amount of the quantity present. If you are ever asked how something behaves in the body and you have no idea of the correct answer, a good guess would be "exponentially." You stand a good chance of being right!

A few examples of exponential behavior in medicine are the following:

1. Radioactivity decays exponentially with time.
2. Growth of bacteria increases exponentially with time.
3. X-rays and gamma rays are absorbed exponentially with distance in the body.
4. Ultrasound is absorbed exponentially with distance in the body.
5. Many substances are removed exponentially with time from the body by the kidneys.
6. Light is absorbed exponentially with distance in the body.

The mathematical equation for exponential behavior is $Y = Y_0 e^{-kx}$, where $e = 2.718$; Y_0 is the initial value of Y when $x = 0$, where x represents

569

distance or time; and k is a constant that is positive for all the examples we consider. This is applicable when substances are decreasing in an exponential manner. A graph of this equation for $Y_0 = 100$ and $k = 1$ is shown in Fig. A.1a on conventional linear graph paper. The same graph is shown on semilog graph paper in Fig. A.1b. The straight line on the semilog graph is characteristic of an exponential function.

The two general forms of the exponential equation that we use are the exponential decrease of some substance with time and the exponential decrease of some substance with distance. They are summarized in Table A.1.

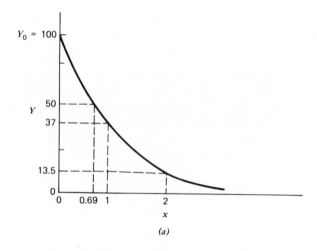

(a)

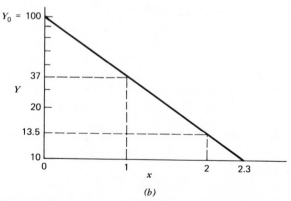

(b)

Figure A.1.   Graphs of the exponential function $Y = 100\ e^{-x}$ (a) on conventional linear graph paper and (b) on semilog graph paper. Note that the semilog graph results in a straight line.

Table A.1. Summary of Exponential Equations Used in This Book

Equation	Constant	Independent Variable	Half-Life or Half-Thickness
$Y = Y_0 e^{-\lambda t}$	λ	t(time)	$t_{\frac{1}{2}} = 0.693/\lambda$
$Y = Y_0 e^{-kx}$	k	x(distance)	$x_{\frac{1}{2}} = 0.693/k$

As an example of exponential behavior with time, consider a radioactive material that is decaying, that is, changing into another element and simultaneously emitting radiation. In this case the amount of radioactivity A is directly related to the amount of the radioactive material. This can be written mathematically as $A = \lambda N$, where A is the activity in disintegrations per second, N is the number of radioactive atoms, and λ is the decay or disintegration constant for the material. If you had 10^6 atoms and $\lambda = 0.01$ sec^{-1}, the activity would be 10^4 disintegrations/sec. Since N is decreasing each second, the activity also decreases with time. After about 69 sec N would be half of its original amount and the activity would also be reduced by half. Thus 69 sec is this radioactive material's half-life. If you continued to measure the activity you would find that after another 69 sec it would again be reduced by half. In fact, it would make no difference when you started measuring, after any 69 sec you would find that half of the atoms had decayed.

The equation $A = \lambda N$ is useful if you know two of the three quantities. A more convenient form for many purposes can be obtained by some mathematical manipulations to relate the activity to the time. This form is $A = A_0 e^{-\lambda t}$, where A_0 is the radioactivity at some arbitrary time $t = 0$. Since the activity is directly proportional to N, this equation can also be written as $N = N_0 e^{-\lambda t}$. However, this last equation is not very useful since N is usually more difficult to measure than A.

A.2. LOGARITHMS

The semilog plot of Fig. A.1b is the logarithm (log) form of the exponential equation. If we take the log of both sides of $Y = Y_0 e^{-kx}$, we get $\log_e Y_0/Y = kx$, where $\log_e$ is the log to the base e, which is also called the natural log, often written as ln. The semilog plot is useful not only because it is linear for exponential behavior, but also because it can be used to plot quantities that cover several orders of magnitude. Quite often descriptions of signal intensity, such as of sound or light, range from 10 to 15 orders of magnitude (i.e., by a factor of 10^{10} to 10^{15}). We usually describe such a signal intensity Y in terms of powers of 10 such as $Y = 10^x$, where x is an exponent (or power) and 10 is the base. Many times we write only the exponent, which is the logarithm of the number Y to the base 10, that is, the power x to which 10 must be raised to give the number Y. This is written as $\log Y = x$. If *log* is used without a subscript you can assume that the base is 10. While it is possible to use any number as the base, for practical purposes the only base used other than 10 is $e = 2.718$. The two types of logarithms are simply related by $\ln Y = 2.3 \log Y$.

We make use of $\log Y$ in this book on a number of occasions, such as in the description of the sensitivity of the ear (Chapter 13). The ear has a maximum sensitivity of about 10^{-16} W/cm² and an upper limit near 10^{-4} W/cm². Since the sensitivity ranges over 12 orders of magnitude, it is usually most practical to plot this sensitivity in terms of $\log R$, where R is a ratio between two sound intensities. We use this ratio to define the bel:

$$\text{bel} = \log\left(\frac{\text{intensity}_1}{\text{intensity}_2}\right)$$

The term intensity$_2$ is often a reference level. Usually a smaller unit, one-tenth of a bel, or a decibel (dB), is used; it is defined

$$dB = 10 \log\left(\frac{\text{intensity}_1}{\text{intensity}_2}\right) = 20 \log\left(\frac{\text{pressure}_1}{\text{pressure}_2}\right)$$

where intensity is proportional to the pressure squared (see Chapter 12, p. 255). If we let intensity$_2$ correspond to the threshold of hearing (10^{-16} W/cm²), then a sound 30 dB (3 bels) above the threshold means that it has an intensity of 10^{-13} W/cm² or that

$$10 \log\left(\frac{\text{intensity}_1}{\text{intensity}_2}\right) = 10 \log\left(\frac{10^{-13}}{10^{-16}}\right) = 10 \log 10^3 = 30 \text{ dB}$$

APPENDIX B

Units

When we wish to speak of the physical quantity involved in a law or principle, we must know how to measure the quantity. There is some arbitrariness in the units used in measurements, although scientists do try to limit this arbitrariness. Scientists consider some physical quantities, which are measured in terms of fundamental (internationally accepted) units, as fundamental quantities; all other quantities are expressed in terms of the fundamental quantities and are called derived quantities.

In the branch of physics known as mechanics, the fundamental quantities are length, mass, and time and the fundamental units are the meter, kilogram, and second. All other physical quantities involved in mechanics can be expressed in terms of these fundamental units; for example, force can be expressed as kilogram meters per second per second. For the other branches of physics we need only use three more fundamental quantities and units; they are temperature (kelvin), electric current (ampere), and luminous intensity (candela).

All other physical quantities and units can be expressed in terms of these six units, which were adopted in 1954 and 1960 and have the formal name of International System, or SI, units (Table B.1). Some derived quantities and units are given in Table B.2.

Unfortunately, the SI system has not been accepted throughout the world (notably in the United States and Canada). In addition, the various branches of medicine use some unique units. In this book we use units other than those derived from SI units when they are in common use in medicine. Some of these non-SI units are given in Table B.3.

Table B.1. International System (SI) Units

Quantity	Unit	Abbreviation
Length	meter	m
Mass	kilogram	kg
Time	second	sec
Current	ampere	A
Temperature	kelvin	K
Luminous intensity	candela	cd

Table B.2. Derived Units

Quantity	Unit	Abbreviation	Dimensions
Force	newton	N	$kg\ m/sec^2$
Pressure	pascal	Pa, N/m^2	$kg/m\ sec^2$
Energy	joule	J, Nm	$kg\ m^2/sec^2$
Power	watt	W, J/sec	$kg\ m^2/sec^3$
Torque	meter-newton	τ, mN	$kg\ m^2/sec^2$
Electric charge	coulomb	C	A sec
Electric potential	volt	V, J/C	$kg\ m^2/sec^3\ A$
Electrical resistance	ohm	Ω, V/A	$kg\ m^2/sec^3\ A^2$
Capacitance	farad	F, C/V, C^2/J	$sec^4\ A^2/kg\ m^2$
Inductance	henry	H, J/A^2, Ωsec	$kg\ m^2/sec^2\ A^2$
Magnetic flux	weber	Wb, J/A, Vsec	$kg\ m^2/sec^2\ A$
Magnetic intensity	tesla	T, Wb/m^2, $Vsec/m^2$	$kg/sec^2\ A$
Frequency	hertz	Hz	sec^{-1}
Disintegration rate	becquerel	Bq	sec^{-1}
Absorbed dose	gray	Gy, J/kg	m^2/sec^2

Table B.3. Non-SI Units

Quantity	Unit	Abbreviation
Mass	gram	g
Length	foot	ft
	centimeter	cm
Volume	liter	—
Time	minute	min
Force	dyne	—
	$pound_{force}$	lb_f
Energy	calorie	cal
	kilocalorie	kcal
Power	kilocalories/minute	kcal/min
Pressure	pounds/inch^2	psi
	millimeter of mercury	mm Hg
	centimeter of water	cm H_2O
	atmosphere	atm
Temperature	fahrenheit	F
	celsius	C

APPENDIX C

Standard Man Data

In medical physics, where we are often concerned with the anatomy and physiology of humans, it is convenient to define a *standard man*. While the standard man is nonexistent, the following somewhat arbitrary values are useful for computational purposes:

Age	30 yr
Height	172 cm (5 ft 8 in.)
Mass	70 kg
Weight	690 N (154 lb)
Surface area	1.85 m²
Body core temperature	37.0°C
Body skin temperature	34.0°C
Heat capacity	0.86 kcal/kg°C
Basal metabolism	38 kcal/m² hr, 70 kcal/hr, 1680 kcal/day
O_2 consumption	260 ml/min
CO_2 production	208 ml/min
Blood volume	5.2 liters
Cardiac output	5 liters/min
Blood pressure	120/80 mm Hg
Heart rate	70 beats/min
Total lung capacity	6 liters
Vital capacity	4.8 liters
Tidal volume	0.5 liter
Dead space	0.15 liter
Breathing rate	15/min
Muscle mass	30,000 g 43% of body mass
Fat mass	10,000 g 14% of body mass

Bone mass	7,000 g	10% of body mass
Blood mass	5,400 g	7.7% of body mass
Liver mass	1,700 g	2.4% of body mass
Brain mass	1,500 g	2.1% of body mass
Mass of both lungs	1,000 g	1.4% of body mass
Heart mass	300 g	0.43% of body mass
Mass of each kidney	150 g	0.21% of body mass
Thyroid mass	20 g	0.029% of body mass
Mass of each eye	15 g	0.021% of body mass

General Bibliography

Ackerman, E., *Biophysical Science,* Prentice-Hall, Englewood Cliffs, N.J., 1962.

Aird, E. G. A., *An Introduction to Medical Physics,* Heinemann, London, 1975.

Benedek, G. B., and F. M. H. Villars, *Physics with Illustrative Examples from Medicine and Biology, Vol. 1: Mechanics,* Addison-Wesley, Reading, Mass., 1973 (See Villars, F. M. H., for Vols. 2 and 3).

Breuer, H., *Physics for Life Science Students,* Prentice-Hall, Englewood Cliffs, N.J., 1975.

Casey, E. J., *Biophysics: Concepts and Mechanisms,* Reinhold, New York, 1962.

Clynes, M., and J. H. Milsum (Eds.), *Biomedical Engineering Systems,* McGraw-Hill, New York, 1970.

Cromer, A. H., *Physics for the Life Sciences,* McGraw-Hill, New York, 1974.

Cromwell, L., F. J. Weibell, E. A. Pfeiffer, and L. B. Usselman, *Biomedical Instrumentation and Measurements,* Prentice-Hall, Englewood Cliffs, N.J., 1973.

Davidovits, P., *Physics in Biology and Medicine,* Prentice-Hall, Englewood Cliffs, N.J., 1975.

Duncan, G., *Physics for Biologists,* Halsted, New York, 1975.

Easton, D. M., *Mechanisms of Body Functions,* Prentice-Hall, Englewood Cliffs, N.J., 1974.

Flitter, H. H., and H. R. Rowe, *An Introduction to Physics in Nursing,* 5th ed., Mosby, St. Louis, 1967.

Geddes, L. A., and L. E. Baker, *Principles of Applied Biomedical Instrumentation,* Wiley, New York, 1968.

Glasser, O. (Ed.), *Medical Physics,* Vol. 1, Year Book Medical, Chicago, 1944.

Glasser, O. (Ed.), *Medical Physics,* Vol. II, Year Book Medical, Chicago, 1950.

Glasser, O. (Ed.), *Medical Physics,* Vol. III, Year Book Medical, Chicago, 1960.

Greenburg, L. H., *Physics for Biology and Pre-Med Students,* Saunders, Philadelphia, 1975.

Gupta, M. L., *Problem and Source Guide for a Quantitative Course in Biophysics,* University Publications, Blacksburg, Va., 1972.

Guyton, A. C., *Textbook of Medical Physiology,* 4th ed., Saunders, Philadelphia, 1971.

Jacobson, B., and J. G. Webster, *Medicine and Clinical Engineering,* Prentice-Hall, Englewood Cliffs, N.J., 1977.

Jensen, J. T., *Physics for the Health Professions,* 2nd ed., Lippincott, Philadelphia, 1976.

577

Kenedi, R. M. (Ed.), *Perspectives in Biomedical Engineering*, University Park Press, Baltimore, 1973.

Krusen, F. H., F. J. Kottke, and P. M. Ellwood, Jr. (Eds.), *Handbook of Physical Medicine and Rehabilitation*, 2nd ed., Saunders, Philadelphia, 1971.

MacDonald, S. G. G., and D. M. Burns, *Physics for the Life and Health Sciences*, Addison-Wesley, Reading, Mass., 1975.

Mountcastle, V. B., *Medical Physiology*, Vol. 1, 13th ed., Mosby, St. Louis, 1974.

Nave, C. R., and B. C. Nave, *Physics for the Health Sciences*, Saunders, Philadelphia, 1975.

Parker, J. G., and V. R. West (Eds.), *Bioastronautics Data Book*, 2nd ed., National Aeronautics and Space Administration, Washington, D.C., 1973.

Plonsey, R., and D. G. Fleming, *Bioelectric Phenomena*, McGraw-Hill, New York, 1969.

Radiological Health Handbook, compiled and edited by the Bureau of Radiological Health and the Training Institute Environmental Control Administration, revised ed., Public Health Service, Rockville, Md., 1970.

Randall, J. E., *Elements of Biophysics*, Year Book Medical, Chicago, 1958.

Richardson, I. W., and E. B. Neergaard, *Physics for Biology and Medicine*, Wiley, New York, 1972.

Rotblat, J., *Aspects of Medical Physics*, Taylor and Francis, London, 1966.

Ruch, T. C., and H. D. Patton (Eds.), *Physiology and Biophysics*, 19th ed., Saunders, Philadelphia, 1965.

Ruch, T. C., and H. D. Patton (Eds.), *Physiology and Biophysics, Vol. II: Circulation, Respiration and Fluid Balance*, 20th ed., Saunders, Philadelphia, 1973.

Ruch, T. C., and H. D. Patton (Eds.), *Physiology and Biophysics, Vol. III: Digestion, Metabolism, Endocrine Function and Reproduction*, 20th ed., Saunders, Philadelphia, 1973.

Sayers, B. McA., S. A. V. Swanson, and B. W. Watson, *Engineering in Medicine*, Oxford University Press, Oxford, England, 1975.

Schwan, H. P. (Ed.), *Biological Engineering*, McGraw-Hill, New York, 1969.

Stacy, R. W., D. T. Williams, R. E. Worden, and R. O. McMorris, *Essentials of Biological and Medical Physics*, McGraw-Hill, New York, 1955.

Stanley, H. E. (Ed.), *Biomedical Physics and Biomaterial Science*, The MIT Press, Cambridge, Mass., 1972.

Stibitz, R., *Mathematics in Medicine and the Life Sciences*, Year Book Medical, Chicago, 1966.

Strong, O., *Biophysical Measurements*, Tektronix, Beaverton, Ore., 1970.

Strother, G. K., *Physics with Applications in Life Sciences*, Houghton Mifflin, Boston, 1977.

Talbot, S. A., and V. Gessner, *Systems Physiology*, Wiley, New York, 1973.

Thompson, D. W., *On Growth and Form*, Cambridge U. P., London, 1961.

Tilley, D. E., and W. Thumm, *College Physics: A Text with Applications to the Life Sciences*, Cummings, Menlo Park, Calif., 1971.

Villars, F. M. H., and G. B. Benedek, *Physics with Illustrative Examples from Medicine and Biology, Vol. 2: Statistical Physics*, Addison-Wesley, Reading, Mass., 1974 (See Benedek, G. B., for Vol. 1).

Villars, F. M. H., and G. B. Benedek, *Physics with Illustrative Examples from Medicine and Biology, Vol. 3: Electricity and Magnetism*, Reading, Mass., 1977.

Solutions

CHAPTER 1

1.11 (a) $\bar{P} = 124$ mm Hg
(b) $\sigma = 8$ mm Hg

CHAPTER 2

2.2 Sum the torques about the point of application of the force W; since F is equal to the weight on the foot, $M \simeq 3F$. W is about $4F$ or four times the weight on the foot!

2.3 Sum the torques about the elbow.
$-4M \cos \alpha + 14H \cos \alpha + 30W \cos \alpha = 0$
or
$M = 3.5H + 7.5W$

2.5 (a) $(20 \times 10^{-4} \text{ m}^2)(3.1 \times 10^7 \text{ N/m}^2) = 6.2 \times 10^4 \text{ N}$
(b) $(\sim 0.75)(6.2 \times 10^4 \text{ N}) = 4.6 \times 10^4 \text{ N}$

2.6 (a) The sum of the torques about $F = 0$.
$0 = Ml_1 - W(l_1 + l_2)$
$Ml_1 = W(l_1 + 3l_1)$
$M = 4W = 400 \text{ N}$
(b) $\dfrac{100 \text{ N}}{0.5 \times 10^{-4} \text{ m}^2} = 2 \times 10^6 \text{ N/m}^2$

2.7 (a) $3W = 5M$
$M = 0.6W = 24 \text{ N}$
$F = W + M = 64 \text{ N}$

(b)   The stress is $64/5 \times 10^{-4}$ or 1.28×10^5 N/m².

(c)   The mass is 66 kg, and the stress is $\dfrac{(66)(9.8)}{5 \times 10^{-4}} = 1.3 \times 10^6$ N/m²,

which is less than 1% of the maximum compression strength.

2.8 $F = \dfrac{\Delta(mv)}{\Delta t} = \dfrac{(50)(4.5)}{0.2} = 1125$ N

2.9 $F = \dfrac{\Delta(mv)}{\Delta t} = \dfrac{(4)(15)}{0.002} = 3 \times 10^4$ N ($\sim$7000 lb)

2.10 $g_{eff} = 4\pi^2 f^2 r = 4\pi^2 (3000/60)^2 (0.22) \simeq 21{,}700$ m/sec² $\simeq 2200\, g$
where g, the acceleration of gravity, is 9.8 m/sec².

CHAPTER 3

3.12 (a)   N(max) = (4 cm²) (10² mm²/cm²) (120 N/mm²) = 48,000 N ($\sim$5 tons)

(b) $\dfrac{\Delta L}{L} = 0.015$ at fracture

$\Delta L = (0.015)(0.35\ \text{m}) = 0.0052\ \text{m} = 0.52$ cm

(c)   Stress $= \dfrac{F}{A} = \dfrac{10^4\ \text{N}}{4 \times 10^{-4}\ \text{m}^2} = 2.5 \times 10^7$ N/m²

$\Delta L = \dfrac{LF}{AY} = \dfrac{(0.35\ \text{m})\,(10^4\ \text{N})}{(4 \times 10^{-4}\ \text{m}^2)(1.8 \times 10^{10}\ \text{N/m}^2)}$
$= 4.9 \times 10^{-4}$ m = 0.49 mm

CHAPTER 4

4.3   A 100°C change gives a 1.8% change in volume. A 1°C change gives a 1.8×10^{-4} fractional change in volume or a change of mercury volume of $(1.8 \times 10^{-4})(10^{-2}) = 1.8 \times 10^{-6}$ cm³. Thus

$\dfrac{\pi d^2}{4}\,(0.5) = 1.8 \times 10^{-6}$ cm³

$d^2 = \dfrac{14.4}{\pi} \times 10^{-6} = 4.6 \times 10^{-6}$

$d = 2.14 \times 10^{-3}$ cm = 2.14×10^{-2} mm

4.10 (a) 2×10^3 °C/min

(b) $(2 \times 10^3$ °C/min$)(t) = (196 + 37) = 233$°C

$t = \dfrac{2.33 \times 10^2}{2 \times 10^3} = 0.116$ min $\simeq 7$ sec

CHAPTER 5

5.3   (a) 10^4 kcal/day

    (b) $\dfrac{10^4}{5} = 2 \times 10^3$ g/day $= 2$ kg/day at BMR

5.5   (a) 3.8 kcal/min $\times \dfrac{20 \text{ km}}{5 \text{ km/hr}} \times 60$ min/hr

       Energy $= 912$ kcal

    (b) $\dfrac{912 \text{ kcal}}{5 \text{ kcal/g}} = 182$ g (0.4 lb)

5.7   External work $= mgh = 70\,(9.8)(45)$ J $\dfrac{\text{kcal}}{4.2 \times 10^3 \text{ J}} = 7.3$ kcal

     Calories needed $= \dfrac{7.3 \text{ kcal}}{\text{efficiency}} = \dfrac{7.3}{0.15} = 49$ kcal

5.9   (a) $mgh = 70\,(9.8)(10^3) = 6.9 \times 10^5$ J work

    (b) $\dfrac{6.9 \times 10^5}{3(3600)} = 64$ W

    (c)   Energy consumed $= (9.6 \text{ kcal/min})(180 \text{ min})(4.2 \times 10^3 \text{ J/kcal})$
                          $= 7.3 \times 10^6$ J

       $\epsilon = \dfrac{6.9 \times 10^5}{7.3 \times 10^6} = 0.094$ or 9.4%

    (d) $(7.3 \times 10^6) - (6.9 \times 10^5) = 6.6 \times 10^6$ J

5.13   (a)   From Fig. 5.5, $K_c = 27.8$ at 5 m/sec
         $H_c = 27.8\,(1.2)(23) = 767$ kcal/hr

    (b) $23.2\,(1.2)(33 - T_a) = 767$ kcal/hr
         $T_a \simeq 5°C$

5.14 $72 = 16.5\,(1.75)(34 - T_w)$
      $T_w = 34 - 2.5 = 31.5°C$

5.15   (a) $H_r = 5(0.9)\,(2) = 9$ kcal/hr
    (b) $H_c = 26.5\,(0.9)\,(2) \simeq 48$ kcal/hr
    (c) $80 + 30 - 48 - 9 - 10 =$ evaporative loss of 43 kcal/hr

CHAPTER 6

6.2 $\dfrac{20 \text{ cm}}{13.6} = 1.47$ cm Hg $= 14.7$ mm Hg

6.4   (a)   1 atm gauge pressure and 2 atm absolute pressure
    (b)   It will be reduced to 3 liters.
    (c)   The eardrum will rupture.

6.6   Pressure of 1 m of blood $= \dfrac{\rho_{\text{blood}}}{\rho_{\text{Hg}}} \times 10^3$ mm

$$= \dfrac{1.04}{13.6} \times 10^3 = 76.5 \text{ mm Hg}$$

Net pressure $= 76.5 - 2 = 74.5$ mm Hg

6.7   (a)   3 atm gauge and 4 atm absolute
       (b)   4 times the rate at sea level

6.8   Net negative pressure $= 100 \text{ mm Hg} - 370 \left(\dfrac{1.0}{13.6}\right) = 73$ mm Hg

Drainage could be accomplished by gravity (siphoning).

6.9   Weight $= mg = (1.3 \times 10^{-3} \text{ g/cm}^3)(980 \text{ cm/sec}^2) = 1.27$ dynes.
       Over 1 cm², this weight would produce a pressure of 1.27 dynes/cm².

$\dfrac{P_{\text{air}}}{P_{\text{Hg}}} = \dfrac{1.3 \times 10^{-3}}{13.6} = 9.6 \times 10^{-5}$

1 atm = 760 mm Hg

1 cm air $= 9.6 \times 10^{-5}$ cm Hg $= 9.6 \times 10^{-4}$ mm Hg

1 cm air $= \dfrac{9.6 \times 10^{-4}}{760} = 1.3 \times 10^{-6}$ atm

6.10 $P = \rho g h = (1.3 \times 10^{-3} \text{ g/cm}^3)(980 \text{ cm/sec}^2)(30 \times 10^2 \text{ cm}) = 3.8 \times 10^3$
       dynes/cm²
       To determine mm Hg, $P = 3.8 \times 10^3$ dynes/cm² $= \rho g h =$
       $(13.6)(980)(h \text{ cm})$
       $h = 0.29$ cm Hg

CHAPTER 7

7.2   22.4 liters contains 6×10^{23} molecules
       4% of 500 cm³ = 20 cm³ = 0.02 liter

$\left(\dfrac{0.020}{22.4}\right) 6 \times 10^{23} \simeq 5 \times 10^{20}$ O$_2$ molecules

7.3 $\dfrac{2500 \text{ cm}^3}{3 \times 10^8 \text{ alveoli}} = \dfrac{8.3 \times 10^{-6} \text{ cm}^3}{\text{alveoli}}$

7.5   (a)   The negative pressure produced by a maximum inspiratory
             effort and the volume of the snorkel tube, which should be as
             small as practical
       (b)   About 90 cm

7.6 $\Delta t \propto D^2$; therefore, $\Delta t = 2^2 = 4$ times longer

7.7 pO_2 = 20% of 447 mm Hg = 89 mm Hg
pN_2 = 80% of 447 mm Hg = 358 mm Hg

7.8 FRC = residual volume + expiratory reserve volume = 1.0 + 1.5 = 2.5 liters

7.9 40% of 2 liters is 0.8 liter He. Let x liters equal total lung capacity; then 10% of (x + 2) must also be 0.8 liter, that is, 0.1 (x + 2) = 0.8 or x = 6 liters.

7.10 1 m³ = 10^3 liters and 1 cm H_2O = 980 dynes/cm² = 980 × 10^4 dynes/m² = 98 N/m²

Thus $\dfrac{0.2 \text{ liter}}{\text{cm } H_2O} = \dfrac{0.2 \times 10^{-3} \text{ m}^3}{98 \text{ N/m}^2} \simeq 2 \times 10^{-6} \text{ m}^3/\text{N m}^{-2}$

7.11 $R_g = \Delta P/\dot{V}$; $\dot{V} = \Delta P/R_g$; R_g = 3 cm H_2O/(liter/sec)
100 mm Hg = 1360 mm H_2O

$\dot{V} = \dfrac{136}{3} = 45.3$ liters/sec

CHAPTER 8

8.4 (80 cm³/beat)(72 beats/min)(1440 min/day) = 8.3 × 10^6 cm³/day

8.6 (No. of cells/mm³)(vol. of a cell in mm³) = fraction of cells
(8000/mm³) [(4/3) π (6 × 10^{-3})³ mm³] $\simeq$ 7 × 10^{-3} = 0.7%

8.7 (3 × 10^5/mm³)[(4/3) π (10^{-3})³ mm³] $\simeq$ 10^{-3} = 0.1%

8.8 (2500 kcal)(4.2 × 10^3 J/kcal) $\simeq$ 10^7 J/day
10 W = 10 J/sec
(10 J/sec)(3600 sec/hr)(24 hr/day) = 8.6 × 10^5 J/day

$\dfrac{8.6 \times 10^5}{10^7} = 8.6 \times 10^{-2}$ or 8.6%

8.11 (See Fig. 8.8.) The pressure difference is equal to the pressure of a column of blood equal to your height. For example, for a height of 180 cm

$\Delta P = \dfrac{(180)(1.04)}{13.6} \simeq 14$ cm Hg

8.12 (a) $\dfrac{(2 \text{ mm})^2}{(3 \text{ mm})^2} \times 50$ cm/sec = 22 cm/sec
(b) no
(c) $\dfrac{1}{2}$ mv² = $\dfrac{1}{2}$ (1.04)(50)² $\simeq$ 1300 ergs/cm³ $\simeq$ 1 mm Hg

8.15 $\dfrac{\text{new rate}}{\text{old rate}} = \dfrac{(40 \ \mu m)^4}{(50 \ \mu m)^4} = 0.41$ or a 59% decrease

CHAPTER 10

10.11 $\Omega = \dfrac{V}{A} = \dfrac{3000 \text{ V}}{20 \text{ A}} = 150 \ \Omega$

CHAPTER 11

11.3   (a) $I = \dfrac{V}{R} = \dfrac{10^3 \text{ V}}{(10^7 + 10^5) \ \Omega} \simeq 10^{-4} \text{ A}$

(b) $I = \dfrac{110 \text{ V}}{10^5 \ \Omega} \simeq 10^{-3} \text{ A}$

(c) $I = \dfrac{10^3 \text{ V}}{(10^6 + 10^5) \ \Omega} \simeq 10^{-3} \text{ A}$

From Fig. 11.1, case b is the most hazardous.

11.4 $I = \dfrac{116}{t^{1/2}}$ or $t = \left(\dfrac{116}{I}\right)^2 = \left(\dfrac{116}{50}\right)^2 = 5.4 \text{ sec}$

CHAPTER 12

12.2 $\lambda = \dfrac{v}{f} = \dfrac{1480 \text{ m/sec}}{1000/\text{sec}} = 1.48 \text{ m}$

12.4 $A = \dfrac{1}{\omega}\left(\dfrac{2I}{Z}\right)^{1/2} = \left(\dfrac{1}{6.28 \times 10^3}\right)\left(\dfrac{2 \times 10^{-7}}{4.3 \times 10^2}\right)^{1/2} \simeq 3.5 \times 10^{-9} \text{ m}$

12.5 $10 \log 1000 = 30 \text{ dB}$

12.7 $\dfrac{R}{A_0} = \dfrac{Z_1 - Z_2}{Z_1 + Z_2} = \dfrac{|(1.33 - 1.64) \times 10^6|}{(1.33 + 1.64) \times 10^6} = 0.10 \text{ or } 10\%$

$\dfrac{T}{A_0} = \dfrac{2 Z_1}{Z_1 + Z_2} = \dfrac{2 (1.33 \times 10^5)}{(1.33 + 1.64) \times 10^5} = 0.90 \text{ or } 90\%$

12.8 $\dfrac{I}{I_0} = e^{-2\alpha x} = e^{-2(0.11)(15)} = 0.037$

CHAPTER 13

13.6 $20 \log 20 = 20 (1.30) = 26 \text{ dB}$

13.11 $125 \text{ mm Hg} = \dfrac{125}{760} = 0.164 \text{ atm}$

For 160 dB the pressure is $(2 \times 10^{-4} \text{ dynes/cm}^2) \times 10^8 = 2 \times 10^4$ dynes/cm²

$2 \times 10^4 \text{ dynes/cm}^2 \simeq \dfrac{2 \times 10^4}{10^6} = 0.02$ atm

CHAPTER 15

15.3 $I = \dfrac{Q}{P} O = \dfrac{0.02 \text{ m}}{150 \text{ m}} (0.3 \text{ m}) = 4 \times 10^{-5} \text{ m} = 40 \ \mu\text{m}$

15.9 $\theta = \dfrac{s}{r}, 2.3 \times 10^{-6} = \dfrac{s}{25 \text{ cm}}$ or $s \simeq 6 \times 10^{-5}$ cm

15.10 $OD = \log \dfrac{I_0}{I} = \log \dfrac{1}{0.5} = 0.3$

15.12 $\dfrac{1}{F} = \dfrac{1}{P} + \dfrac{1}{Q}$. For $P = 0.15$ m and $Q = 0.02$ m, $\dfrac{1}{F} = 56.7$ D

without glasses and $\dfrac{1}{F} = 55.7$ D with glasses.

$55.7 = \dfrac{1}{P} + \dfrac{1}{0.02}$ or $P \simeq 0.18$ m, the near point with glasses.

15.13   For distance vision:

$\dfrac{1}{\infty} + \dfrac{1}{0.02} = 50$ D. An accommodation of 3 D gives 53 D at the near point P_0.

$\dfrac{1}{P_0} + \dfrac{1}{0.02} = 53$ or $P_0 = 1/3$ m

15.14   Let the near point be P_0 without glasses and P_g with glasses.

At $P_g = 25$ cm, $\dfrac{1}{F} = \dfrac{1}{0.25} + \dfrac{1}{0.02} = 54$ D.

Without glasses the person would have a maximum of 52 D.

$52 \text{ D} = \dfrac{1}{P_0} + \dfrac{1}{0.02}$ or $P_0 = 0.5$ m.

CHAPTER 16

16.5 $P = IV = (3 \times 10^{-3} \text{ A})(8 \times 10^4 \text{ V}) = 240$ W

16.6 $10^{-9} ZV = 10^{-9} (74)(8 \times 10^4) = 5.9 \times 10^{-3} = 0.59\%$

16.9   3600 rpm = 60 rps

0.01 sec = 0.6 revolutions or 3.8 radians

16.12 $\mu = \dfrac{0.693}{3 \text{ mm}} = 0.231 \text{ mm}^{-1}$

CHAPTER 17

17.7   (a) $T_{1/2} = \dfrac{0.693}{\lambda} = \dfrac{0.693}{0.001 \text{ days}^{-1}} = 693$ days
  (b) $\tau = 1.44\, T_{1/2} = 1.44\,(693 \text{ days}) = 1000$ days
  (c) For small λ
    $A = A_0(1 - \lambda t) = 10[1 - (0.001)1] = 9.99$ MBq

17.10 $N_{\text{net}} = N_g - N_b = 3700 - 800 = 2900$ counts
    $\sigma_{\text{net}} = \sqrt{3700 + 800} \approx 67$ counts

CHAPTER 19

19.11 $D^2 = \Gamma\, N/I_\gamma$
    $= \dfrac{(8.4 \times 10^{-6} \text{ R m}^2/\text{MBq hr})(10 \text{ MBq})}{0.01 \text{ R/hr}}$
    $= 8.4 \times 10^{-3} \text{ m}^2$
  $D = \sqrt{8.4 \times 10^{-3} \text{ m}^2} \approx 9.2 \times 10^{-2} \text{ m} = 9.2$ cm

19.12 One half-value layer, 1.2 cm Pb

19.13 $I_\gamma = \Gamma\, N/D^2$
    $= \dfrac{(1.9 \times 10^{-6} \text{ R m}^2/\text{MBq hr})(500 \text{ MBq})}{(0.3 \text{ m})^2}$
    $= \dfrac{9.5 \times 10^{-4}}{9 \times 10^{-2} \text{ R/hr}} \approx 0.01 \text{ R/hr} = 10 \text{ mR/hr}$

INDEX

587

position of, during normal respiration, 289
Vocal sounds, production of, 289
Vocal tract, model of, 289
  modification of, during speech, 291
  sound transmission characteristics of, 291
Voice, frequencies of, 303
  female, fundamental frequency of, 290
  male, fundamental frequency of, 290
  power of, 121
  sound levels of, 303
Volume, of blood in body, 154
  of heart pump, 154
Vomiting, 109
Von Bekesey, George, 296, 297
Vought, C., 319
Vowel sounds, power of, 292

Walking, oxygen use in, 93
Waller, A. C., 198
Warts, treatment of, 84
Water, viscosity of, 167
Water-head (hydrocephalus), 107, 314
Water molecule, electric dipole in, 242
Weed, Lawrence, 564
Weight, apparent change of with heartbeat, 28
  determination of, 9
  loss of by dieting and exercise, 92
Weightlessness, 16, 32
  physiological effects of, 32
Well counter, 465, 550
Wheatstone stereoscope, 360
Whiplash injury, 28
White blood cells, *see* Blood cells, white
White count, 156
White radiation, x-ray, 393
Whole body rectilinear scanner, 476
Wind chill, 100, 101
Window of PHA, 458
Windpipe, 122, 289
Wind speed, effect of, on heat loss, 100
Work, aerobic and anaerobic, 96
  of breathing, 147, 148
  capacity of body, 96
  of climbing, 94
  external, 87, 93
  of heart, 159, 160, 1
  internal, 94
  mechanical, 93

and power of body, 93-96
of speaking, 121

Xenon, gas of, for ionography, 433
  light source, 330
Xenon 133 (^{133}Xe), 478
Xeroradiography, 73, 432
X-ray, absorbers of, 395
  attenuation of, 395
  automatic exposure meter for, 415
  biological effects of, 486, 523
  cancer treatment with, 486
  characteristic, 393, 445
  of chest, 390, 415
  damage from, 416
  dangers of, 386, 523
  efficiency of producing, 390
  exposures from, 408
    table of, 416
  fluorescent light from, 386, 420
  hard, 396
  intensifying screens, for, 415
  laws controlling medical use of, 521
  of lower spine, exposure from, 539-540
  monoenergetic, 396
  scattering of, 395
  slices of body with (tomography), 425-432
  soft, 395
  spectrum of, 394
  wavelengths of, 315
X-ray beam, equivalent energy of, 397
  measuring attenuation of, 395
  production of, 388-395
X-ray film, latitude of, 415
  processing of, 415
  reading of, 355
  speed of, defined, 414
X-ray image, electrostatic technique for, production of, 432
  production of, 407
  resolution in, 415
  subtraction technique in, 406
X-raying women, policies on, 541
X-ray photons, maximum energy of, 390
X-ray safety, responsibility for, 521
X-ray technician, training of, 388
X-ray tube, components of, 388
  heat produced in anode of, 390
  power produced in, 390